Hot and Cold Water Supply
Third Edition

Prepared by

R.H. Garrett, EngTech, MIPHE RP
*Past President of the National Association of
Plumbing Teachers
Plumber of the Year 1988*

for

British Standards Institution

A John Wiley & Sons, Ltd., Publication

This edition first published 2008
First edition published 1991
Second edition published 2000
© 2008 British Standards Institution

Blackwell Publishing was acquired by John Wiley & Sons in February 2007. Blackwell's publishing programme has been merged with Wiley's global Scientific, Technical, and Medical business to form Wiley-Blackwell.

Registered office
John Wiley & Sons Ltd, The Atrium, Southern Gate, Chichester, West Sussex, PO19 8SQ, United Kingdom

Editorial offices
9600 Garsington Road, Oxford, OX4 2DQ, United Kingdom
2121 State Avenue, Ames, Iowa 50014-8300, USA

For details of our global editorial offices, for customer services and for information about how to apply for permission to reuse the copyright material in this book please see our website at www.wiley.com/wiley-blackwell.

Wiley also publishes its books in a variety of electronic formats. Some content that appears in print may not be available in electronic books.

Designations used by companies to distinguish their products are often claimed as trademarks. All brand names and product names used in this book are trade names, service marks, trademarks or registered trademarks of their respective owners. The publisher is not associated with any product or vendor mentioned in this book. This publication is designed to provide accurate and authoritative information in regard to the subject matter covered. It is sold on the understanding that the publisher is not engaged in rendering professional services. If professional advice or other expert assistance is required, the services of a competent professional should be sought.

Library of Congress Cataloging-in-Publication Data

Garrett, R.H. (Robert H.)
  Hot and cold water supply / prepared by R.H. Garrett for British Standards Institution. – 3rd ed.
        p.   cm.
  Includes bibliographical references and index.
  ISBN 978-1-4051-3002-8 (pbk. : alk. paper) 1. Plumbing. 2. Water-supply engineering. 3. Hot water. I. British Standards Institution. II. Title.
  TH6521.G37 2008
  696′.12–dc22   2008022792

A catalogue record for this book is available from the British Library.

Set in 10/12 pt Sabon by Graphicraft Limited, Hong Kong
Printed in Singapore by Fabulous Printers Pte Ltd

1   2008

Contents

Acknowledgements   viii

Introduction   x

1 General considerations   **1**
  1.1 Scope of the standard   1
  1.2 Water Regulations   2
  1.3 Building Regulations   5
  1.4 The Health and Safety at Work etc. Act 1974   11
  1.5 Definitions   12
  1.6 Graphical symbols   17
  1.7 Materials   21
  1.8 Initial procedures   26

2 Cold water supply   **30**
  2.1 Drinking water   30
  2.2 Cold water systems   31
  2.3 Storage cisterns   36
  2.4 Valves and controls   50
  2.5 Water systems outside buildings   57
  2.6 Water revenue meters   58
  2.7 Boosted systems   66
  2.8 Water treatment   73

3 Hot water supply   **79**
  3.1 System choice   79
  3.2 Instantaneous water heaters   81
  3.3 Water-jacketed tube heaters (primary stores)   85
  3.4 Storage type water heaters and boiler heated systems   88
  3.5 Primary circuits   105
  3.6 Secondary hot water distributing systems   123
  3.7 Components for hot water systems   125
  3.8 Hot water provision for the less able   147

4 Prevention of bursting   **155**
  4.1 Energy control and safety devices   155
  4.2 Pressure and expansion control   167
  4.3 Control of water level   172

5 Pipe sizing   **174**
  5.1 Sizing procedure for supply pipes   176
  5.2 Tabular method of pipe sizing   184
  5.3 Sizing cold water storage   188

	5.4	Sizing hot water storage	189
	5.5	*Legionella* – implications in sizing storage	193
6		**Preservation of water quality**	**194**
	6.1	Materials in contact with water	194
	6.2	Stagnation of water and *Legionella*	197
	6.3	Prevention of contamination by cross connection	198
	6.4	Backflow protection	201
	6.5	Backflow prevention devices	203
	6.6	Secondary or zone backflow protection	220
	6.7	Application of backflow prevention devices	222
7		**Frost precautions**	**234**
	7.1	Protection from frost	234
	7.2	Protection of pipes and fittings	236
	7.3	Draining facilities	240
	7.4	Insulation against frost damage	243
8		**Water economy and energy conservation**	**247**
	8.1	Water economy	247
	8.2	Grey water and recycled rainwater	257
	8.3	Energy conservation	258
	8.4	Building Regulations and energy conservation	259
	8.5	Insulation to meet Building Regulations	263
9		**Noise and vibration**	**267**
	9.1	Flow noises	267
	9.2	Water hammer noise	270
	9.3	Other noises	272
	9.4	Noise transmission and reduction	273
10		**Accessibility of pipes and water fittings**	**275**
	10.1	Pipes passing through walls, floors and ceilings	275
	10.2	Stopvalves	279
	10.3	Water storage cisterns	281
11		**Installation of pipework**	**282**
	11.1	Steel pipes	282
	11.2	Copper pipes	286
	11.3	Stainless steel pipes	304
	11.4	Plastics pipes	305
	11.5	Iron pipes	320
	11.6	Asbestos cement pipes	326
	11.7	Lead pipes	327
	11.8	Connections between pipes of different materials	327
	11.9	Connections to cisterns and tanks	329
	11.10	Branch connections for buildings	332
	11.11	Contamination of mains	335
	11.12	Laying underground pipes	335
	11.13	Pipework in buildings	345
	11.14	Electrical earthing and bonding	350

	11.15	Jointing of pipework for potable water	350
	11.16	Testing	353
	11.17	Identification of valves and pipes	353
12	**Commissioning and maintenance of pipelines, services and installations**		**356**
	12.1	Inspections	356
	12.2	Testing for soundness	357
	12.3	Testing methods	358
	12.4	Flushing and disinfection	361
	12.5	Commissioning hot water and heating systems	365
	12.6	Maintenance	366
	12.7	Locating leaks	369
	12.8	Disconnection of unused pipes and fittings	372
	12.9	Occupier information	372
13	**Firefighting systems**		**373**
	13.1	Fire safety	373
	13.2	Openings for pipes	375
	13.3	Fire mains within buildings	376
	13.4	Certification and accreditation of fire protection installations	377
14	**Sprinkler systems for domestic and residential premises**		**378**
	14.1	Scope of BS 9251	378
	14.2	Terms and definitions	379
	14.3	Consultation	380
	14.4	Sprinkler systems and water supply methods	380
	14.5	Pressure requirements and system flow rates	385
	14.6	System components	386
	14.7	Installation	390
	14.8	Commissioning sprinkler systems	391
	14.9	Maintenance	392
	14.10	Documentation	392
	14.11	Hydraulic calculations for sprinkler systems	393
British Standards relevant to this book			397
References and further reading			405
Index			407

Acknowledgements

BSI wishes to thank Frank Young for his assistance with the first edition and the following companies/organizations for their kind permission to use material.

Braithwaite Engineers Ltd
British Coal Corporation
British Gas PLC
Caradon Mira Ltd
Cistermiser Ltd
Construction Industry Training Board
Copper Development Association
Council for the Registration of Gas Installers (CORGI)
Danfoss Ltd
Department for Communities and Local Government
Department of the Environment, Transport and the Regions
Ecowater Systems Ltd
F W Talbot & Co Ltd
George Fischer Sales Ltd
Gledhill Water Storage Ltd
Globe Fire Sprinkler Corporation
Glow-worm Ltd
Grundfoss Pumps
Heatrae Sadia Heating
Honeywell Control Systems Ltd
H Warner & Son Ltd
IBP Conex Limited
IMI Range Cylinders Ltd
IMI Santon Ltd
IMI Yorkshire Copper Tube Ltd
Lancashire Fittings
Meter Options Ltd
North West Water Ltd
Pegler Limited
Polystel Ltd
Polytank Ltd
Reliance Water Controls
South Northants District Council
Talon Manufacturing Ltd
The Architectural Press Ltd
The Institute of Plumbing
Tyco Thermal Controls

Unifix Ltd
Water Research Centre
Water Training International
Worcester Bosch Group
Yorkshire Fittings Ltd

Introduction

The development, construction, installation and maintenance of hot and cold water supply systems are vital areas of concern for public health. Water quality is by turns a political, environmental and technical issue. It is governed by legislation, Water Regulations, Building Regulations and technical standards intended to safeguard quality.

This book is a thorough introduction and guide to hot and cold water supply. It is based on a British Standard, BS 6700, which is a specification for design, installation, testing and maintenance of water supply services for domestic buildings. It also includes information on the law and on Water and Building Regulations. It is an invaluable work of reference for engineers, designers, installers and contractors who work on water supply installations. It is essential reading for teachers and students of plumbing, central heating and other water related occupations. It is especially useful to those who are working towards S/NVQ, BTech or other water related qualifications. It can also provide an important aid to technical staff of water undertakers, building control officers, architects, surveyors, building contractors and property services managers.

At present it is not generally a legal requirement that anyone installing or repairing domestic water services should be properly qualified although installers are required under both Building and Water Regulations to be competent. Notification to water suppliers and to building control bodies is a requirement for an increasing number of work operations. Under various competent person schemes, installers who can prove their competence may be permitted to self-certify their work. By failing to comply with the regulations, the installer or repairer could be prosecuted for offences under both Water and Building Regulations or could be prosecuted under civil law by a householder.

With increasing pressure to maintain the highest standards of health and safety, to prevent waste, misuse or contamination of water, and to maintain more stringent standards in energy efficiency, it is important that technical knowledge is up to date. There is also a strong movement to make water supply practice consistent throughout Europe and new EN standards are being developed to achieve this aim. This book helps practitioners to come to terms with these changes.

Hot and Cold Water Supply was written for the British Standards Institution by Bob Garrett, and has benefited from the assistance and comments of other plumbers, designers and teachers. A principle of *Hot and Cold Water Supply* is to support technical information with examples and illustrations that make the text easier to understand. It is an immensely practical book, which makes easy the job of turning theory into action.

This edition has been updated to take account of BS 6700:2006 and makes extensive reference to the latest Building Regulations, including Part L (energy conservation) and Part P (electrical safety). It includes valuable information on the Water Supply (Water Fittings) Regulations 1999 and new or revised British and European Standards including BS EN 805 (pipes below ground). It also considers the needs of the less able and in particular the provision of hot water that is safe to the user. Additionally, this edition introduces two new chapters on fire protection and sprinkler systems in domestic and residential premises.

Chapter 1
General considerations

1.1 Scope of the standard

This book is based primarily on BS 6700 but also contains information from other standards and regulations. BS 6700:2006 is a revision of BS 6700:1997. It specifies requirements and gives recommendations for the design, installation, testing and maintenance of services supplying water for domestic use within buildings and their curtilages. It covers systems in domestic and non-domestic buildings for the supply of water for drinking and culinary use, domestic laundry, ablutionary, cleansing and sanitary purposes.

The following are not included although, in parts, the standard may apply to them: hot water systems whose temperature exceeds 100°C; central heating systems; firefighting; and water for industrial purposes. However, because hot water is increasingly linked with central heating, and aspects of fire protection are included in Building Regulations, these subjects have been taken into account in this book.

BS 6700 does not give detailed specifications for the design and installation of large-scale systems below ground. These come within the scope of BS EN 805 *Water supply – Requirements for systems and components outside buildings*.

It is expected that BS 6700 will eventually be replaced by BS EN 806 *Specification for installations inside buildings conveying water for human consumption*. This European standard will be in five parts but, at the time of writing, only parts 1, 2 and 3 have been published.

Note on the scope of BS EN 805

BS EN 805 specifies general requirements for water supply systems and their components outside buildings (see figure 1.1). It includes specifications for the design, installation and testing of potable water mains and service pipes, service reservoirs and other facilities, but excludes treatment works and water resources development. The requirements of BS EN 805 are applicable to:

- all water installations within the supply system outside buildings, irrespective of ownership or responsibility for pipes or other apparatus;
- the design and construction of new water supply systems; and
- significant extension, modification and/or rehabilitation of existing water supply systems.

It is not intended that existing water supply systems be altered to comply with this standard, provided that there are no significant detrimental effects on water quantity, security, reliability and adequacy of the supply.

This book takes into account the requirements of BS EN 805 in respect of water systems outside buildings but is concerned only with supply pipes within premises that are not the responsibility of the water supplier.

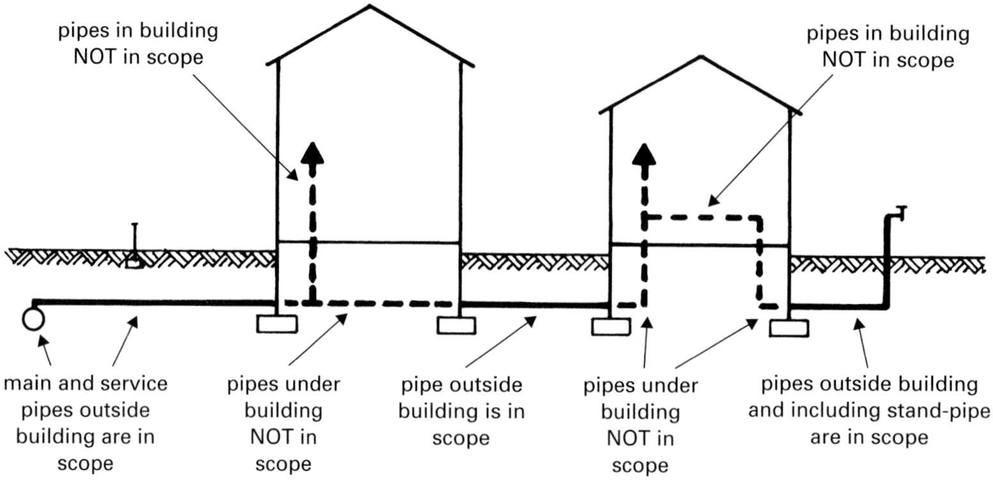

pipes in building NOT in scope

pipes in building NOT in scope

main and service pipes outside building are in scope

pipes under building NOT in scope

pipe outside building is in scope

pipes under building NOT in scope

pipes outside building and including stand-pipe are in scope

Pipes and fittings outside buildings are in scope of BS EN 805 irrespective of ownership.

Stopvalves and other fittings not shown.

Figure 1.1   Application of BS EN 805

1.2   Water Regulations

In England and Wales, water supply was for many years governed by water byelaws, made by water undertakers under Section 17 of the Water Act 1945. Byelaws have now been replaced by the Water Supply (Water Fittings) Regulations 1999 for the prevention of waste, misuse, undue consumption, contamination or erroneous measurement of water supplied by a water undertaker.

Made by the Secretary of State for the Department of the Environment, and the Secretary of State for Wales, the Water Fittings Regulations came into force on 1 July 1999 and apply to any water fitting in premises to which water is supplied by a water undertaker.

Water Fittings Regulations, applicable to England and Wales, are made under Sections 74, 84 and 213 of the Water Industry Act 1991. The requirements are similar in content to the previous byelaws, but have been amended to take account of the latest technical advancement and innovation. In addition to the 14 regulations, there are two schedules, which set out technical requirements in greater detail. Unlike previous legislation, the Water Fittings Regulations could only be made after consultation, through the European Commission, with other Member States of the European Community. Similar legislation applies to water installations in Scotland and Northern Ireland.

In Scotland, the Scottish Water Byelaws 2004 are made under Section 70 of the Water (Scotland) Act 1980 and came into force on 30 August 2004. In Northern Ireland, Water Regulations are made under Article 40 of the Water and Sewerage Services (Northern Ireland) Order 1973. Legislation in Northern Ireland is being revised to make Water Regulations technically similar to those in the rest of the UK.

A building owner or occupier can demand a supply of water for domestic purposes provided the relevant requirements of the Water Industry Act 1991 have been complied with and the installation satisfies the requirements of the Water Fittings Regulations.

Whilst it is the duty of the water supplier to accede to the owner's demand, the supplier must also uphold the requirements of the Water Regulations and has the right to refuse connection to the mains of any new installation which is not in compliance with them.

To avoid unnecessary dispute with water undertakers, and perhaps lengthy legal proceedings, it is advisable to consult water undertakers about their regulations at an early stage, and particularly their requirements arising from local water or soil characteristics.

Although there is no legal requirement for a person installing or repairing water services to be suitably qualified, the work should be done in a 'workmanlike manner', which means conforming to the regulations and following the recommendations of relevant British or European Standards.

Enforcement of the Water Fittings Regulations is the responsibility of water undertakers for their area of supply. It is a statutory requirement that they are notified in advance of any installation, alteration or disconnection of water fittings. Anyone who carries out work of this kind without the consent of the water undertaker can be prosecuted for an offence against current Water Regulations.

It should be noted that householders also commit an offence under the regulations if they 'use' a fitting which does not comply with the regulations. Knowledge of the regulations is therefore desirable for both installer and user.

Particular note should be taken of Regulation 5 which requires notice in writing before commencing work on any of the following:

(1)   The erection of a building or other structure, not being a pond or swimming pool.
(2)   The extension or alteration of a water system on any premises other than a house.
(3)   A material change of use of any premises.
(4)   The installation of:
    (a)   a bath having a capacity, measured to the centre line of overflow, of more than 230 l;
    (b)   a bidet with an ascending spray or flexible hose;
    (c)   a single shower unit (which may consist of one or more shower heads within a single unit) not being a drench shower installed for reasons of safety or health, connected directly or indirectly to a supply pipe which is of a type specified by the regulator;
    (d)   a pump or booster drawing more than 12 l/min, connected directly or indirectly from the supply pipe;
    (e)   a unit which incorporates reverse osmosis;
    (f)   a water treatment unit which produces a waste discharge or which requires the use of water for regeneration or cleaning;
    (g)   a reduced pressure zone valve assembly or other mechanical device for protection against a fluid which is in fluid-risk category 4 or 5;
    (h)   a garden watering system unless designed to be used by hand; or
    (i)   any water system laid outside a building and either less than 750 mm or more than 1350 mm below ground level.
(5)   The construction of a pond or swimming pool with a capacity greater than 10 000 l which is designed to be replenished by automatic means and is to be filled with water supplied by a water undertaker.

Under Regulation 7 any person who on summary conviction is found guilty of Water Regulations contraventions is liable to a fine, for each separate offence, not exceeding level three on the standard scale. At the time of writing this means a fine of up to £1000 for each offence. In Scotland the fine is level four on the standard scale.

Water undertakers are encouraged under the Water Fittings Regulations to set up approved contractors' schemes within their area of supply. The scheme requires approved contractors to certify to the undertaker that water fittings installed are in compliance with the regulations. Approved contractors will be excused from prior notification for some work operations, but by no means all of them. Work which may be self-certified by approved contractors includes:

- the extension or alteration of a water system in premises other than a house (extensions or alterations in houses need not be notified);
- a bidet with an ascending spray or flexible hose;
- an RPZ valve or other mechanical device used to protect against a fluid risk category 4 or 5.

The scheme will be of benefit to consumers who, in any proceedings against them, can show 'that the work was carried out by, or under the direction of, an approved contractor, and that the contractor certified to them that the water fittings complied with the requirements of the Regulations'.

Measures for water conservation permit the use of dual-flushing cisterns, and include the use of non-siphonic flushing methods for WCs. Maximum WC flushing volumes are 6 l for a single flush and 6 l/4 l for dual-flush cisterns.

A Secretary of State Specification for WC Suite Performance requires extensive testing by manufacturers for all WCs whether they are siphonic or non-siphonic.

Pressure flushing valves are permitted (but not in a house) for use in the flushing of both WCs and urinals which may be connected directly to either a supply pipe or a distributing pipe, but they must have backflow protection fitted.

Backflow protection requirements have been revised. The new regulations introduce, in Schedule 1, five fluid-risk categories (rather than the three under previous byelaws) to bring us into line with European practices. At the same time backflow prevention devices are categorized differently in a new Secretary of State specification.

We have been used to safety devices for unvented hot water heaters for a long time. Thermostatic control and temperature relief valves have, of course, been required by the Building Regulations for a good many years. The Water Fittings Regulations now duplicate this requirement.

In Schedule 2 to Regulation 4(3), Paragraph 18 it says 'Appropriate vent pipes, temperature control devices and combined temperature pressure relief valves shall be provided to prevent the water temperature within a secondary hot water system from exceeding 100°C'.

Water systems are required under Paragraph 13 of Schedule 2 to be tested, flushed and, where necessary, disinfected before use. This is not new, but more emphasis is given to the need for these important aspects of water installations.

It should be added that the Water Fittings Regulations are *not* made for the specific protection of people or property, but solely for the prevention of waste, undue consumption, misuse and contamination of water supplied by water undertakers:

'waste'   means water which flows away unused, e.g. from a dripping tap or hosepipe left running when not in use, or from a leaking pipe.

'undue consumption'   means water used in excess of what is needed, e.g. full bore tap to wash hands when half or quarter flow will suffice, or automatic flushing cistern that flows even when urinals are not in use (at night).

'misuse' means water used for purposes other than that for which it is supplied, e.g. use of garden sprinkler when paying only for domestic use or taking supply from domestic premises for industrial or agricultural use.

'contamination' means pollution of water by any means, e.g. by cross connection between public and private supplies or by backflow through backsiphonage.

'erroneous measurement' means incorrect meter reading, e.g. connections made which may not be detected by the meter.

Guidance and approval of water fittings

Guidance on particular Water Regulation matters may be sought from local water under-takers. Their inspectors are trained in the application of Water Regulations.

The Water Regulations Advisory Scheme (WRAS) has published a guide to the applica-tion and interpretation of the Water Fittings Regulations. It contains useful information and background knowledge for those concerned with water services. WRAS also operate a voluntary national scheme for the testing and approval of water fittings. Fittings which pass the Centre's tests are listed in its publication *Water Fittings and Materials Directory*, together with the names and addresses of manufacturers and any applicable installation requirements. Thus, the connection and use of any listed fitting carry with them virtual certainty of acceptance by water undertakers.

1.3 Building Regulations

Building Regulations in England and Wales are made under the Building Act 1984. Their purpose is to control building work for the health and safety of occupants and users of buildings and to conserve energy and power. The responsibility for enforcement of build-ing work under Building Regulations lies with local authorities and their building control officers (building inspectors) or other approved inspectors such as the National House Building Council (NHBC). Local authorities and approved inspectors have powers to inspect work during construction. Any disputes that arise may have to be decided ultimately in a court of law. To contravene Building Regulations is a criminal offence.

It should also be noted that building work generally involves more than one Statutory Instrument and any person undertaking water installations or other works should be aware of the relevant requirements in such statutory or guidance documents as Building Regula-tions, Water Fittings Regulations, Gas Regulations and Electrical Wiring Regulations.

In Scotland, the Building Standards (Scotland) Regulations are made under the Building (Scotland) Act 1959.

In Northern Ireland, the Building Regulations (Northern Ireland) are made under the Building (Northern Ireland) Order 1979.

Building Regulations include a number of water-related provisions. These are:

- pipes passing through compartment walls and floors (B3);
- sprinkler systems (B4) and buildings fitted with fire mains (B5);
- provision for wash basins with hot and cold water in the vicinity of water closets (WCs), and provision for these to be effectively cleaned (Gl);

- provision for baths and showers and the supply of hot and cold water to them (G2);
- requirements for the safety of hot water systems and, in particular, prevention of explosion (G3);
- provision for the removal of condensate from high-energy boilers (H1);
- the use of grey water (H2) and the re-use of rainwater (H3);
- provisions for the control of air supply to combustion appliances (J1) and the discharge of the products of combustion (J2);
- protection of the building from combustion appliances (J3) and the positioning of liquid fuel storage to reduce risk of fire (J5);
- requirements for the control of energy in heating and hot water systems and the insulation of hot water pipes and storage vessels (L1 and L2);
- safe access to and use of buildings, and, in particular, sanitary provision for the less able (Part M); and
- electrical safety (P), which is increasingly applicable to water systems, e.g. controls to boilers and hot water/heating systems.

Building Regulation 7 requires building work to be carried out:

- using adequate and proper materials; and
- in a workmanlike manner.

Materials

Materials used in building work are required to be suitable for the purpose and conditions in which they are used and should not have any adverse affect on the health and safety aspects of building work.

Methods to establish fitness of materials are set out in Section 1 of the Approved Document to support Regulation 7. There are a number of ways in which compliance can be established:

- The material complies with a relevant and up to date British or European Standard. Most materials will conform to a British Standard and the material or its packaging will carry the BS Kitemark (see figure 1.2(a)). However, many of these are currently being revised to become European Standards. This transposition of the full range of standards will take place over a number of years. An example of this can be seen in copper tube which formerly complied with BS 2871 but is now manufactured to BS EN 1057 (see figure 1.2(c)).
- The material conforms to a technical specification of another European Member State providing its level of performance is at least equal to that of a British Standard (see figure 1.2 (d)).
- The material has CE marking which gives a presumption of conformance with legal minimum requirements as set out in the Construction Products Regulations 1991 (see figure 1.2(b)).
- Materials may be approved through one of a number of independent certification bodies within the UK many of which will be tested and approved through UKAS (United Kingdom Accreditation Service).

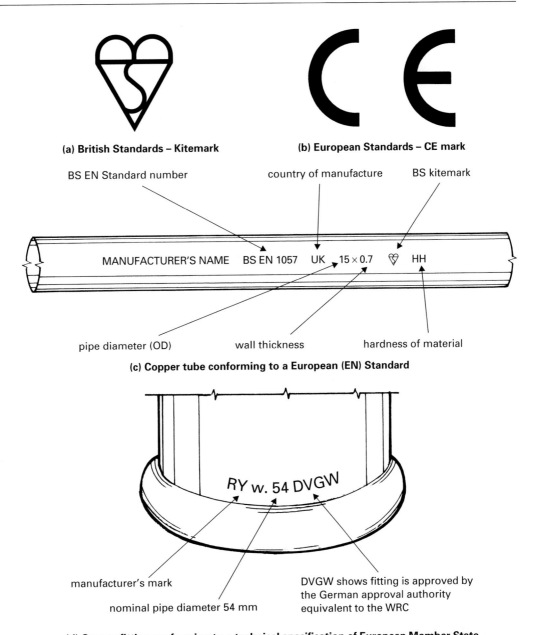

(a) British Standards – Kitemark

(b) European Standards – CE mark

BS EN Standard number

country of manufacture

BS kitemark

MANUFACTURER'S NAME   BS EN 1057   UK   15 × 0.7   ♡   HH

pipe diameter (OD)

wall thickness

hardness of material

(c) Copper tube conforming to a European (EN) Standard

RY w. 54 DVGW

manufacturer's mark

nominal pipe diameter 54 mm

DVGW shows fitting is approved by
the German approval authority
equivalent to the WRC

(d) Copper fitting conforming to a technical specification of European Member State

Figure 1.2   Material fitness approval marks

- Additionally materials may be accepted if it can be shown by tests, by calculation, or by past experience that the material is capable of performing its intended function.
- The material is covered by a national or European certificate issued by a European Technical Approvals Body, e.g. the British Board of Agrément. Unvented hot water storage appliances, for example, are required to be approved under these procedures. See chapter 3.

Workmanship

Good workmanship implies that the correct choice of materials is made, they are properly fixed and installed, and they adequately perform the functions for which they are intended. The adequacy of workmanship can be established in a number of ways. The work will be deemed to be satisfactory if:

- it is carried out in accordance with the recommendations of a current British Standard Code of Practice; or
- it conforms to a technical specification of a European Member State;
- it is covered by a national or European certificate issued by a European Technical Approvals Body; or
- workmanship can be established by following quality management schemes complying with BS EN ISO 9000 many of which will be accredited in the UK by UKAS.

Note   Building Regulations Approved Document 7 refers to BS 8000 *Workmanship on building sites*, the section relevant to the application of this book being Part 15: 1990 *Code of practice for hot and cold water services (domestic scale)*. It is unfortunate that this standard has not been updated to take account of the Water Supply (Water Fittings) Regulations 1999. Nor does it take account of recent amendments to Building Regulations, e.g. conservation of energy and electrical safety. For this reason Part 15 of BS 8000 should be read with care and direct reference made to current Water and Building Regulations.

Notification and self-assessment

Regulations 12, 13 and 14 set out provisions for self-certification schemes, which in turn are listed in Schedule 2A of the Regulations.

Under the above-mentioned Building Regulations, installers of hot water or heating installations, classed as 'building work', are required to notify a building control body that the work is to be carried out. This may be done:

- By making a 'full plans' application to the local authority or an approved inspector in accordance with Regulation 14. For the majority of hot water installations 'full plans' notification is not likely to be needed unless the installation is part of a larger construction project that itself requires a 'full plans' application. Most hot water installations are carried out as replacement or renewal of existing work and in these cases a building notice will normally suffice.
- By giving a building notice to the local authority or an approved inspector, detailing the work that is to be carried out along with information on the location of the work.

Following acceptance of the plans the installer is required to inform the building control body that he intends to commence the work, at least 2 days before the work is begun. Additionally, the control body should be notified within 5 days of completion of the work, that the work has been finished.

There are, however, exceptions to the above rules. A building notice or a full plans submission will not be required where the work is carried out on a 'controlled service or fitting' by a 'competent person' who is 'approved' under the regulations to carry out the work. In such cases the competent person may 'self-certificate' the work.

Note A controlled service or fitting means a service or fitting in relation to which Part G, H, J, L or P of Schedule 1 imposes a requirement.

Competent persons schemes have been introduced by government to permit individuals and businesses to self-certify that their work complies with Building Regulations as an alternative to submitting a building notice or using an approved inspector.

To be included in a scheme an operative or company will need to be assessed to ensure they have the required knowledge and expertise in the prescribed area of work. Once competency can be proved the operative will be permitted to issue a certificate to the customer that the work complies with relevant Building Regulations. He will also be required to send a similar notice to a building control body, within 2 days of completion, that the work has been finished.

A list of 'building work' that may be self-certificated is given in Schedule 2A of the Building Regulations along with details of the persons deemed competent to carry out the work. It should be noted that at the time of writing some of the building work listed in Schedule 2A does not have self-assessment schemes in place but schemes will be introduced as and when relevant parts of Schedule 1 are amended and re-written. Currently self-certification schemes are operating fully within the scope of Part L and Part P. (More schemes for other parts of the Regulations are in the pipeline.)

Part L: Conservation of fuel and energy in dwellings (L1 and L2)

Part L of the Building Regulations provides for the control of energy and conservation of fuel. Under the 2006 edition of Part L, provision must be made in both new and existing buildings to achieve improved energy efficiency compared with what was previously required. A proportion of these requirements apply particularly to hot water (and heating) installations.

Guidance as to how compliance can be achieved is given in Part L, which is completely re-written and published in four separate documents:

- Approved Document L1A *Conservation of fuel and power in new dwellings*
- Approved Document L1B *Conservation of fuel and power in existing dwellings*
- Approved Document L2A *Conservation of fuel and power in new buildings other than dwellings*
- Approved Document L2B *Conservation of fuel and power in existing buildings other than dwellings*

This book will look primarily at the documents L1A and L1B, which deal with work in dwellings.

The Domestic Heating Compliance Guide

To support the guidance given in Approved Documents L1A and L1B, the ODPM (Office of the Deputy Prime Minister) has published a second-tier document entitled the *Domestic Heating Compliance Guide*. Approved Documents L1A and L1B refer to the Guide which sets out minimum provisions for compliance with the hot water and space heating energy efficiency requirements of the Regulations. Figure 1.3 shows how the *Domestic Heating Compliance Guide* fits into the legislative process.

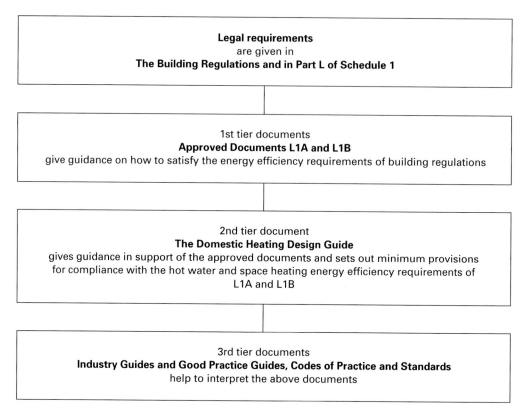

Figure 1.3   Status of the Domestic Heating Compliance Guide

Part M: Access to and use of buildings

Part M of the Building Regulations applies mainly to non-domestic buildings but there are some exceptions including the provision of WCs in the entrance storey of dwellings. The object of Part M is to encourage the provision of facilities within buildings for those people who for various reasons have difficulty in using 'normal' conveniences and fittings. It calls for reasonable access in and around buildings to enable all people to make better and easier use of the building's facilities.

Part M provisions were originally aimed towards those with some sort of disability, e.g. wheelchair users or ambulant disabled, but has been extended in the 2004 edition to accommodate the needs of *all* people and includes anyone with a limited physical ability, including people with babies and those encumbered with luggage.

The scope of this book in relation to Part M is concerned with sanitary facilities in buildings and in particular the safe supply and delivery of hot water, and the provision of taps, valves and mixers that are easy to use by everyone. See chapter 3.

Part P: Electrical safety

Electrical work in dwellings (including gardens and outbuildings) is controlled under Building Regulations Part P *Electrical safety*.

Electrical work, which includes installation, inspection and testing, is required to be notified to a local building control office, or to an approved inspector, at least 2 days before work is commenced, except when:

- The work is to be carried out by a 'competent person', who is listed on an approved register and is therefore capable of 'self-certifying' his or her own work; or
- The work is of a 'minor' nature and not installed in a 'special installation' or in a 'special location' or in a kitchen.

'Minor work' may include the replacement of electrical accessories such as socket outlets, control switches and ceiling roses. It includes the addition of socket outlets or lighting points to existing circuits but does NOT include the provision of a new circuit.

'Special installations or locations' include rooms with a bath or shower, swimming pools or paddling pools.

Within 30 days of completion of the work the competent person is required to complete a 'Certificate of Compliance' stating that the work complies with Building Regulations. This should be given to the person who has ordered the work and a further copy sent to a building control body.

Alternative routes to Part P compliance are shown in figure 1.4.

All electrical work is required to conform to the current edition of BS 7671 *Requirements for electrical installations.*

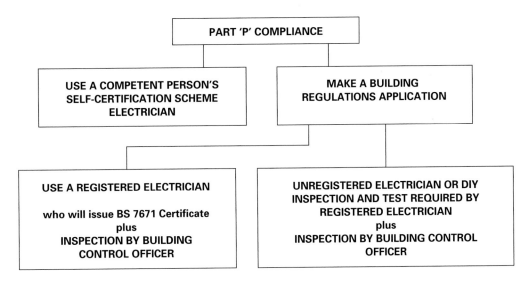

Figure 1.4   Routes to Part P compliance

1.4   The Health and Safety at Work etc. Act 1974

This Act provides for securing the health, safety and welfare of persons at work, for controlling the use and storage of dangerous substances and for the control of certain emissions into the atmosphere. The Health and Safety at Work etc. Act 1974 is an enabling act under which many safety regulations are made. Some of those relevant to this book include:

- The Workplace (Health, Welfare and Safety) Regulations which regulate the provision of drinking water and sanitary accommodation in the workplace.
- The Gas Safety (Installation and Use) Regulations, used to control the installation of gas appliances, provide for the registration of gas installers and require that all gas-fitting operatives shall be competent in the area of gas work that they do.
- The Control of Asbestos at Work Regulations.
- Control of Substances Hazardous to Health Regulations.

1.5   Definitions

Definitions given here are taken from BS 7600, the Water Fittings Regulations and from Building Regulations. Definitions are important to the understanding of the document in which they are given. For this reason, different interpretations are sometimes given to the same term depending on which document they are taken from. For example, 'building' is given a different interpretation under BS 6700 than that shown under Building Regulations.

From BS 6700

backflow   means flow upstream, that is in a direction contrary to the intended direction of flow, within or from a water fitting

building   means a structure (including a floating structure) of a permanent character or not, and movable or immovable, connected to the water supplier's mains

cavity wall   means a structural or partition wall, formed by two upright parts of similar or dissimilar building materials, suitably tied together with a gap formed between them, which might be (but need not be) filled with insulating material

chase   means a recess cut into an existing structure

cistern   means a fixed, vented container for holding water at atmospheric pressure

combined feed and expansion cistern   means a cistern for supplying cold water to a hot water system without a separate expansion cistern

composite fitting   means a combination of fittings or valves incorporated into one body

contamination   means a reduction in chemical or biological quality of water due to a change in temperature or the introduction of polluting substances

cover   means a panel or sheet of rigid material fixed over a chase, duct or access point, of sufficient strength to withstand surface loadings appropriate to its position. *NOTE: Except where providing access to joints or changes of direction (i.e. at an inspection access point), a cover may be plastered or screeded over*

direct hot water supply system   means a hot water supply system in which the water supplied to draw-off points is heated by a direct source of heat

distributing pipe   means a pipe (other than a warning, overflow or flushing pipe) conveying water from a storage cistern, or from hot water apparatus supplied from a cistern and under pressure from that cistern

duct means an enclosure designed to accommodate water pipes and fittings and other services, if required. *NOTE: Ideally, a duct should be constructed so that access to the interior can be obtained either throughout its length or at specified points by removal of a cover or covers*

dwelling means premises, buildings or part of a building providing accommodation, including a terraced house, a semi-detached house, a detached house, a flat in a block of flats, a unit in a block of maisonettes, a bungalow, a flat within any non-domestic premises, a maisonette in a block of flats, or any other habitable building and any caravan, vessel, boat or houseboat that can accommodate a single family unit connected to the water supplier's mains

expansion valve means a pressure-activated valve designed to release expansion water from an unvented water heating system

flushing cistern means a cistern provided with a valve or device for controlling the discharge of the stored water into a water closet pan or urinal

indirect hot water supply system means a hot water supply system in which the water supply to draw-off points is heated by an indirect cylinder or calorifier

inspection access point means a position of access to a duct or chase whereby the pipe or pipes therein can be inspected by removing a cover which is fixed by removable fastenings but does not necessitate the removal of surface plaster, screed or continuous surface decoration

overflow pipe means a pipe from a cistern in which water flows only when the level in the cistern exceeds a predetermined level

pressure relief valve means a pressure-activated valve which opens automatically at a specified pressure to discharge fluid

primary circuit means an assembly of water fittings in which water circulates between a boiler or other source of heat and a primary heat exchanger inside a hot water storage vessel, and includes any space heating system

removable fastenings means fastenings that can be removed readily and replaced without causing damage, including turn buckles, clips, magnetic or touch latches, coin-operated screws and conventional screws, but not nails, pins or adhesives

RPZ valve means a type BA backflow protection device that conforms to BS EN 1717. *NOTE: 'RPZ valve' stands for 'verifiable backflow preventor with reduced pressure zone'*

secondary circuit means an assembly of water fittings in which water circulates in supply pipes or distributing pipes of hot water storage system

secondary system means an assembly of water fittings comprising the cold feed pipe, any hot water storage vessel, water heater and pipework from which hot water is conveyed to all points of draw-off

servicing valve means a valve for shutting off, for the purpose of maintenance, the flow of water in a pipe connected to a water fitting or appliance

sleeve means an enclosure of tubular or other section of suitable material designed to provide a space through an obstruction to accommodate a single water pipe and to which access to the interior can be obtained only from either end

stopvalve   means a valve, other than a servicing valve, used for shutting off the flow of water in a pipe

storage cistern   means a cistern for storing water for subsequent use, not being a flushing cistern

tap size designations   means numbers directly related to the nominal size of the thread on the inlet of the tap, which in turn is unchanged from the nominal size in inches before metrication (e.g. $\frac{1}{2}$ nominal size tap means a tap with an inlet having a G $\frac{1}{2}$ thread)

tank   means a closed vessel holding water at greater than atmospheric pressure

temperature relief valve   means a valve which opens automatically at a specified temperature to discharge fluid

terminal fitting   means a water outlet device

tundish   means a funnel for catching overflow or discharge

vent pipe   means a pipe, open to the atmosphere, which exposes the system to atmospheric pressure

walkway or **crawlway**   means an enclosure similar to a duct, but of such size as to provide access to the interior by persons through doors or manholes and which will accommodate water pipes and fittings and other services if required

warning pipe   means a pipe from a cistern in which water flows only when the level in the cistern is about to exceed the predetermined overflow level to warn of impending overflow

From Water Regulations

A further list of definitions relating to Schedule 2 of the Water Supply (Water Fittings) Regulations 1999:

backflow   means flow upstream, that is in a direction contrary to the intended normal direction of flow, within or from a water fitting

cistern   means a fixed vented container for holding water at atmospheric pressure

combined feed and expansion cistern   means a cistern for supplying cold water to a hot water system without a separate expansion cistern

combined temperature and pressure relief valve   means a valve capable of performing the function of both a temperature relief valve and a pressure relief valve

concealed water fitting   means a water fitting that:
(a)   is installed below ground;
(b)   passes through or under any wall, footing or foundation;
(c)   is embedded in any wall or solid floor;
(d)   is enclosed in any chase or duct; or
(e)   is in any other position which is inaccessible or renders access difficult

contamination   includes any reduction in chemical or biological quality of water due to raising its temperature or the introduction of polluting substances

distributing pipe   means any pipe (other than a warning, overflow or flush pipe) conveying water from a storage cistern, or from hot water apparatus supplied from a cistern and under pressure from that cistern

expansion cistern   means a cistern connected to a water heating system which accommodates the increase in volume of that water in the system when the water is heated from cold

expansion valve   means a pressure-activated valve designed to release expansion water from an unvented water heating system

overflow pipe   means a pipe from a cistern in which water flows only when the water level in the cistern exceeds its normal maximum level

pressure flushing cistern   means a WC flushing device that utilizes the pressure of water within the cistern supply pipe to compress the air and thus increase the pressure of water available for flushing a WC pan

pressure relief valve   means a pressure-activated valve which opens automatically at a specified pressure to discharge fluid

primary circuit   means an assembly of water fittings in which water circulates between a boiler or other source of heat and a primary heat exchanger inside a hot water storage vessel

secondary circuit   means an assembly of water fittings in which water circulates in supply pipes or distributing pipes to and from a hot water storage vessel

secondary system   means that part of any hot water system comprising the cold feed pipe, any hot water storage vessel, water heater and flow and return pipework from which hot water is conveyed to all points of draw-off

servicing valve   means a valve for shutting off the flow of water in a pipe connected to a water fitting for the purpose of maintenance or service

spill-over level   means the level at which the water in a cistern or sanitary appliance will first spill over if the inflow of water exceeds the outflow through any outflow pipe and any overflow pipe

stopvalve   means a valve, other than a servicing valve, for shutting off the flow of water in a pipe

supply pipe   means so much of any pipe as is not vested in the water undertaker

temperature relief valve   means a valve which opens automatically at a specified temperature to discharge water

terminal fitting   means a water discharge point

unvented hot water storage vessel   means a hot water storage vessel that is not provided with a vent pipe but is fitted with safety devices to control primary flow, prevent backflow, control working pressure and accommodate expansion

vent pipe   means a pipe open to the atmosphere which exposes the system to atmospheric pressure at its boundary

warning pipe   means an overflow pipe whose outlet is located in a position where the discharge of water can be readily seen

From Building Regulations

A selection of terms and definitions from Regulation 1 is as follows:

building   means any permanent or temporary building but not any other kind of structure or erection, and a reference to a building includes a reference to part of a building

building work   has the meaning given in Regulation 3(1). *NOTE: Building work includes: the provision or extension of a controlled service or fitting in or in connection with a build-ing and the material alteration of a controlled service or fitting. For a full definition readers should consult Regulation 3(1)*

controlled service or fitting   means a service or fitting in relation to which Part G, H J, L or P of Schedule 1 imposes a requirement

domestic hot water   means water that has been heated for ablution, culinary and cleansing purposes. The term is used irrespective of the type of building in which an unvented hot water storage system is installed

dwelling   includes a dwelling-house and a flat

dwelling-house   does not include a flat or a building containing a flat

energy rating   of a dwelling means a numerical indication of the overall energy efficiency of that dwelling obtained by the application of a procedure approved by the Secretary of State under regulation 16(2) of the Regulations

European Technical Approval issuing body   means a body authorized by a Member State of the European Economic Area to issue European Technical Approvals (a favourable technical assessment of the fitness for use of a construction product for the purposes of the Construction Products Directive (a))

final certificate   means a certificate given under Section 51 of the Act

flat   means separate and self-contained premises constructed or adapted for use for residential purposes and forming part of a building from some other part of which it is divided horizontally

full plans   means plans deposited with a local authority for the purposes of Section 16 of the Act in accordance with Regulations 12(2)(b) and 14

initial notice   means a notice given under section 47 of the Act

package   means an unvented hot water storage system having the safety devices described in paragraph 3.3 or 3.4 factory-fitted together with a kit containing other applicable devices, supplied by the package manufacturer, to be fitted by the installer

unit   means an unvented hot water storage system having the safety devices described in paragraph 3.3 or 3.4 and all operating devices factory-fitted by the manufacturer

unvented hot water storage system   means an unvented vessel for either:
a)   storing domestic hot water for subsequent use; or
b)   heating domestic water that passes through an integral pipe or coil (e.g. water jacketed tube heater/combi boiler)
and fitted with safety devices to prevent water temperatures exceeding 100°C and other applicable operating devices to control primary flow, prevent backflow, control working pressure and accommodate expansion

1.6   Graphical symbols

See figure 1.5.

Symbols used in this book are, where possible, based on those given in BS 1192: Part 3. However, there are many components not included in BS 1192, so symbols from other sources have been used.

Symbol	Description	ref. BS 1192 EN 806-1	Application
▶	draw-off tap (valve port)	7.3 7.4 6.4.1	
	shower head		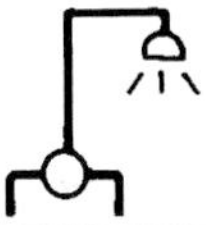
△	sprinkler head (spray outlet)	7.212	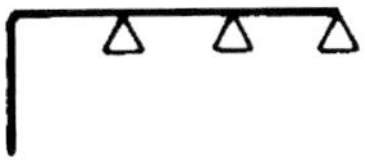
	float-operated valve (ballcock)	7.309	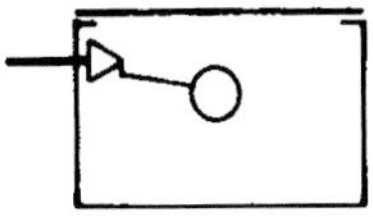
	float switch (hydraulic type)		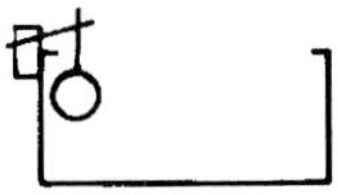
	float switch (magnetic type)		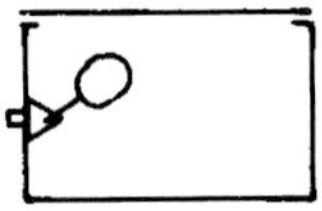
	filter or screen		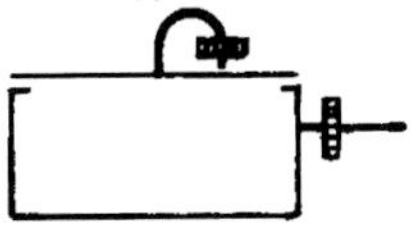
 ⋈	supply stopvalve (SV) (valve)	7.301 6.3.1	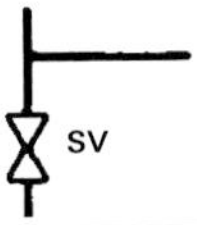
 ▶◀	servicing valve (SV)		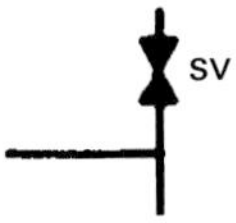
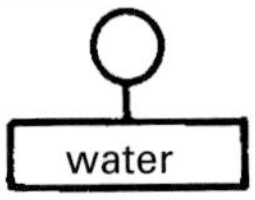 water	water meter	7.226	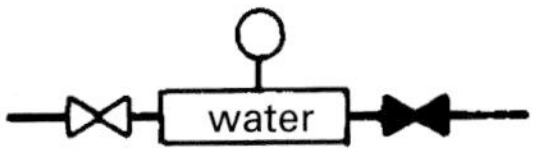

Figure 1.5   Graphical symbols and abbreviations used in this book

Symbol	Description	ref. BS 1192 EN 806-1	Application
⋈	draining valve (BS 1192) (drain valve) (drain cock)	7.305	
⊏	hose connection (used in this book as a draining valve)	7.3E8	
⊤	line strainer	7.218	
⋈	pressure reducing valve (small end denotes high pressure)	7.307	
⊖	expansion vessel	7.610	
⋈	pressure relief valve (expansion relief valve)	7.316	
▷	check valve or non-return valve (NRV)	7.306 6.6.4	nrv
▷▷	double check valve assembly		
⬓	combined check and anti-vacuum valve (check valve and vacuum breaker)		
△	air inlet valve		

Figure 1.5   continued

Symbol	Description	ref. BS 1192 EN 806-1	Application
CWSC	cold water storage cistern (storage and feed cistern) (feed cistern)		CWSC
F&ExC	feed and expansion cistern		F&ExC
HWC	hot water storage cylinder or tank (plan)		HWC
HWC	direct hot water storage cylinder or hot store vessel (elevation)		HWC
HWC	indirect hot water storage cylinder or hot store vessel (elevation)		HWC
	boiler (elevation)		
	temperature relief valve (spring loaded safety valve)	6.6.11	
Y	tundish		Y

Figure 1.5   continued

Symbol	Description	ref. BS 1192 BSEN 806-1	Application
	pump (centrifugal)	7.609	
	pump (circulating)	7.608	
	automatic air vent	7.224	
	insulation	2.209 2.210	
	sink (elevation)		
	wash basin (elevation)		
	bidet (elevation)		
	bath (elevation)		
	water closet (WC) (elevation)		
	urinal bowl (elevation)		

Figure 1.5   continued

BS 1192 is now withdrawn and is replaced by a number of EN standards. Unfortunately it has proved extremely difficult to find a range of EN symbols that are suitable for this book. The author has for the present retained the original symbols and made reference to EN symbols where appropriate.

1.7 Materials

Selection and use of materials are dealt with in greater depth in chapter 11.

Choice of material for a particular water installation may be determined by the following factors:

- effect on water quality;
- cost, service life and maintenance needs;
- internal and external corrosion (particularly from certain waters);
- compatibility of materials;
- ageing, fatigue and temperature effects, especially in plastics;
- mechanical properties and durability;
- vibration, stress or settlement;
- internal water pressures;
- internal and external temperatures;
- permeation.

The water supplier should be consulted at an early stage, particularly about the choice of materials in relation to the character of the water supply and ground conditions. Some waters are aggressive to certain pipes as are certain types of soil. The water supplier should be able to advise on local conditions, and give guidance on the suitability and application of proposed materials.

Pipes and fittings should be used only within the limits stated in relevant British or European Standards and in accordance with any manufacturers' recommendations, and the requirements of Water or Building Regulations should be met.

Manufacturers should identify pipes and components, either on the product or its packaging, with the following information:

- a product standard number, i.e. EN XXX;
- the manufacturer and site of production;
- the year of manufacture;
- the certification body, if any;
- its class (or grade), where applicable; and
- where applicable, suitability for use with potable water.

Installations should be capable of operating effectively under the conditions they will experience in service. Pipes, joints and fittings should be capable of withstanding sustained temperatures as shown in table 1.1, without damage or deterioration.

Pipes, joints and fittings of dissimilar metals should not be connected together unless precautions are taken to prevent corrosion. This is particularly important in below ground installations where conditions are often conducive to corrosion.

Table 1.1   Temperature limits for pipes and fittings

Cold water installations	40°C
Hot water installations	95°C (with occasional short-term excursions up to 100°C)
Discharge pipes connected to temperature or expansion relief valves in unvented hot water systems	114°C

Table 1.2 gives a selection of pipes of various materials along with a range of comparative internal and external diameters.

Detailed information on pipes and fittings can be seen in chapter 11.

Pipe sizes

Since about 1970 many pipe sizes have gradually changed from imperial to metric measurement. Pipes in some materials were metricated quickly, e.g. copper, while others even now retain their imperial identification.

With metrication came a move away from designating pipe sizes by the inside diameter towards the use of the outside diameter.

Currently European Standards are being developed and are replacing many of our existing British Standard specifications.

Table 1.2 provides a comparison between the sizes of pipes of different materials, using inside diameters as a base, because it is the inside diameter which determines the water-carrying capacity of the pipe.

Polyethylene tube, low density to BS 1972 and high density PE tube to BS 3284, has been widely used in the past for applications in farming and agriculture. It was also extensively used for supply pipes to dwellings and other buildings. In the mid 1980s blue medium density tube to BS 6572 and black medium density tube to BS 6730 were introduced. Two main changes occurred at this time: pipe sizes were changed from internal dimensions to external measurement and blue polyethylene tube was introduced to provide ready identification of pipes below ground.

Thousands of pipelines using all of these materials have been installed over the years and will remain in use for many years to come. It is important that information on these pipes is still available because connections and repairs will still need to be made to many of them in the future.

More recently standards for polyethylene pipes have changed again. The above standards are expected to be replaced by BS EN 12201 Parts 1 to 5, which sets out requirements for pipes, valves and fittings. However, for the time being BS 6572 and BS 6730 are still current.

BS EN 805 specifies two series of nominal sizes for pipes and components for water supply. These cover sizes within the range 20 mm to 4000 mm and should be designated by the use of DN. One relates to internal diameter (DN/ID) and the other to external diameter (DN/OD) but for this book extracts are shown up to size 200.

- DN/ID: 20, 30, 40, 50, 60, 65, 80, 100, 125, 150, 200
- DN/OD: 25, 32, 40, 50, 63, 75, 90, 110, 125, 160, 180, 200

Table 1.2 Equivalent pipe sizes

Rigid pipes

Comparative internal size in millimetres	Copper* to BS EN 1057 nominal size ID mm	Copper OD mm	Stainless Steel to BS EN 10312 Series 1 nominal size (DN) ID mm	Stainless Steel OD mm	Steel (screwed) (galvanized or black)* to BS 1387 (medium grade) nominal size ID mm	Steel OD mm	Steel thread designation BS 21 inches	Grey iron to BS 4622 nominal diameter (DN) ID mm	Grey iron OD mm	Ductile iron to BS EN 545 nominal diameter (DN) ID mm	Fibre cement* to BS EN 512: 19 nominal diameter (DN) ID mm	Comparative internal size in inches
4.5	4.8	6	4.8	6	–	–	–	–	–	–	–	$\frac{1}{8}$
6	6.8	8	6.8	8	–	–	–	–	–	–	–	$\frac{1}{4}$
8	8.8	10	8.8	10	8	13.6	$\frac{1}{4}$	–	–	–	–	$\frac{1}{4}$
10	10.8	12	10.8	12	10	17.1	$\frac{3}{8}$	–	–	–	–	$\frac{3}{8}$
13	13.6	15	13.6	15	15	21.4	$\frac{1}{2}$	–	–	–	–	$\frac{1}{2}$
15	16.4	18	16.6	18	–	–	–	–	–	–	–	
20	20.2	22	20.6	22	20	26.9	$\frac{3}{4}$	–	–	–	–	$\frac{3}{4}$
25	26.2	28	26.4	28	25	33.8	1	–	–	–	–	1
32	32.6	35	33	35	32	42.5	$1\frac{1}{4}$	–	–	–	–	$1\frac{1}{4}$
40	39.6	42	39.8	42	40	48.4	$1\frac{1}{2}$	–	–	40	–	$1\frac{1}{2}$
50	51.8	54	51.6	54	50	59.3	2	–	–	50	50	2
63	64.6	67	64.3	66	65	80.1	$2\frac{1}{2}$	–	–	60	60	$2\frac{1}{2}$
75	73.1	76.1	73.1	76.1	80	88.8	3	80	98	80	80	3
100	105	108	105	108	100	113.9	4	100	118	100	100	4
125	130	133	130	133	125	139.6	5	–	–	125	125	5
150	155	159	155	159	150	165.1	6	150†	170	150†	150†	6

Note Some intermediate sizes have been omitted.
* In some materials only one grade is shown.
† Larger sizes have been excluded.

Table 1.2   continued

Flexible pipes

Comparative internal size in millimetres	Unplasticized (PVC-U) to BS 3505 (Class E) nominal size ID inches	OD mm	ID mm	Propylene copolymer to BS 4991 nominal size ID inches	OD mm	Polyethylene to BS 6572 medium density BLUE ID mm	nominal size OD mm	to BS 6730 medium density BLACK ID mm	nominal size OD mm	to BS 1972 low density (withdrawn) Class C OD mm	nominal size ID inches	Comparative internal size in inches
4.5	–	–	–	–	–	–	–	–	–	–	–	⅛
6	–	–	–	–	–	–	–	–	–	–	–	¼
8	–	–	–	¼	13.6	–	–	–	–	–	–	¼
10	–	–	–	⅜	17.1	–	–	–	–	17.1	⅜	⅜
13	⅜	17.1	15.3	½	21.3	–	–	–	–	21.3	½	½
15	½	21.3	17.5	–	–	15.1	20	15.1	20	–	–	–
20	¾	26.7	21.7	¾	26.7	20.4	25	20.4	25	26.7	¾	¾
25	1	33.5	28.3	1	33.5	26	32	26	32	33.5	1	1
32	1¼	42.2	34.8	1¼	42.2	–	–	–	–	42.3	1¼	1¼
40	1½	48.2	41.0	1½	48.3	40.8	50	40.8	50	48.3	1½	1½
50	2	60.3	51.3	2	60.3	51.4	63	51.4	63	60.3	2	2
63	–	–	–	2½	75.3	–	–	–	–	–	–	2½
75	3	88.9	76.5	3	88.9	–	–	–	–	–	–	3
100	4	114.3	97.7	4	114.3	–	–	–	–	–	–	4
125	5	140.2	120	–	–	–	–	–	–	–	–	5
150	6†	168.2†	144†	6†	168.3†	–	–	–	–	–	–	6

Note Some intermediate sizes have been omitted.
* In some materials only one grade is shown.
† Larger sizes have been excluded.
BS 3505 is to be replaced by BS EN 1452.
BS 6572 and BS 6730 are to be replaced by BS EN 12201.

Table 1.2 continued

Flexible pipes (continued)

Comparative internal size in millimetres	Polybutylene (PB) to BS 7291: Parts 1 and 2				Cross-linked polyethylene (PE-X) to BS 7291: Parts 1 and 3				Chlorinated polyvinyl chloride (PVC-C) to BS 7291: Parts 1 and 4		Comparative internal size in inches
	Consistent with BS 5556	nominal size	Consistent with BS 2871	nominal size	Consistent with BS 5556	nominal size	Consistent with BS 2871	nominal size		nominal size	
	ID mm	OD mm	ID mm	OD mm	ID mm	OD mm	ID mm	OD mm	ID mm	OD mm	
6	6.7	10	6.7	10	6.7	10	6.7	10	–	–	$\frac{1}{4}$
8	8.7	12	8.7	12	8.7	12	8.7	12	8.5	12	$\frac{3}{8}$
12	12.7	16	12.2	15	12.7	16	12.2	15	12	16	$\frac{1}{2}$
15	15.9	20	14.2	18	15.9	20	14.2	18	15.8	10	–
20	20.1	25	17.7	22	20.1	25	17.7	22	19.9	25	$\frac{3}{4}$
25	26.1	32	22.5	28	26.1	32	22.5	28	25.6	32	1
32	–	–	28.3	35	–	–	28.3	35	32.2	40	$1\frac{1}{4}$
40	–	–	–	–	–	–	–	–	40.3	50	$1\frac{1}{2}$
50	–	–	–	–	–	–	–	–	50.85	63	2

1.8   Initial procedures

Hot and cold water installations should be designed to conform with Water Regulations (byelaws in Scotland) and with relevant parts of Building Regulations. Consideration should be given to economic maintenance of the installation throughout its working life. Particular attention should be given to the prevention of bacterial contamination in both hot and cold water services. Bacterial contamination is more likely to occur in buildings of multiple occupation. Systems should be designed to avoid trapping of air and formation of air locks during operation. Temperatures of both hot and cold water should be controlled to avoid conditions which encourage bacterial growth. Guidance on the preservation of water quality is given in chapter 6.

Attention should also be given to the insulation and control of temperatures for the conservation of energy.

Preliminary design factors are as follows:

- flow requirements and estimated likely consumption;
- location of available water supply;
- quality, quantity and pressure of available supply;
- quantity of storage required;
- requirements of Building Regulations and Water Regulations and particularly the need for notification of work;
- ground conditions, e.g. subsidence/contamination of site;
- liaison with other parties;
- the specific requirements of the water supplier;
- the likelihood of transient and surge pressures during system operation.

The design should include provision for appliances likely to be added later.

Availability of water supplies

Water supplies are available from:

(1)   Nearby public water main at cost to owner or within contract price. If not readily available, the water undertaker must provide mains at the expense of the owner or applicant. It is important that the water supplier is consulted at an early stage.
Note   Mains may not be laid until road line and kerb level are permanently established.

(2)   Suitable and available supply pipe (not favoured by water suppliers who generally insist that each premises has a separate supply pipe).

(3)   Private source. The condition and purity of the water should be considered. Chemical and bacterial analyses are advisable before the source is used. Approval is needed from the local public health authority for drinking water supplies, and a licence to abstract may be required from the water authority.
Note 1   If private supply and public supply are taken to a single property, the water undertaker must be informed and regulations complied with. Water from a private source must not be connected to a supply pipe from a water supplier's main.
Note 2   The water supplier may require the supply to be metered.

(4)   Water supplies may be supplemented by the use of reclaimed waste water or rain-water collected from roofs provided that precautions are taken for the prevention of contamination.

Water suppliers will need full details of non-domestic water supply requirements to assess the likely demand and effects on water mains and other users in the locality (see figure 1.6).

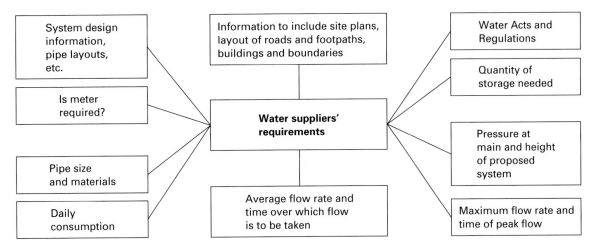

Figure 1.6   Water suppliers' requirements

Ground conditions to be considered

(1)   Likelihood of contamination of site (local authorities may provide information).
(2)   Likelihood of subsidence and other soil movement from:
    (a)   mining;
    (b)   vibration from traffic (consider increased depth);
    (c)   moisture swelling and contraction;
    (d)   building settlement.

Laying pipes outside buildings

When laying underground service pipes make allowances for pipe materials, jointing and methods of laying. Consider methods of passing pipes through walls. Pipes must be free to deflect.
    As far as possible underground pipes:

• should be laid at right angles to the main;
• should be laid in straight lines to facilitate location for repairs, but with slight deviation to adjust to minor ground movement;
• should not be laid under surfaced footpaths or drives;
• should be laid at a minimum depth of 750 mm to avoid frost and other damage, and not deeper than 1350 mm to permit reasonable access.

Service pipe diameters should be determined by hydraulic calculation. Flow velocities should preferably be kept within the range 0.5 m/s to 2 m/s but in exceptional circumstances a maximum velocity of 3.5 m/s may be allowed. To avoid undue stresses in the pipeline due to thermal movement, consideration should be given to the temperature of the pipeline when installed compared with the expected temperature when filled and in use. Provision should be made for the flushing and removal of air and for the testing of the pipeline. Where applicable the pipeline should be sterilized and samples taken for chemical and bacterial analyses.

The use of external pipes above ground should be avoided. Where unavoidable, pipes should be lagged with water-proofed insulation and provision made to drain pipes as a precaution against frost.

Liaison and consultation

From an early stage in the design process, the designer should consult with others involved in the design, installation and use of the system (see figure 1.7).

Where work is to be carried out in a public highway, the highway authority and all private/public utility undertakers should be notified and all relevant notices completed and lodged with those likely to be affected. These notices should include any drawings and other details of work to be done. Particular attention should be given to the submission of notices to water undertakers and to building control bodies, to whom notice should be given before commencement of the work. Designers and installers should be aware of requirements under both Water and Building Regulations for the notification of work, and of conditions relating to the self-certification of work.

The designer or installer should provide full working drawings, including precise location of all pipe runs, method of ducting, description of all appliances, valves and other fittings, methods of fixing, protections and precautions.

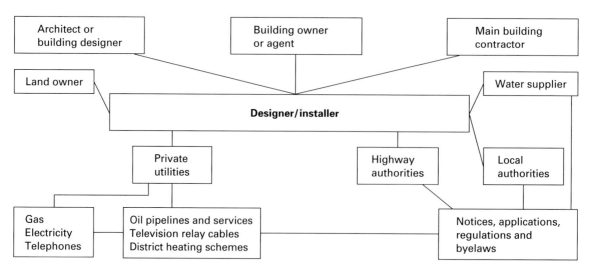

Figure 1.7   Designers'/installers' liaisons and consultations

Water pipes of metal should be arranged to permit equipotential bonding to the main electrical earth.

The programme of work should consider:

- method of construction;
- sequence of events including handover to owner;
- coordination of services;
- time needed for construction and services works;
- size and position of incoming services.

Water services to buildings should be coordinated with other services and laid in an orderly sequence and at a line and level that will readily permit maintenance at a later date. On new sites the recommendations of the *National Utilities Group Publication No 6* should be followed.

Chapter 2
Cold water supply

2.1 Drinking water

Under the Water Supply (Water Fittings) Regulations 1999 every dwelling is required to have a wholesome water supply, and this should be provided in sufficient quantities for the needs of the user, and at a temperature below 20°C. There may be occasions during summer months when mains water could rise as high as 25°C. Precautions should be taken to make sure that this temperature is not exceeded. The most important place to provide drinking water in dwellings is at the kitchen sink (see figure 2.1). However, because there is a likelihood that all taps in dwellings will be used for drinking, they should all be connected in such a way that the water remains in potable condition. This means that all draw-off taps in dwellings should either be connected direct from the mains supply, or from a storage cistern that is 'protected'. To avoid confusion and to protect disabled and blind persons, the hot tap should be positioned on the left and the cold on the right, as viewed by the user.

Drinking water supplies should also be provided in suitable and convenient locations in places of work such as offices and commercial buildings, particularly where food and drink are prepared or eaten. If no such locations exist, drinking water should be provided near but not in toilets. However, drinking water fountains may be installed in toilet areas, provided they are sited well away from WCs and urinals and comply with the requirements of BS 6465: Part 1.

Where drinking water taps are fitted in places of work along with taps for other purposes, all taps should be labelled 'drinking water' or 'non-drinking water' as appropriate.

To avoid stagnation drinking water points should not be installed at the ends of long pipes where only small volumes of water are likely to be drawn off, but rather be fitted where there is a high demand downstream.

As far as possible pipe runs to drinking water taps should not follow the routes of space heating or hot water pipes or pass through heated areas. Where this is unavoidable, both hot and cold pipes should be insulated.

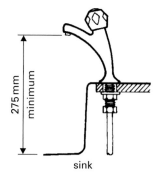

Positioned so that outlet is at least 275 mm above bottom of bowl, to allow buckets and other utensils to be filled.

Outlets designed to make hosepipe connections difficult.

Hot tap on the left, cold on the right.

Figure 2.1 Tap at kitchen sink

In tall buildings where drinking water is needed above the pressure limit of the mains water supply, it may be necessary for the water to be pumped to the higher level. In such cases any drinking water must be supplied from a 'protected' drinking water cistern or from a drinking water header (see later in figures 2.36 and 2.37).

Where a water softener of the base exchange type is installed, it is recommended that connections for drinking water are made upstream of the softener so they do not receive softened water. The Department of Health recommends that unsoftened drinking water should be available for infant feed preparation and to assist recommendations to reduce sodium intake in the general population.

2.2   Cold water systems

Systems in dwellings

Water may be supplied to cold taps either directly from the mains via the supply pipe or indirectly from a protected cold water storage cistern. In some cases a combination of both methods of supply may be the best arrangement.

A supply 'direct' from the mains is preferred because water quality from storage cannot be guaranteed. However, pressure reliability of the mains supply should be considered especially where connections are made near to the ends of distributing mains. Where constant supply pressure may be a problem, storage should be considered.

Factors to consider when designing a cold water system should take into account the available pressure and reliability of supply, particularly where any draw-off point is at the extreme end of a supply pipe or situated near the limit of mains pressure.

Figures 2.2 and 2.3 give characteristics of 'direct' and 'indirect' cold water systems and should be useful when considering a system for a particular application.

Systems in buildings other than dwellings

In the case of small buildings where the water consumption is likely to be similar to that of a dwelling, the characteristics in figures 2.2 and 2.3 should be considered.

For larger buildings such as office blocks, hostels and factories it will usually be preferable for all water, except drinking water, to be supplied indirectly from a cold water storage cistern or cisterns. Drinking water should be taken directly from the water supplier's main wherever practicable (see figure 2.4).

Pumped systems

Where the height of the building lies above statutory levels or when the available pressure is insufficient to supply the whole of a building and the water supplier is unable to increase the supply pressure in the supplier's mains, consideration should be given to the provision of a pump. Where the pump delivers more than 0.2 l/s the water undertaker must be notified and written consent obtained before the pump is fitted.

It is important that the water supplier is consulted and written consent received before fitting any pump. The water supplier will wish to ensure that Water Regulations are complied with in respect of backflow risk, and that any pump will not have adverse effects on the mains and other users.

Cistern omitted where hot water is supplied from: (a) an unvented hot water system; or (b) mains-fed instantaneous water heater.

Higher pressure supply is more suitable for instantaneous type shower heaters, hose taps and mixer fittings used in conjunction with a high pressure (unvented) hot water supply.

Expensive dual-flow mixer fittings required if used in conjunction with a low pressure (vented) hot water supply.

All taps are supplied under mains pressure and quality is assured for drinking and food preparation.

Pumped supply may be needed where mains pressures are very low.

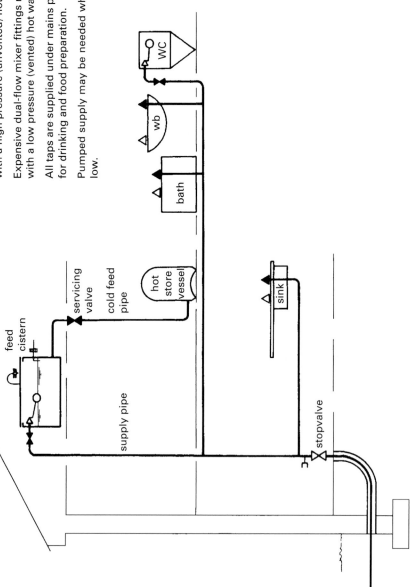

Figure 2.2 Characteristics of cold water supply direct from supply pipe

Risk of frost damage in roof space.

Reserve supply of water available in case of mains failure.

Cistern provides additional protection against contamination of mains.

Pressure available from storage may not be sufficient for some types of tap or shower, e.g. power shower.

Cistern must be continuously protected against the entry of any contaminant; cistern may need to be replaced occasionally.

Space occupied and cost of cistern, structural support and additional pipework must be considered.

Constant low supply pressure reduces the risk and rate of leakage and is suitable for supply to mixer fittings for low pressure (vented) hot water supply.

At least one tap supplied directly from supply pipe for drinking and cooking purposes.

Reduced risk of water hammer and noise from outlets, but additional noise may be generated by the float-operated valve controlling the supply into the cistern.

Drinking water quality at kitchen sink without the need to pass through storage cistern.

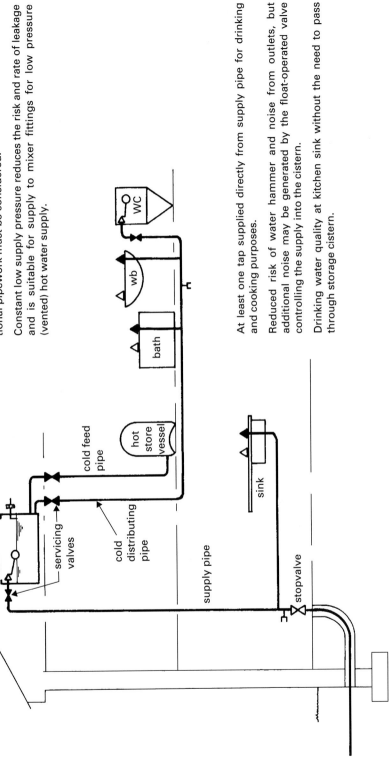

Figure 2.3 Characteristics of cold water supply via storage cistern

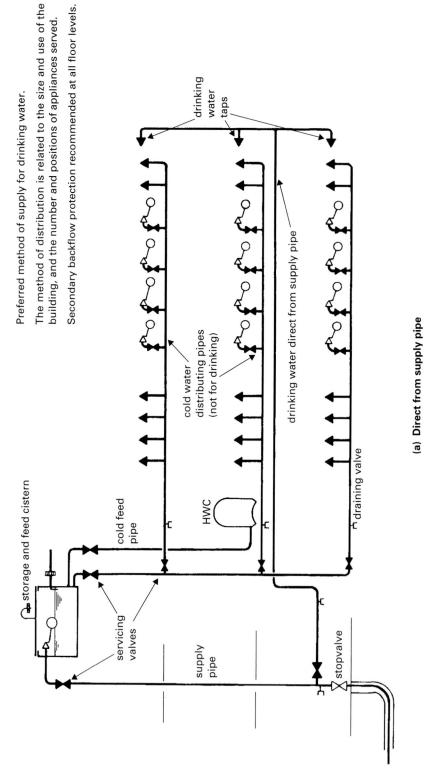

Preferred method of supply for drinking water.

The method of distribution is related to the size and use of the building, and the number and positions of appliances served.

Secondary backflow protection recommended at all floor levels.

drinking water taps

cold water distributing pipes (not for drinking)

drinking water direct from supply pipe

(a) Direct from supply pipe

storage and feed cistern

cold feed pipe

HWC

servicing valves

supply pipe

draining valve

stopvalve

Figure 2.4 Drinking water supply

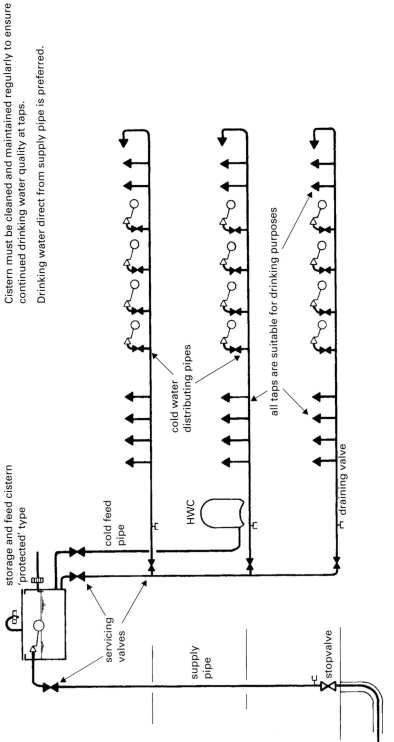

For use where a separate drinking water supply cannot be provided.

Secondary backflow protection recommended at all floor levels.

All taps provide water of drinking quality.

Cistern must be cleaned and maintained regularly to ensure continued drinking water quality at taps.

Drinking water direct from supply pipe is preferred.

storage and feed cistern 'protected' type

cold feed pipe

HWC

all taps are suitable for drinking purposes

cold water distributing pipes

draining valve

servicing valves

supply pipe

stopvalve

(b) **Alternative system from storage**

Figure 2.4   continued

2.3 Storage cisterns

Drinking water storage cisterns and lids, which include all cisterns used for domestic purposes, should not impart taste, colour, odour or toxicity to the water, nor promote microbial growth.

Cisterns for use with cold water should be listed in the *Water Fittings and Materials Directory* or conform to one of the following British Standards specifications:

- BS 417 Galvanized low carbon steel cisterns, cistern lids, tanks and cylinders. Part 2: Metric units.
- BS 1563 Cast iron sectional tanks (rectangular).
- BS 1564 Pressed steel sectional rectangular tanks.
- BS 1565 Specification for galvanized mild steel indirect cylinders, annular or saddle-back type.
- BS 4213 Specification for cold water storage and combined feed and expansion cisterns (polyolefin or olefin copolymer) up to 500 l capacity used for domestic purposes.

Cistern requirements

Cistern materials should be suitable for maintaining potable water quality and must not deform unduly in use. Cisterns should be resistant to corrosion or coated with a material that is suitable for use with potable water. All cisterns and pipes should be situated away from heat and insulated against the effects of frost or heat.

Domestic cisterns and those supplying drinking water should be 'protected' to prevent contamination of water and should comply with the requirements of the Water Fittings Regulations (see figure 2.5). Cisterns should be supplied from a supply pipe or pumped from a cistern that is 'protected'.

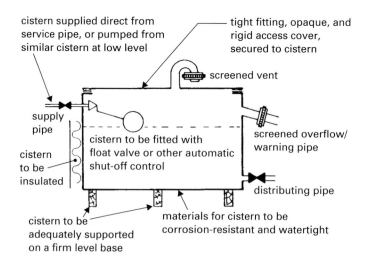

(a) General requirements for cisterns

Figure 2.5   Cistern requirements

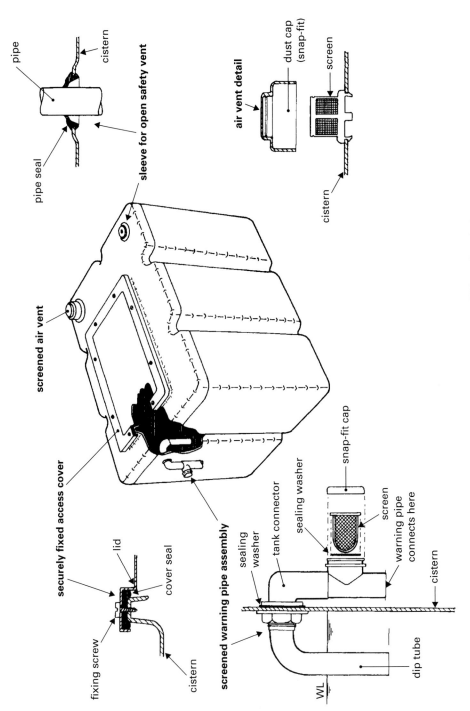

pipe

cistern

pipe seal

sleeve for open safety vent

air vent detail

dust cap (snap-fit)

screen

cistern

screened air vent

securely fixed access cover

lid

cover seal

fixing screw

cistern

screened warning pipe assembly

sealing washer

tank connector

sealing washer

snap-fit cap

screen

warning pipe connects here

cistern

WL

dip tube

(b) Cistern to meet the requirements of Water Regulations

Figure 2.5   continued

Cisterns should be fitted with warning pipes and overflow pipes as appropriate, and a vent to permit air movement within the cistern. These should be screened to prevent insects from entering the cistern to contaminate the water.

Construction of cisterns should permit easy access to the interior of the cistern and inlet valve for inspection, maintenance and cleansing. Lids should be rigid, close fitting and securely fixed, and should fit closely around any vent pipe or air inlet pipe.

All water cisterns for domestic purposes must be of the 'protected' type, on the grounds that water is likely to be drunk from all taps in dwellings. This is a departure from past practice which should greatly improve cisterns hygienically.

All cisterns should be supported on a firm level base capable of withstanding the weight of the cistern when filled with water to the rim. Where cisterns are located in the roof space, the load should be spread over as many joists as possible (see figure 2.6).

Occasionally large cisterns are buried or sunk in the ground. In these cases measures need to be taken to detect leakage and to protect the cistern from contamination (see figure 2.7).

(a) Flexible cisterns   **(b) Rigid cisterns**

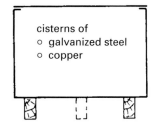

plastics cisterns, e.g.
   o polypropylene
   o polyethylene
   o glass reinforced
     polyester

cisterns of
   o galvanized steel
   o copper

Continuous support needed over whole base area.

Two or more timber joists depending on size of cistern.

No connections to be made to base of plastics cisterns.

Continuous support not needed, and is undesirable for galvanized steel cisterns.

Figure 2.6   Support for cisterns

Cistern capacities

When calculating cold water storage capacities, the designer should take account of:

- the need to prevent stagnation by ensuring that water is stored for as short a time as possible;
- the requirements of any associated water-using fittings and appliances, particularly where supply interruptions could cause damage to property or inconvenience to the consumer;
- the likely pattern of water use (draw-off rates and their durations);
- the requirements of the water undertaker, who should be consulted before finalizing cistern capacities, particularly to non-domestic premises. The water undertaker should be able to advise on any local conditions such as low mains pressures that are likely to affect cistern refilling at times of peak demand.

The installation should include:
- o an electrical sump pump to automatically remove any water that might build up in the chamber due to the cistern overflowing, or through ingress of surface water;
- o an audible or visual warning device to indicate when the cistern water reaches overflowing level;
- o an audible or visual warning to show if cistern water or surface water is building up in the chamber.

Depending on ground conditions, cisterns may need to be anchored to prevent them lifting if the chamber becomes flooded and the cistern is empty or partially full.

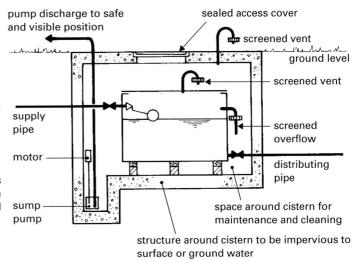

Figure 2.7   Sunken cistern

Where storage is a requirement under the Water Industry Act 1991, a maximum capacity of 80 l per person normally resident will probably prove satisfactory. A larger capacity based on 130 l per person would be appropriate where cistern refilling normally takes place only during the night hours.

Note   Under the 1991 Act the water undertaker may require storage to be provided in premises:

- • where constant pressure cannot or need not be maintained; or
- • where the water is delivered at a point higher than 10.5 m below the draw-off level of the water supplier's reservoir or tank.

In general the following storage recommendations should suffice:

Smaller houses	cold water outlets only	100 l to 150 l
	hot and cold outlets	200 l to 300 l
Larger houses	per bedroom	80 l where adequate supply pressures are guaranteed
		130 l where refilling only takes place at night

See section 5.3 for more detailed cistern sizing.

Cistern control valves

Every pipe supplying water to a cistern should be fitted with a float-operated valve or some other equally effective device to control the inflow of water and to maintain it at the required level.

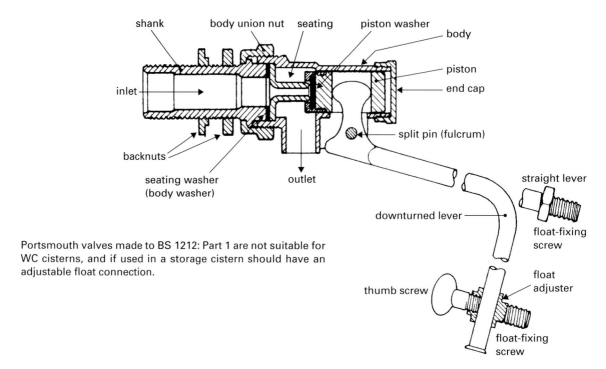

Portsmouth valves made to BS 1212: Part 1 are not suitable for WC cisterns, and if used in a storage cistern should have an adjustable float connection.

Figure 2.8   Portsmouth type float valve to BS 1212: Part 1 showing alternative lever arms

Float valves should comply with BS 1212: Parts 1 to 4 (see figures 2.8, 2.9 and 2.11) and be used with a float complying with BS 1968 or BS 2456 of the correct size corresponding to the length of the lever arm and the water supply pressure.

Float valves should be clearly marked with the water pressure, temperature and other characteristics for which they are intended to be used.

This is the preferred type for use in water storage cisterns.

May be used in any situation that requires the use of a float-operated valve.

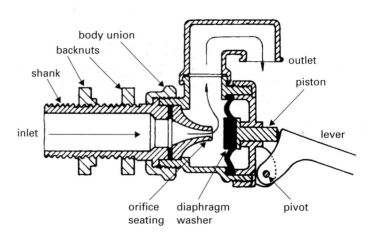

Figure 2.9   Diaphragm float valve to BS 1212: Part 2 (brass) and Part 3 (plastics)

Every float-operated valve or other inlet device should be securely fixed to the cistern and, where necessary, braced to prevent any movement of the float or cistern wall which might affect the inflow to the cistern, or cause noise.

Where a non-British Standard float-operated valve or other level control device is used, it must meet the requirements of Water Regulations and be listed in the *Water Fittings and Materials Directory* produced by the Water Regulations Advisory Scheme (WRAS) or other similar fittings approval scheme. Figures 2.10 and 2.12 illustrate two types of non-British standard valves.

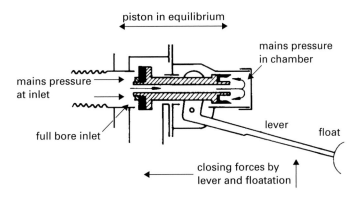

Figure 2.10   Portsmouth equilibrium float valve

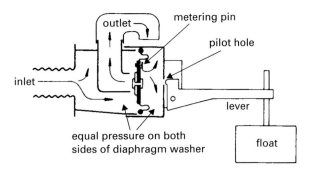

BS 1212: Part 4 covers this type of valve which is designed
primarily for use in WC flushing cisterns.

Figure 2.11   Diaphragm equilibrium float valve

Used to provide a full flow of water at all times, it completely eliminates 'dribble conditions' often associated with a conventional ball valve; this valve is particularly applicable to automatic pumping and booster systems and water treatment. 'Arclion' is a registered trade name of H. Warner & Son Ltd.

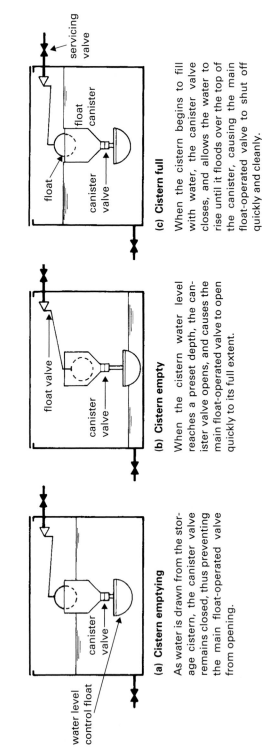

(a) Cistern emptying

As water is drawn from the storage cistern, the canister valve remains closed, thus preventing the main float-operated valve from opening.

(b) Cistern empty

When the cistern water level reaches a preset depth, the canister valve opens, and causes the main float-operated valve to open quickly to its full extent.

(c) Cistern full

When the cistern begins to fill with water, the canister valve closes, and allows the water to rise until it floods over the top of the canister, causing the main float-operated valve to shut off quickly and cleanly.

Overflow/warning pipes not shown.

Figure 2.12 Arclion® delayed action float valve

Warning/overflow pipe must be large enough to carry away leakage under worst conditions.

Dimensions are in millimetres.

Cistern must comply with Water Regulations if it is used for domestic purposes.

Float valve connection should be braced to prevent movement by water pressure thrust.

Dimension 'A'. Distributing pipe connection should be as low as is practically possible.

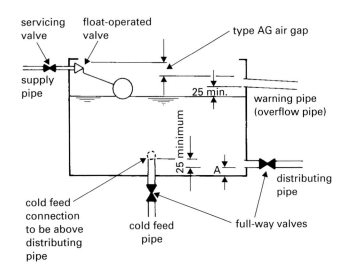

(a) Small cisterns (up to 1000 l capacity)

Washout connection flush with base of cistern at its lowest point.

Washout pipe to be plugged when not in use and must discharge to open air at least 150 mm above any drain.

Overflow pipe must be large enough to carry away leakage under worst conditions.

Cold feed and cold distributing pipes to be fitted with corrosion-resistant strainers.

Cold feed pipe to be above distributing pipe.

Dimensions are in millimetres.

Cistern must comply with Water Regulations if it is used for domestic purposes.

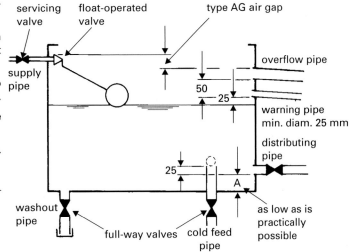

(b) Large cisterns (over 1000 l capacity)

Figure 2.13 Connections to cisterns

Connections to cisterns

Connections to cisterns will typically be similar to that shown in figures 2.13 and 2.14. It is advised that all connections be made at prescribed distances above the base of cisterns to permit debris particles to settle out rather than pass into the water system and possibly cause problems in tap seatings or other fittings with moving parts.

To avoid *Legionella* cisterns should:
 o be small enough to ensure a rapid turnover and thus prevent stagnation;
 o have float valves arranged to open and close together;
 o have inlet and outlet connections at opposite ends;
 o be regularly inspected and maintained in clean condition;
 o conform to the requirements of Water Regulations.

The diagram shows separate overflow pipes. A common overflow pipe or warning pipe may be fitted provided cisterns are linked to form one storage unit.

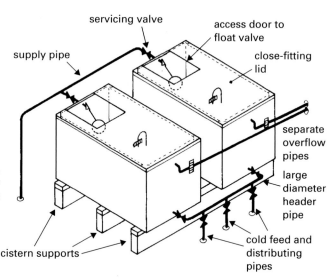

(a) Cisterns connected in parallel

This method may be preferred for prevention of *Legionella.*

To take cistern 1 out of commission for cleansing:
 o fit temporary connection between link (b) to distributing pipe;
 o remove link (a) and cap off close to cisterns.

To take cistern 2 out of commission for cleansing
 o fit temporary connection to float valve in cistern 1 and disconnect branch pipe to float valve in cistern 2;
 o remove links (a) and (b) and cap off near cisterns.

Sterilize any pipes and fittings used before they are fitted.

Sterilize cistern and pipes before putting back into service.

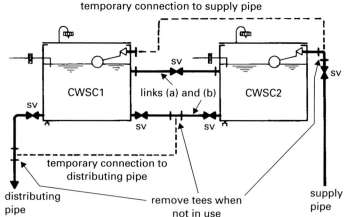

(b) Cisterns connected in series

Figure 2.14   Linked cisterns

It is thought that to reduce the risk of contamination with *Legionella* and other similar water-borne diseases, outlet connections should be positioned as near to the bottom of the cistern as is reasonably practical to enable small particles to pass through the system rather than to cause unhealthy accumulations at the base of cisterns. This should not, however, encourage connections of pipes to the underside of any cistern, particularly those of flexible materials that need to be fully supported over the whole base area.

Outlet connections to hot water apparatus, for example cold feed pipes, should be positioned 25 mm above connections to cold water fittings. This is especially important where mixer fittings and showers are fitted and will, if the storage cistern should fail, enable the hot water to shut off before the cold and thus reduce the risk of scalding.

Pipework to and from cisterns

Supply and distributing pipes to and from cisterns should comply with the following recommendations.

(1) All pipework should be insulated to reduce heat losses and gains, to minimize frost damage, and to prevent condensation.
(2) Pipes should preferably be laid to a fall to reduce the risk of air locks and to facilitate filling and draining.
(3) Pipes should be closely grouped for neatness, but not so close as to gain heat from one another.
(4) Pipes should be securely fixed and adequately supported.
(5) A cold feed pipe should not be used other than to supply the hot water apparatus for which it is intended.

Linked cisterns

On occasion, to provide large quantities of water storage, or because of space restrictions, two or more cold water storage cisterns may need to be linked. Figure 2.14 shows two methods which permit cisterns to be cleaned, repaired or replaced without interruption of supplies to the building.

Cisterns mounted outside buildings

Whether fixed to the building itself or supported on an independent structure, cisterns outside buildings, see figure 2.15, should be enclosed in a well-ventilated, yet draught-proof housing. This should be constructed to prevent the entry of birds, animals and insects, but provide access to the interior of the cistern for maintenance. Ventilation openings should be screened by a corrosion-resistant mesh with a maximum aperture size of 0.65 mm.

Large cisterns

Generally these provide over 5000 l storage and are often made up of preformed panels or are constructed of concrete. They should preferably be divided into two or more

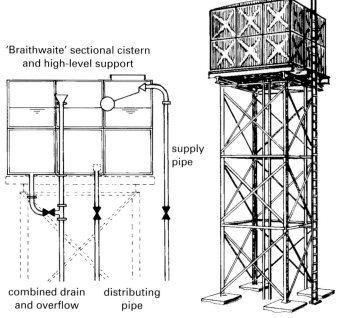

'Braithwaite' sectional cistern
and high-level support

supply
pipe

Washout fitted flush with base of
cistern, and base of cistern laid to
fall to assist draining and cleanup.

combined drain   distributing
and overflow   pipe

Cistern and pipes to be enclosed and insulated against frost.

Enclosure to be ventilated, but draughtproof, and arranged to
prevent entry by birds, animals and insects.

Overflow/drain pipe must terminate in a conspicuous position
above ground level.

Figure 2.15   Typical exterior storage cistern

compartments to avoid interruption of the water supply when carrying out repairs or
maintenance to the cistern.

Warning and overflow pipes

Warning and overflow pipes (see figure 2.16) serve two purposes:

(1)   to give warning that inlet valve to cistern has failed to close,
(2)   to remove safely from the buildings any water which leaks from the inflow pipe.

On cisterns of less than 1000 l capacity, one pipe will serve both as an overflow pipe and a
warning pipe, but on larger cisterns it may be necessary to have a separate pipe for each
function (see figures 2.16–2.18).

Note   For Water Regulations purposes cistern capacities are measured to the level at
which the water starts to flow into the overflow pipe.

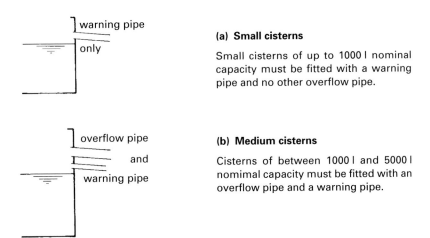

(a) **Small cisterns**

Small cisterns of up to 1000 l nominal capacity must be fitted with a warning pipe and no other overflow pipe.

(b) **Medium cisterns**

Cisterns of between 1000 l and 5000 l nomimal capacity must be fitted with an overflow pipe and a warning pipe.

Cisterns of more than 5000 l nominal capacity may have other warning devices fitted in place of the warning pipe (see figures 2.17 and 2.18).

Figure 2.16 Warning and overflow pipes for cisterns

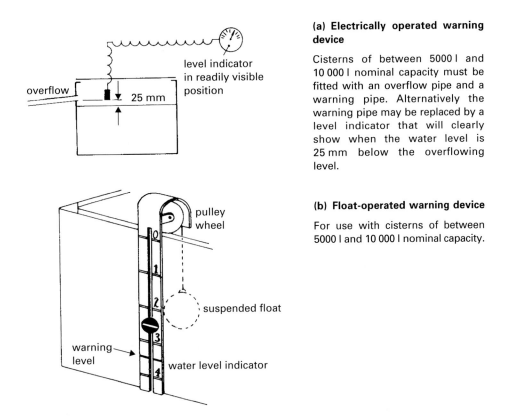

(a) **Electrically operated warning device**

Cisterns of between 5000 l and 10 000 l nominal capacity must be fitted with an overflow pipe and a warning pipe. Alternatively the warning pipe may be replaced by a level indicator that will clearly show when the water level is 25 mm below the overflowing level.

(b) **Float-operated warning device**

For use with cisterns of between 5000 l and 10 000 l nominal capacity.

Figure 2.17 Overflow and warning arrangements for large cisterns

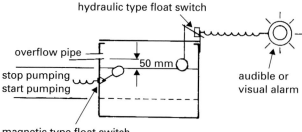

hydraulic type float switch

overflow pipe

50 mm

stop pumping

start pumping

audible or
visual alarm

magnetic type float switch

(a) Electrically operated alarm

Cisterns of more than 10 000 l nominal capacity must be fitted with an overflow pipe and a warning pipe. Alternatively the warning pipe may be replaced by an audible or visual alarm that will clearly show when the water level is 50 mm below the overflowing level.

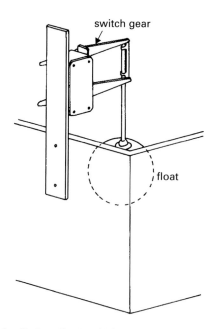

switch gear

float

(b) Hydraulic type float switch

Float switch can be used to operate an audible or visual alarm. Also used for water level control.

Figure 2.18   Overflow and warning arrangements for very large cisterns

The following recommendations for warning pipes and overflow pipes are noted (see figures 2.19–2.21).

(1)   Overflow and warning pipes should be made of rigid, corrosion-resistant material.
(2)   Overflow and warning pipes fitted to feed and expansion cisterns must be of metal, cross-linked polyethylene, polybutylene or chlorinated PVC, and be able to resist heat.

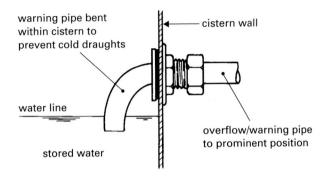

Figure 2.19   Overflow/warning pipe

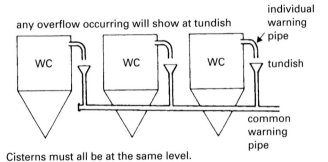

Cisterns must all be at the same level.

Individual warning pipes must terminate over tundish before connection to common pipe, so that any discharge is readily visible.

Common warning pipe must terminate in a conspicuous position.

Figure 2.20   Common warning pipes to WCs

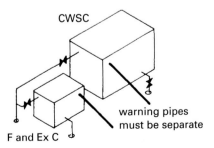

Figure 2.21   Warning pipes from feed and expansion cisterns

(3)   The overflow pipe or pipes should be able to carry away all water which is discharged into the cistern in the event of the inlet control becoming defective, without the water level reaching the spill-over level of the cistern, or submerging the discharge orifice of the inlet pipe or valve. Every overflow pipe shall discharge immediately the water in the cistern reaches the overflowing level (invert of overflow pipe).

(4)   Warning and overflow pipes must fall continuously to points of discharge.

(5)   Where overflow and warning pipes discharge through the external wall of a building they should be arranged to prevent cold draughts by turning down the warning pipe into the cistern and below the water line (see figure 2.19).

(6)   Every warning pipe should discharge in a conspicuous position, preferably outside the building where this is appropriate. In some circumstances warning pipes may be permitted to terminate into a special bath overflow fitting.

(7)   For cisterns at or above ground the overflow/warning pipe should terminate at least 150 mm above the surface of the ground.

(8)   For cisterns below ground level, the outlet of any warning pipe or combined warning and overflow pipe should be arranged to give warning of an overflow. Cisterns sunk in the ground shall have special measures to detect leakage, e.g. electronic warning devices.

(9)   Warning pipes from more than one similar cistern may be linked together to one common outlet providing they cannot discharge one into the other, and as long as the source of any overflow can be readily identified (see figure 2.20).

(10) Warning pipes from feed and expansion cisterns must be separate from those from storage cisterns.

(11) When a single overflow pipe or warning pipe is used, it should have a minimum diameter of 19 mm, or be at least one size larger than the cistern inlet pipe, whichever is the greater.

2.4   Valves and controls

Water Regulations require the use of three types of control valve on water installations within buildings: stopvalves, servicing valves and isolating valves.

Stopvalves

A stopvalve must be provided on the supply pipe to any premises occupied as a dwelling, or in every building, or part of a building, which is separately chargeable. The stopvalve must control the whole of the supply to those premises without shutting off the supply to any other premises. It must be:

* installed within the building or premises to which it is supplied;
* in an accessible position above floor level; and
* near to the point of entry of the pipe.

Where separate premises are supplied with water from a remote or communal cistern the rules are equally applied to the distributing pipe. Stopvalves must be readily accessible to users of the premises and should not be installed in any part of a building to which the users do not have access. Suitable types of stopvalve are illustrated in figure 2.22.

Stopvalves to BS 1010 are for above ground use only.

Stopvalves to BS 5433 for underground use are of heavier quality and made of corrosion-resistant material such as gunmetal or DZR brass and have an enclosed (shrouded) washer.

Washer plates should be 'fixed' so as to open mechanically when valve spindle is screwed to open position.

Illustration shows stopvalve to BS 1010 with type A compression joint for copper.

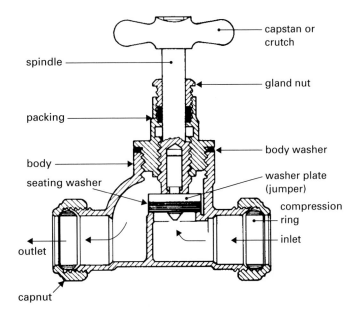

(a) Screwdown stopvalve to BS 1010 and BS 5433

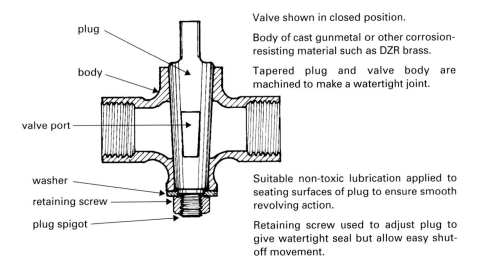

Valve shown in closed position.

Body of cast gunmetal or other corrosion-resisting material such as DZR brass.

Tapered plug and valve body are machined to make a watertight joint.

Suitable non-toxic lubrication applied to seating surfaces of plug to ensure smooth revolving action.

Retaining screw used to adjust plug to give watertight seal but allow easy shut-off movement.

Open area of valve port to coincide with waterway in valve body to give full flow with minimum resistance.

Valve connections shown are for threaded joint to BS 21.

Valves are available with ends for other materials, i.e. copper, polyethylene.

(b) Underground plug cock to BS 2580

Figure 2.22   Stopvalves

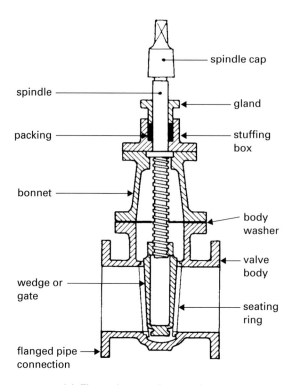

spindle cap

spindle

gland

packing

stuffing box

bonnet

body washer

valve body

wedge or gate

seating ring

flanged pipe connection

(c) Flanged gate valve to BS 5163

Figure 2.22   continued

Stopvalve positions and supply pipe arrangements

It is advisable that every separate premises or building that is supplied with water, be fitted with a separate isolating stopvalve close to its branch connection so that the whole of each branch pipe is fully controlled.

It should be possible to turn off the supply to premises in an emergency or for repair without affecting the supply to any other premises.

There are various methods of supply to premises that allow proper control of supply pipes and these are shown in figures 2.23–2.25.

It should be borne in mind that the water undertaker may not permit the use of joint supply pipes. If these are planned the water undertaker's advice should be sought.

When a stopvalve is installed on an underground pipe it should be enclosed in a pipe guard under a surface box (see figures 10.5 and 11.63–11.67).

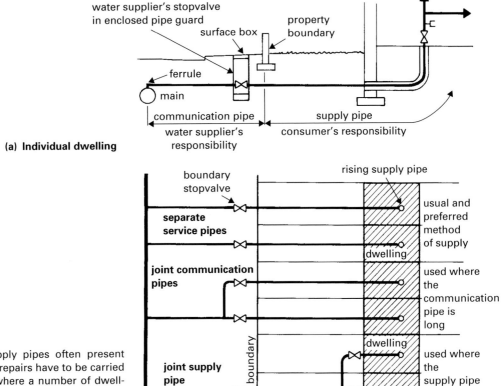

(a) **Individual dwelling**

Note Joint supply pipes often present problems when repairs have to be carried out, especially where a number of dwellings are supplied. For this reason, joint supply pipes are not favoured by water authorities, who will usually insist on separate service pipes.

(b) **Methods of supplying more than one dwelling**

Figure 2.23   Stopvalves to dwellings

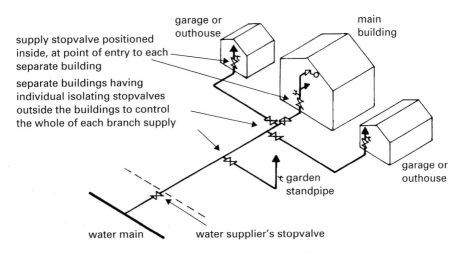

Figure 2.24   Stopvalves for premises having separate buildings

A similar arrangement may be necessary where separate premises are supplied from a common distributing pipe from a common cold water storage cistern at high level (see figure 6.23b).

Backflow protection may be required at each floor level.

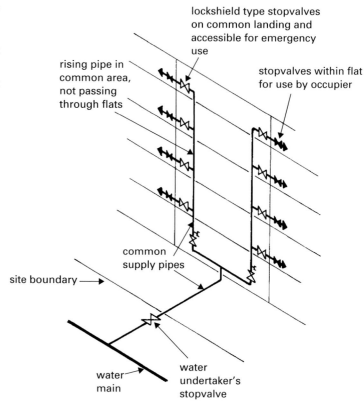

lockshield type stopvalves on common landing and accessible for emergency use

rising pipe in common area, not passing through flats

stopvalves within flat for use by occupier

common supply pipes

site boundary ⟶

water main

water undertaker's stopvalve

(a) Internally separated system

Figure 2.25   Common supply pipes to flats

Servicing valves

Servicing valves must be provided in accessible positions to enable the flow of water to individual or groups of appliances to be controlled and to limit inconvenience caused during maintenance and repair (see figures 2.26 and 2.27).

Traditionally gate valves have been used for this purpose rather than screwdown valves because of their lower hydraulic resistance. In recent years spherical plug valves have been introduced and are particularly suited to control individual fittings or appliances.

Servicing valves must be fitted:

- immediately before every float valve connected to a service pipe or distributing pipe;
- to every pipe carrying water from a hot water storage cylinder, tank or cistern;
- on every cold feed and distributing pipe from any feed cistern or storage cistern of more than 18 l capacity.

However, cold feed pipes to primary hot water circuits should not be fitted with a servicing valve.

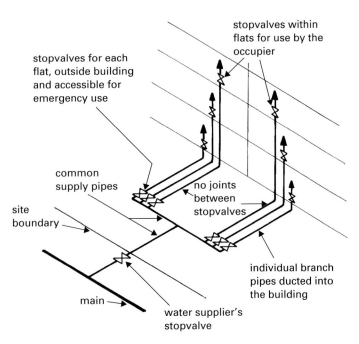

stopvalves within flats for use by the occupier

stopvalves for each flat, outside building and accessible for emergency use

stopvalves within flats for use by the occupier

common supply pipes

no joints between stopvalves

site boundary

individual branch pipes ducted into the building

main

water supplier's stopvalve

(b) Externally separated system

The externally separated system is the method preferred by water authorities who may also insist on a separate mains connection for each flat

Figure 2.25   continued

Stopvalves

Stopvalves that meet the requirements of Water Regulations.

Up to 50 mm diameter:
  o screwdown stopvalves to BS 5433
  o plug cock to BS 2580
  o screwdown stopvalves to BS 1010-2
     (above ground use only)

Above 50 mm diameter:
  o flanged gate valve to BS 5163-2
     (may be used above or below ground)

Servicing valves and isolating valves

  o any of the stopvalves listed above, or
  o wheel operated (gate) valve to BS 5154
  o lever-operated, spherical plug valve to
     BS 6675
  o lever-operated, spherical plug valve to
     BS 6675
  o screwdriver-operated (slot type),
     spherical plug valve to BS 6675

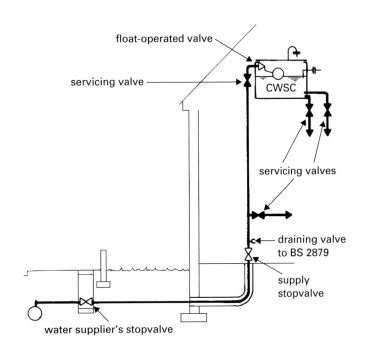

float-operated valve

servicing valve

CWSC

servicing valves

draining valve to BS 2879

supply stopvalve

water supplier's stopvalve

Figure 2.26   Control valves within dwellings

(a) Gate valve to BS 5154
Can be used on distributing pipe outlets from cisterns.

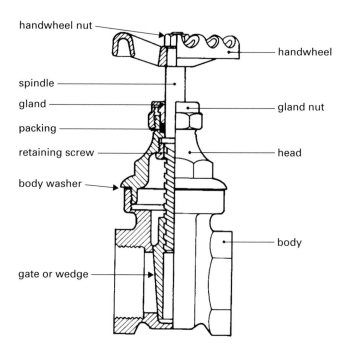

(b) Spherical plug valves to BS 6675

Suitable for use near to single outlet fittings and appliances.

Available in sizes up to 25 mm diameter.

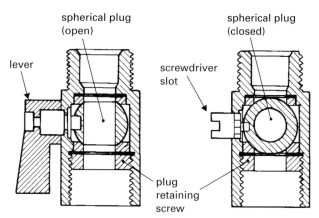

Figure 2.27 Types of servicing valve

Isolating valves

Isolating valves are required by Water Regulations 'to be installed to isolate parts of the system'. These will be useful for maintenance and for closing down parts of the supply when leakages are taking place.

On larger systems control valves should be fitted:

- to isolate pipework on different floors;
- to isolate various parts of an installation at the same level;
- to isolate branch pipes that lead to a range of appliances.

Stop valves and draining valves

Water Regulations require that 'sufficient number of stopvalves and draining valves be fitted to minimize the discharge of water when water fittings are maintained or replaced'.

Draining valves should be fitted at all low points on supply pipes and distributing pipes so that systems can be drained for frost protection and for convenience when carrying out repairs. Draining valves should only be used for draining purposes. See also chapter 7.

2.5   Water systems outside buildings

Pipeline systems should conform to national or European standards as appropriate. The system should be designed to ensure that the quality of potable water in the system does not deteriorate. Precautions should be taken to prevent risk of contamination through backflow, stagnation, cross connections or any other cause. Systems outside buildings should generally be designed for a life of at least 50 years although replacement or renovation of components such as pumps, meters and controls may need earlier attention.

Service pipe diameters should be determined by hydraulic calculation after consultation with the consumer and the water supplier. This is particularly important in the case of non-domestic users where design requirements may create greater demands on the available supply. Calculations should also take into account any future supply needs that might be expected. In premises where water for firefighting purposes is required, the local fire service should be consulted and relevant regulations complied with.

Service pipes should be sized to meet the maximum design flow rate at times of peak demand. Account should be taken of available pressures and flow rates from the water supplier's main and the effects of frictional resistances in pipelines and fittings. Consideration should also be given to the velocity of flow. Excessively high velocities can cause noise pollution, and very low velocities may require overlarge pipes. Flow velocities between 0.5 m/s and 2.0 m/s will normally be suitable but in exceptional circumstances a maximum velocity up to 3.5 m/s may be accepted.

Pipelines should be designed to accommodate thermal movement. Particular attention should be given to the difference in temperature during installation compared to the expected temperature when filled and in use. Considerable stress can occur in pipelines where insufficient expansion joints are inserted. This can be a particular problem when plastics pipes are pulled through the ground by mole plough when stresses caused by 'stretching' are combined with those of contraction through cooling.

Water pressure in pipes exerts considerable outward forces. Uneven forces occur at valves, changes in direction and diameter, branches and blank ends. These forces can move pipes and even push joints apart. Uneven forces should be compensated by the use of restrained joints, thrust blocks or other anchorages. Where thrust blocks are to bear against the soil, the safe bearing capacity of the soil should be determined so that the size and strength of the thrust block can be adequately calculated. See also chapter 11.

Pipes and pipelines should be protected against potential damage from aggressive soils and pollutants and the water supplier should be contacted regarding any corrosive aspects of the water supply before selecting the pipe material.

The design layout for below ground water systems will depend on local circumstances, but in all cases consideration should be given to the following:

- adoption of shortest practical route to allow reasonable access for maintenance;
- minimum depth of cover for frost penetration and the maximum depth of cover for ease of repair;
- location of other services, buildings and structures. Minimum distances should be maintained between water, other services and buildings (see figure 11.59);
- provision and location of valves, air valves, washouts and hydrants where appropriate;
- pipe materials and corrosion protection systems in aggressive or contaminated soils;
- adverse ground conditions and difficult terrain, earth loads and traffic loads;
- risk of damage to and from trees and tree roots;
- risk of damage to and from other utilities, works and apparatus.

Valves

Isolating valves should be fitted so as to allow individual branch pipes to be shut off in an emergency or for repairs without disrupting the supply to the whole of the premises. Consideration should be given to the number of buildings served by the system and the likely effects on people and industrial or commercial activities that might be disrupted (see figure 2.28).

Hydrants are required for firefighting purposes. They may also be used for filling, draining, venting and flushing of the main. The location and type of hydrants should be decided after consultation with the owner/designer of the premises and the local fire service. Hydrants should be positioned so that stagnation cannot occur.

Washout valves should be provided for draining or flushing of large diameter pipes. The size of these should be related to the volume of water to be drained, the time available and the capacity of surrounding draining area. Washout valves should have a maximum discharge diameter of DN 200.

Air valves. Provision should be made for the release of air at high flow rates when the pipeline is being filled and to permit the entry of air when pipes are being drained. In long runs of large diameter pipes this can be achieved with large orifice air valves fitted at high points in the system or alternatively hydrants or washouts could be used. For the release of accumulations of air during normal operation, small orifice air release valves can be used.

Valve chambers should be constructed to avoid inflow of external water to chambers that might create a contamination risk.

2.6   Water revenue meters

This section should be regarded as for information only.

Water revenue meters are normally fitted by the water undertaker although there may be occasions when they might approve an installation by the consumer's contractor. In any event the meter will be supplied by, and will remain the property of, the water supplier. The water undertaker should consult and agree with the consumer on siting and installation details.

The preferred position for the meter is at the boundary of the premises at the end of the communication pipe so that it will register the whole supply (see figure 2.29).

In the case of premises with multiple occupations, e.g. flats, and where underground installations are not practicable, an internal installation may be acceptable provided the whole supply is registered.

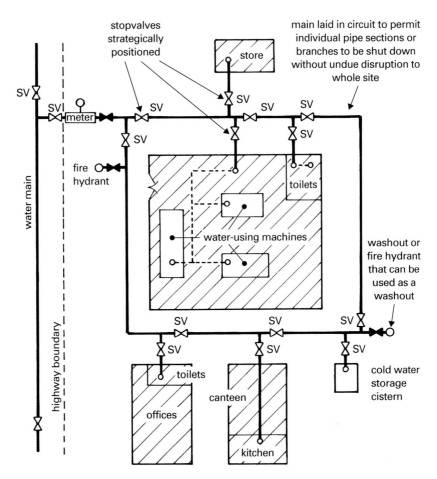

Figure 2.28   Positioning of stopvalves to permit emergency shut-off without disruption of whole supply

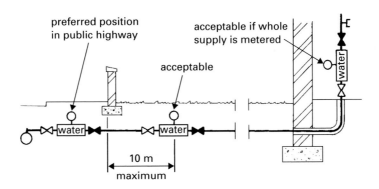

Figure 2.29   Meter positions

Continuity bonding

BS 7671 *Requirements for electrical installations – IEE Wiring Regulations* does not allow the use of water pipes as an electrode for earthing purposes. However, any metal water supply pipe must be bonded to the electrical installation main earth terminal as near as possible to its point of entry into the building.

A suitable conductor should be installed between pipes on either side of the meter and stopvalves, to protect the installer against electrical fault, and for the maintenance of the earth connection during use, and particularly when the meter is being replaced.

The continuity bond should have a cross-sectional area of at least 10 mm².

External meter installations

- Meters should be positioned below the ground, in a suitable chamber that will permit ample space for meter removal or replacement.
- For smaller installations of up to 3.5 l/h the chamber may be constructed of glass reinforced plastics or PVC; larger installations should have chambers of brick or concrete.
- The chamber should be well constructed, with a cover marked 'water meter' and fitted with slots or lifting eyes.
- The clear opening of the surface box (cover) should be equal to the internal dimensions of the chamber. Covers should not be of concrete.
- A meter below ground should be installed in a horizontal position.
- The meter should be fitted with isolating valves on both inlet and outlet to facilitate meter changing.
- The meter should be supported so as not to cause any differential load on its connecting pipework.
- Pipes should be flexible enough, with space around, to permit meter exchanges, but at the same time secured against movement of any mechanical joint.
- Pipes, cables and drains other than meter pipework must not pass through the meter chamber.

See table 2.1 and figure 2.30 for installation details.

Table 2.1   Recommended internal dimensions for meter chambers

Size of meter	Size of chamber			Remarks
	Length	Breadth	Depth	
15 to 20	430	280	To suit pattern of meter but not less than 750	Or 380 circular
25	600	600		
40 to 50	900	600		
80*	1900	750		
100*	2000	750		
150*	2150	750		

* Dimensions for these sizes take account of compound assemblies but do not provide for isolating valves

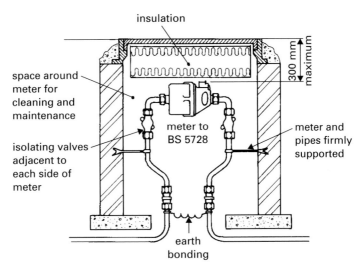

Pipes passing through meter chamber wall or floor should have clearance space to permit changing of meter and to prevent damage through ground movement.

Stopvalves shown are not in the most accessible positions.

Figure 2.30   Below ground meter installation

Note   Whilst BS 6700 and many water authorities favour raising the meter as shown in figure 2.30 to make meter reading and changing easier, the author would prefer the meter to remain at the same level as the service pipe, as in figure 2.31, in order to comply with the Water Regulations requirement of 750 mm depth of cover to avoid frost damage.

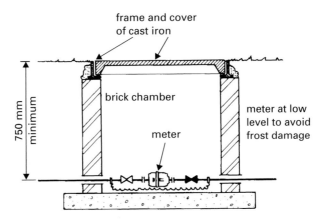

Figure 2.31   Meter installation to avoid frost damage

Internal meters

Internal meters may be fixed horizontally or vertically provided the dial is not more than 1.5 m above floor level and readily visible for reading (see figure 2.32).

Pipework to be adequately supported.

Meter to be protected from frost, especially in exposed positions, e.g. garage.

Meter in cupboard may be brought forward to within 300 mm of the front of cupboard for ease of reading, providing it is properly supported.

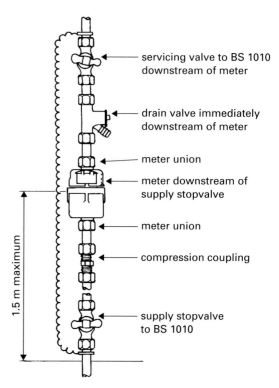

servicing valve to BS 1010 downstream of meter

drain valve immediately downstream of meter

meter union

meter downstream of supply stopvalve

meter union

compression coupling

supply stopvalve to BS 1010

1.5 m maximum

Figure 2.32   Above ground meter installation

Non-revenue meters

Installation should be as described for revenue meters but the water undertaker will not need to be consulted about the meter's position. The recommendations of BS EN 14154 should be followed. Information on selection of a suitable meter with regard to pressure and flow requirements can be obtained from the manufacturer.

Patent meter connections

These save space and installation costs (see figure 2.33).

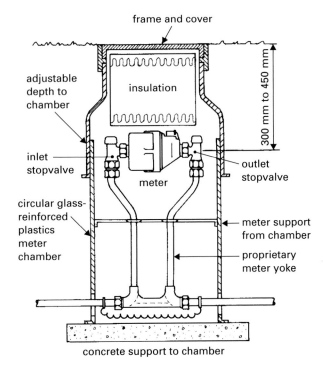

(a) **Below ground, using meter yoke**

Figure 2.33   Patent meter connections

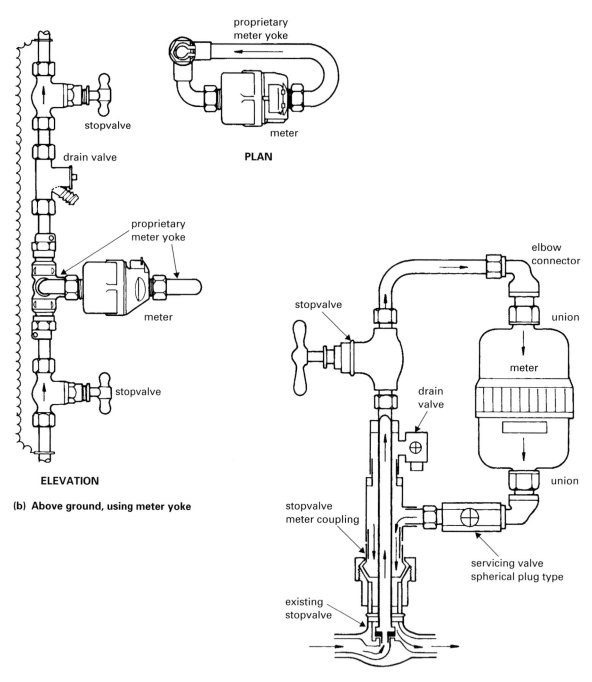

proprietary
meter yoke

PLAN

meter

stopvalve

drain valve

proprietary
meter yoke

meter

stopvalve

ELEVATION

(b) Above ground, using meter yoke

elbow
connector

stopvalve

union

meter

drain
valve

union

stopvalve
meter coupling

servicing valve
spherical plug type

existing
stopvalve

(c) Above ground, using stopvalve body for connection

Figure 2.33   continued

Meter readings

A typical domestic meter reading is shown in figure 2.34.

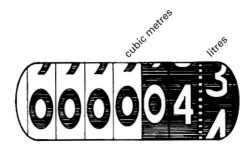

meter readings   above – 0.043 m³ or 43 litres
below – 1234.56 m³

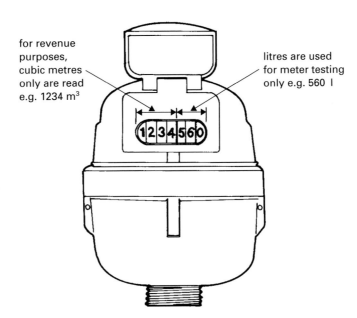

for revenue
purposes,
cubic metres
only are read
e.g. 1234 m³

litres are used
for meter testing
only e.g. 560 l

Figure 2.34 Meter reading – typical domestic meter

2.7   Boosted systems

Water supplies to buildings vary greatly in pressure and quantity available. In some cases this may give rise to intermittent supplies and in others, especially high-rise developments, parts of the building may be above the pressure limit of the mains supply. In these situations there is a case for the use of pumps. Boosted supplies will normally be pumped from a break cistern but, where water pressure in the supply pipe is insufficient and the demand is less than 0.2 l/s, or if the demand is greater and the water undertaker approves, drinking water may be pumped directly off the supply pipe.

There are two ways that pumps can be used to deal with the above problems. These are by direct boosting and indirect boosting. Indirect systems are more common than direct systems; the latter are rarely permitted by water suppliers because they reduce the mains pressure available to other consumers and can increase the risks of backflow. Under no circumstances should any pump be connected directly to a supply pipe without first obtaining the written consent of the water undertaker.

Booster pumps can cause excessive aeration. Although this does not cause deterioration of water quality, the 'milky' appearance can cause concern amongst consumers.

It is desirable that sample taps are fitted at the outlet of pumps to provide for periodic sampling to ensure the continued quality of the pumped water.

Basic systems

The basic systems are as follows:

- simple direct boosting, see figure 2.35;
- direct boosting to header and duplicate storage cisterns, see figure 2.36;
- indirect boosting to storage cistern, see figure 2.37;
- indirect boosting with pressure vessel, see figure 2.38.

The *pneumatic pressure vessel*, as seen in figure 2.39, contains both compressed air and water. As the water is drawn off, the water level drops, the air expands with a resulting loss of water, and a float switch starts the pumps. The water level rises to a predetermined level at which the float switch stops the pumps. The sequence of events will continue to supply the building until the water level falls and the cycle begins again. As and when air is lost from the pressure vessel by absorption into the water, pressure will fall, and the air is replaced via the air compressor.

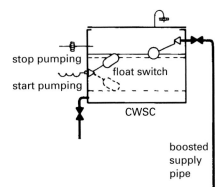

stop pumping

float switch

start pumping

CWSC

boosted
supply
pipe

When used for drinking water, the storage vessel should be of the 'protected' type.

Pump control provided by level switch or similar device in the high-level storage cistern.

Pumps switch on when the level of water drops to a pre-determined depth (normally about half the depth of the cistern) and they should switch off when the water level rises to about 50 mm below the shut-off level of the float-operated valve.

The frequency with which the pumps switch on and off should be limited to reduce wear on them, but the frequency of operation depends on the quantity of water used and stored, and on the pump rating.

Where the water supplier permits, pumps are connected to the incoming supply pipe to enable the pressure head to be increased.

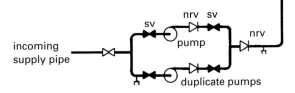

incoming
supply pipe

Figure 2.35 Simple direct boosting

System used for large and high-rise buildings.

Cisterns at high-level supply non-drinking water.

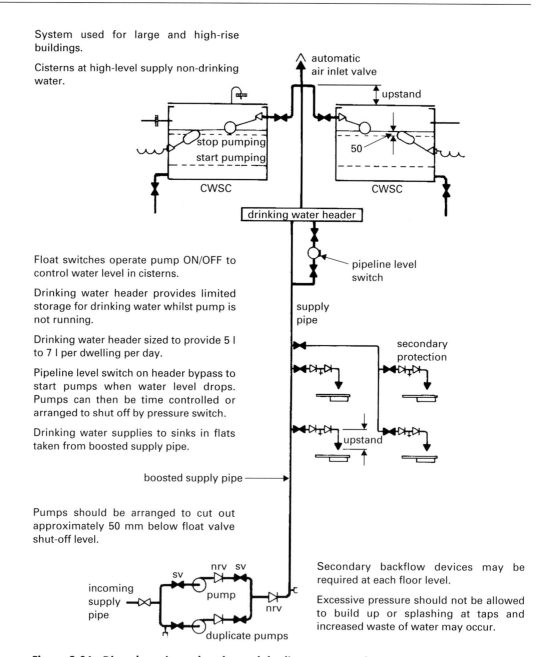

Float switches operate pump ON/OFF to control water level in cisterns.

Drinking water header provides limited storage for drinking water whilst pump is not running.

Drinking water header sized to provide 5 l to 7 l per dwelling per day.

Pipeline level switch on header bypass to start pumps when water level drops. Pumps can then be time controlled or arranged to shut off by pressure switch.

Drinking water supplies to sinks in flats taken from boosted supply pipe.

Pumps should be arranged to cut out approximately 50 mm below float valve shut-off level.

Secondary backflow devices may be required at each floor level.

Excessive pressure should not be allowed to build up or splashing at taps and increased waste of water may occur.

Figure 2.36   Direct boosting to header and duplicate storage cisterns

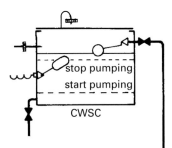

Float switch in storage cistern to operate pumps as water level in cistern rises and falls.

Pumps should cut out as water level rises to approximately 50 mm below float valve shut-off level.

For drinking water, protected cisterns should be used.

Break cistern should have effective capacity equivalent to at least 15 minutes pump output but should not be over-sized as this could increase risk of stagnation of water.

Float switch in break cistern to shut off pumps when water level drops to approximately 225 mm above pump suction connection. This will ensure that pump does not run dry.

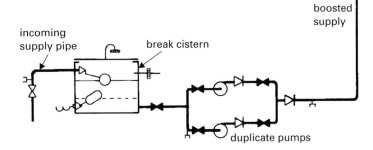

Figure 2.37   Indirect boosting to storage cistern

For use in buildings where a number of storage cisterns are supplied at various levels and where it is not practicable to control pumps by level switches.

Normally pressure vessel, pumps, air compressor and control equipment are purchased as a packaged pressure set.

Floors above limit of mains pressure supplied via break cistern and pumps.

Unboosted supply to floors within limit of mains pressure.

Drinking water cisterns in dwellings must be protected from contamination.

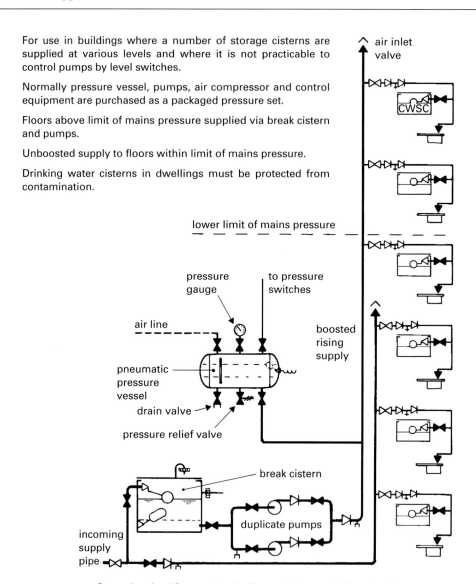

Secondary backflow protection is required at each floor level where separately chargeable premises are supplied.

Figure 2.38   Indirect boosting with pressure vessel

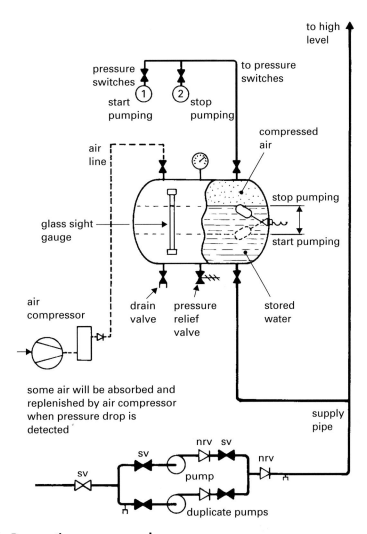

Figure 2.39 Pneumatic pressure vessel

Some systems incorporate a sealed pressure vessel containing a flexible membrane to separate the air from the water. This will reduce air loss and eliminate the need for a permanent compressor. However, air pressures should be checked at intervals and replenished if necessary. Some units contain nitrogen instead of air, and it is important that these are topped up with nitrogen, not air or some other gas. A multi-storey installation is illustrated in figure 2.40.

Pumps and equipment

Pumps and other associated equipment are usually located within the building being served, preferably as near as possible to the point of entry of the incoming pipe.

Electrically driven centrifugal pumping plant is normally used; pumps and other equipment should be duplicated.

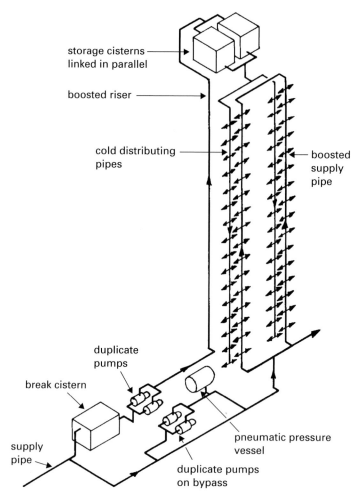

storage cisterns linked in parallel

boosted riser

cold distributing pipes

boosted supply pipe

duplicate pumps

break cistern

supply pipe

pneumatic pressure vessel

duplicate pumps on bypass

Drinking water supply direct from mains using pneumatic pressure vessel.

Indirect boosting for storage cisterns.

Control valves and backflow prevention devices omitted for clarity.

Figure 2.40   Typical installation to multi-storey building

Pumps may be either horizontal or vertical types, directly coupled to their electric motors. A solid foundation is essential for all motors and pumps and anti-vibration mountings should normally be specified.

Automatic control of pumping plant is essential and pressure switches, level switches, or high-level and low-level electrodes should give reliable control. Other methods of control, both mechanical and electrical, could be considered. Pumping should be controlled using a pump selector switch and an ON/OFF/AUTO control. Motor starters should incorporate overload protection.

Pumps should be installed in duplicate and sized so that each pump is capable of overcoming the static lift plus the friction losses in both pipework and valves. Where pumps are connected directly to the service pipe, allowance should be made for the minimum operating pressure in the service pipe, since the pump head is added to this and does not cancel out any existing pressure.

Care should be taken in pump and pipe sizing to minimize the risk of water hammer due to surge when pumps are started and stopped.

Transmission of pump and motor noise via pipework can be reduced by the use of flexible connections. Small-power motors of the squirrel cage induction type are suitable for most installations. Low-speed pumps are preferable to promote a long efficient life and reasonably quiet operation. The fitting of motors with sleeve type super-silent bearings should be considered for quiet running.

All pipework connections to and from pumps should be adequately supported and anchored against thrust to avoid stress on pump casings and to ensure proper alignment.

Most small air compressors used for charging pneumatic pressure vessels are of the reciprocating type, either air-cooled or water-cooled. A water-cooled after-cooler for the condensation and extraction of oil and moisture from the compressed air should be installed. The air to be compressed should be drawn from a clean cool source and should be protected from contamination. Check or non-return valves should offer a minimum of frictional loss and should be non-concussive.

The pump room (see figure 2.41) should be of adequate size to accommodate all the plant and also provide adequate space for maintenance and replacement of parts; it should be dry, ventilated and protected from frost and flooding. Entry of birds and small animals must be prevented. Access should be restricted to authorized persons.

Provision should be made for the pumps to be supplied by an alternative electricity supply in the event of mains failure.

Maintenance and inspection

A responsible person should be appointed to oversee the proper installation of any pumped water scheme, and the user should arrange for regular maintenance and inspection of the pumps and plant.

All work carried out and inspections made should be recorded in a suitable log book which should be kept in the plant room.

2.8   Water treatment

Water treatment is considered only briefly in BS 6700, its main concerns being the risk to health should backflow occur, and energy efficiency in hot water heating circuits.

Water softeners

The main purpose of a water softener is to reduce scale formation in hot water systems and components. Modern softeners are predominantly controlled by electricity, but hydraulically powered units that do not need electricity are available.

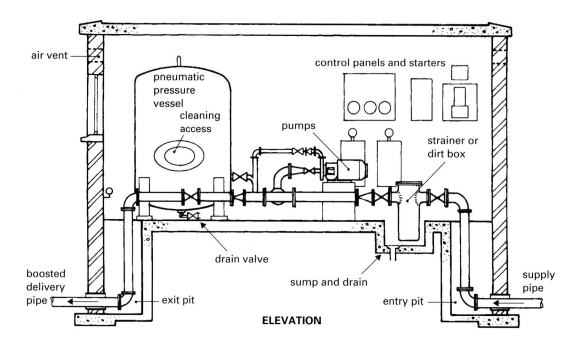

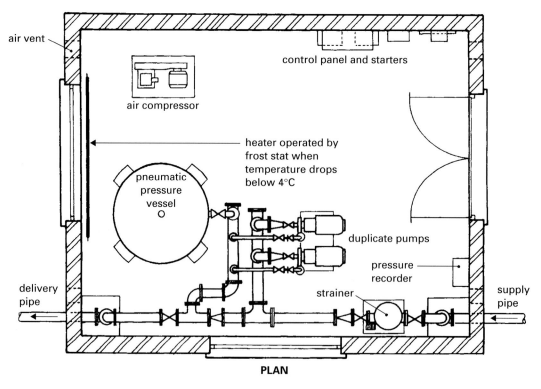

Figure 2.41 Typical pump room

Base exchange (ion exchange) water softeners work by passing the hard water through an enclosed tank containing resin particles (beads). The resin attracts and absorbs the hardness salts, mainly calcium and magnesium, from the water, and at the same time replaces them with sodium from the resin. After a while the resin becomes saturated with hardness salts and needs to be regenerated using a brine (salt) solution to put sodium back into the resin. The hardness salts are given up from the resin and washed to waste down the drain.

Most modern water softeners are electronically controlled to recharge the resin, either at timed intervals or as the resin becomes saturated, and to automatically flush the residual hardness salt to waste. The only maintenance requirement is the occasional replacement of salt in the brine tank (see figure 2.42).

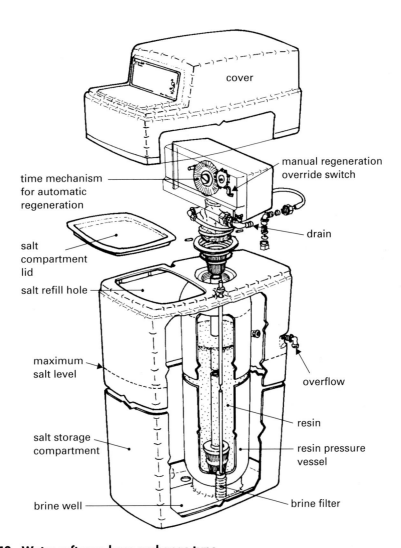

Figure 2.42   Water softener, base exchange type

Advantages

Some of the advantages of water softeners are as follows:
(1) Savings in soap and reduction of scum, resulting in:
  - reduced expenditure on soap purchase (small savings);
  - easier cleaning of appliances;
  - cleaner crockery from dishwashers.
(2) Smooth, gentle feel of bath and shower water.
(3) Scale reduction in appliances and components, resulting in:
  - longer life for cylinders, immersion heaters and other components;
  - less maintenance, e.g. shower outlets less likely to become clogged with fur;
  - improved heat transfer in boilers and heat exchangers resulting in savings in fuel and energy;
  - reduction in scale deposits in hot water cylinders which in turn will lessen risk of legionella.
(4) Soft water **may** be beneficial to sufferers of *some* skin complaints.

Disadvantages

(1) Additional installation costs (manufacturers claim a six-year payback).
(2) Cost of running the unit, e.g. electricity supply and salt.
(3) Drain needed for brine rinse.
(4) User must add salt periodically.
(5) Where water softeners are installed to serve existing hot water systems, the softened water can, over time, dissolve any scale previously formed in the system. In some cases, this could result in the release of particulate lead associated with the scale from lead pipes or from copper pipes with lead solder joints downstream of the softener.

Installation

Water softeners should be sited near the incoming supply pipe and where drain access is available (see figure 2.43).

Electricity supply may be required for automatic control and operation.

Water softeners of the salt regenerated type installed in dwellings are considered to be a fluid category 2 backflow risk and as such a check valve should be installed on the supply pipe before the softener connection.

Drinking water supply should be taken off before the softener, and upstream of any check valves fitted.

In dwellings a single check valve is required under Water Regulations to protect the water supply from backflow.

A softener located other than in a dwelling requires the use of a double check valve arrangement or a combined check and anti-vacuum valve.

Other forms of treatment

As previously mentioned, BS 6700 deals briefly with water softeners, and does not refer to other forms of treatment which are shown in figures 2.44 and 2.45. It is not the intention here to recommend their use, rather to point out that there are such treatments readily available, and that they can in certain circumstances have a limited application.

Before any of the chemical devices are used, advice should be sought from local water undertakers, particularly regarding toxicity.

Backflow protection shown for domestic softener is a single check valve.

Non-domestic softeners require a double check valve assembly.

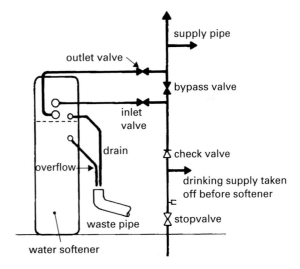

Figure 2.43   Installation of base exchange water softener for domestic use

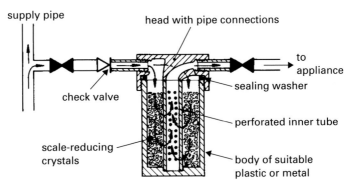

Useful for supplies to instantaneous water heaters.

Figure 2.44   Pipeline dispenser

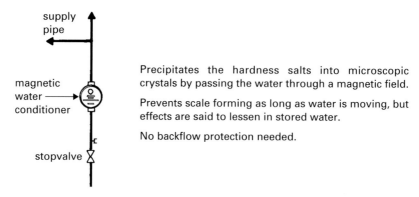

Figure 2.45 Magnetic water conditioner

Polyphosphate dosing

This form of treatment can be used to reduce scale deposits, and in certain conditions can help protect pipes from corrosion. It usually consists of a small dosing chamber containing polyphosphate beads or crystals that dissolve as the water passes through (see figure 2.44).

In stagnant systems excessive concentrations of phosphates can build up to make the water less palatable, and could cause blockages due to the crystals reforming.

Physical conditioners

Physical conditioners may be of a number of types, using a variety of methods to reduce the effects of scaling. These may be electronic, electrolytic, magnetic (figure 2.45), electro-magnetic, or ionic devices that either have a direct water connection or are strapped in some way to the water pipe. The advantage of these is that they may give the effect of softened water without the addition or loss of mineral substances to the water. It should be said, however, that these devices do not soften the water, rather they inhibit the hardness particles so that they do not readily form scale deposits.

Physical conditioners are particularly suited to instantaneous type water heaters, but their effects may be lost in stored hot water.

Chapter 3
Hot water supply

Hot water is an essential requirement of all dwellings and most working environments. Buildings are required under Building Regulations (G1) to be provided with sanitary conveniences. These should sufficient in number, and suitable for the sex, age and physical ability of the persons using the building. Rooms containing water closets (WCs) are required to have wash basins with hot and cold water, sited in or adjacent to the room containing the WC.

Under G2 of the Regulations every dwelling (house, flat or maisonette) is required to have at least one bathroom with a fixed bath or shower which must be supplied with an adequate supply of hot and cold water.

Hot water supply cannot be considered entirely in isolation from central heating because systems commonly combine both functions. This book, like BS 6700, is primarily concerned with hot water supply but with reference to central heating in combined hot water and heating systems up to 44 kW output.

3.1 System choice

There are many methods by which hot water can be supplied, ranging from a simple gas or electric single point arrangement for one outlet, to the more complex centralized boiler systems supplying hot water to whole buildings (see figure 3.1).

The following factors should be considered in the selection and design of hot water systems:

- type and use of the building and the needs of the user;
- quantity of hot water required;
- quality of water supplied and the prevention of *Legionella*;
- temperature in storage and at outlets;
- cost of installation and maintenance;
- fuel energy requirements and running costs;
- conservation of water and energy;
- safety of the user.

BS 6700 recommends that where user requirements are not specified, or not known, e.g. in speculative housing, the user's needs may be assessed on the basis of the size and type of building, experience and convention.

Where a dwelling has only one bathroom it may be assumed that immediately after filling a bath, some hot water will be required for kitchen use, but a second bath will not be required within 20 to 30 minutes. Where a dwelling has two or more bathrooms, it can be assumed that all the baths will be filled in succession and hot water will immediately be required for kitchen use.

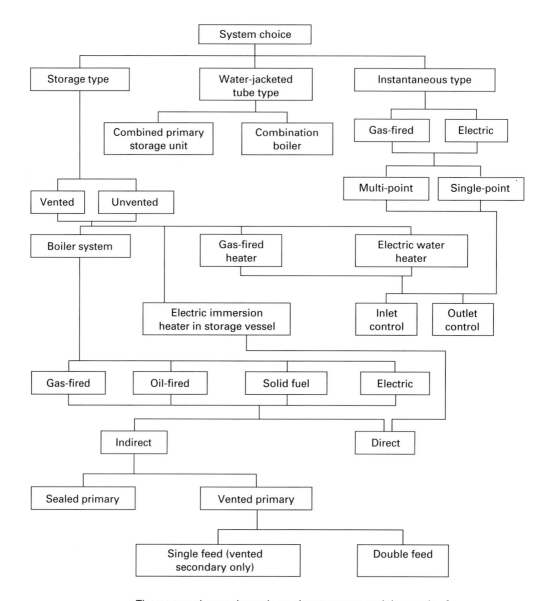

The system chosen depends on circumstances and the needs of the user, and may require the use of one method or a combination of two or more.

Figure 3.1 Alternatives for hot water supply

3.2   Instantaneous water heaters

Instantaneous water heaters (see figures 3.2–3.6) should be chosen with the following considerations in mind.

(1) Some of these heaters have relatively high power ratings (up to 28 kW if gas fired) so it is important that adequate gas or electricity supplies are available.

(2) The water in instantaneous water heaters is usually heated by about 55°C at its lowest flow rate, and its temperature will rise and fall inversely to its flow rate.

(3) Where constant flow temperature is important, the heater should be fitted with a water governor at its inflow. Close control of temperature is of particular importance for showers.

(4) To attain constant temperatures on delivery, water flow and pressure must also be constant. Variations in pressure can cause flow and temperature problems when the heater is in use, and when setting up or adjusting flow controls.

(5) The use of multi-point heaters for showers should be avoided, except where the heater only feeds a bath with a shower over it.

(6) Gas-fired heaters fitted in bathrooms must be of the room-sealed type. Room-sealed types are preferred in other locations.

(7) Gas supply installations are required to comply with the Gas (Installation and Use) Regulations and follow the recommendations of BS 5546 and BS 6899.

(8) Flueing and ventilation of gas appliances should comply with BS 5440-1 and BS 5440-2.

(9) Electrically powered heaters in bathrooms must be protected against the effects of steam.

(10) Electrical work is required to comply with Part P of the Building Regulations and conform to BS 7671 *Requirements for electrical installations – IEE Wiring Regulations*. Electrical instantaneous water heaters should conform to BS EN 60335-2-35.

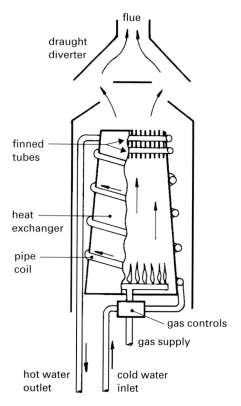

Position heater near most used appliance, usually the kitchen sink.

Where pipe runs from a multi-point heater are likely to be long consider using a number of single-point heaters.

Multi-point heaters operate most satisfactorily when only one outlet is used at any one time.

Flow rate is variable and heater is generally not suited for use with a shower.

Figure 3.2   Typical instantaneous water heater, gas-fired with conventional flue

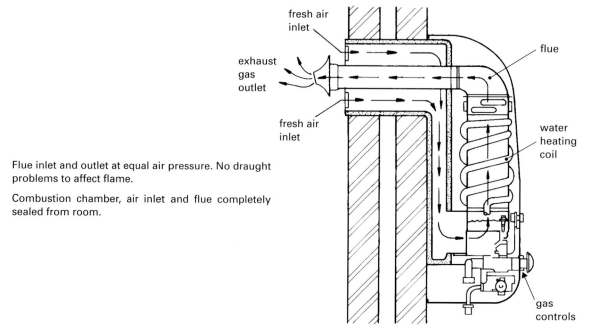

Flue inlet and outlet at equal air pressure. No draught problems to affect flame.

Combustion chamber, air inlet and flue completely sealed from room.

Figure 3.3   Typical instantaneous water heater, gas-fired with balanced flue

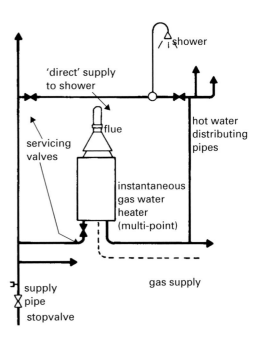

(a) Directly supplied heater

Constant flow rate needed to maintain 55°C temperature difference between feed water and heated water.

Pressure and flow variations will affect temperatures at outlets.

Showers are not recommended because of possible loss of constant temperature control and pressure.

Use only thermostatically controlled shower mixer.

The usual arrangement is direct from the supply pipe as shown here because installation cost is lower. However, supply from storage will give constant flow.

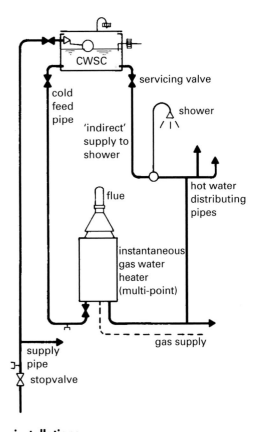

(b) Indirectly supplied heater

High installation cost compared with mains-fed system.

Constant pressure from storage for shower and other fittings gives more stable temperature control.

Figure 3.4　Centralized gas-fired instantaneous water heater installations

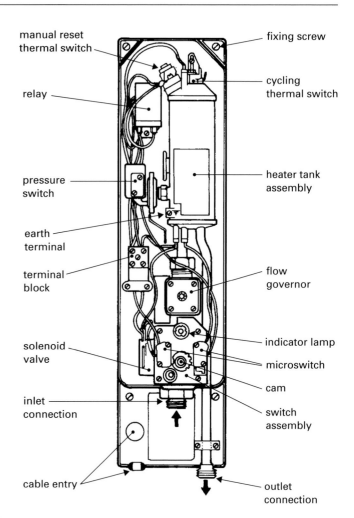

Heater shown is made for direct mains connection. Some types suited to storage-fed supplies.

Heater may be rated up to 8 kW. Need to ensure that electricity supply is adequate.

Flow governor will compensate for pressure variations.

Figure 3.5   Instantaneous electric shower heater

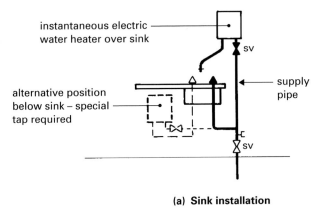

(a) Sink installation

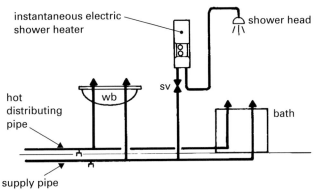

(b) Shower installation

Where shower unit is fitted it will need a minimum head of 10.5 m.

Flexible shower outlet may be a contamination risk if nozzle can become submerged (see figure 3.52b).

Figure 3.6 Typical uses for instantaneous electric water heaters

3.3 Water-jacketed tube heaters (primary stores)

The water drawn off for use passes through a heat exchanger in a reservoir of primary hot water (see figure 3.7). The size of this reservoir and its heat input determine the volume and rate of flow of hot water that can be provided without an unacceptable temperature drop. The cold water feeds to the heater may be from the mains or from a storage cistern.

Insulation for primary store vessels is required to be of a higher standard than for normal storage cylinders because they work at high temperatures of up to 82°C. Primary stores designed for gas, oil or solid fuel systems should be insulated to the standards set out in sections 4.3.1 and 4.3.2 of the Water Heater Manufacturers Association *Performance specification for thermal stores*. Where primary stores are designed for heating by electricity, the insulation should give a 15% lower standing heat loss than that specified above for heaters designed for use with gas, oil and solid fuel.

Water-jacketed tube heater is a form of instantaneous heater.

Primary circuit may be vented system or sealed system.

Heat exchanger warms secondary supply water as it passes through.

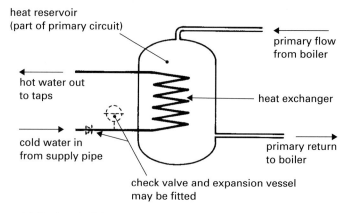

(a) Basic principles

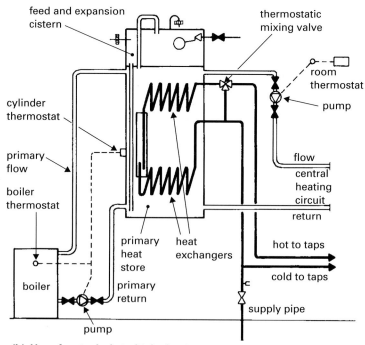

(b) Use of water-jacketed tube heater

Note 1 Primary water from the boiler flows to the heat store as programmed by the cylinder thermostat. Hot water is pumped to the radiator heating circuits and returns to the heat store. Cooler water from the heat store then returns to be reheated in the boiler.

Note 2 Cold water under mains pressure from the supply pipe enters the lower heat exchanger to be partially heated. It then passes through the upper heat exchanger where it is fully heated before being distributed to the taps.

Drawing shows the 'Boilermate' system using a combination unit.

Figure 3.7   Water-jacketed tube heater

Combination boilers

Combination boilers, more commonly known as combis, provide hot water on demand and full central heating all from one centralized heating source (see figures 3.8 and 3.9). Hot water is heated indirectly using the principle of the water-jacketed tube heater in its water-to-water heat exchanger.

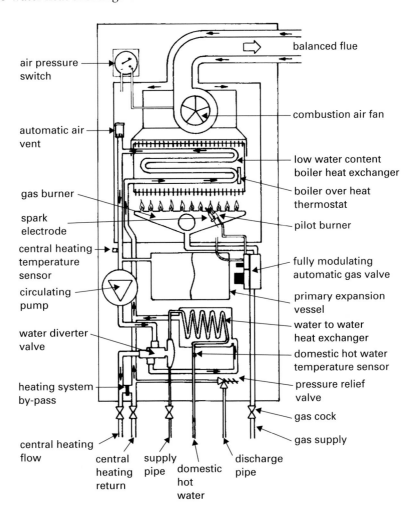

Figure 3.8 Typical combination boiler

BS 6700 does not discuss combination boilers although a large proportion of the heating and hot water market favours this type of system for both new and replacement work.

Features of the combination boiler include the following:

- It provides instant hot water on demand at a constant temperature.
- Water is only heated as and when it is needed.
- Cold water storage is eliminated, saving on materials and installation costs, but there is no reserve water provision.

- Electronic modulating regulation provides for a wide range of heating loads.
- It provides a constant flow rate of 8 l to 13.5 l per minute depending on boiler output rating and on the adequacy of the incoming water supply.
- The sealed system of the central heating primary circuit gives priority to hot water when needed.

System is relatively simple and cheap to install.

Heating circuit usually of the 'sealed' type.

No system controls or valves shown.

Temporary fill point must be disconnected from supply pipe after filling or topping up.

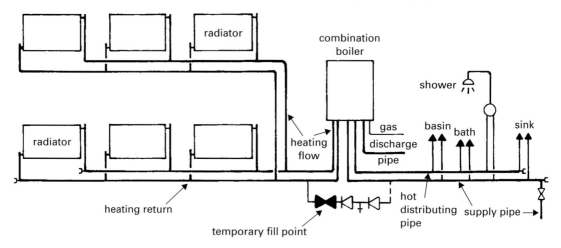

Figure 3.9   System using combination boiler

Combined primary storage unit

This appliance is described in the Energy Efficiency Best Practice booklet *Domestic heating by gas: boiler systems* as a special category of 'Combi' boiler and is designed to provide both hot water and space heating. It contains a large store of heated water (at least 70 l) within the primary circuit to reduce boiler cycling. It will provide a quick warm-up to radiators and is capable of supplying hot water to the tap at a high flow rate.

3.4   Storage type water heaters and boiler heated systems

Domestic hot water supply installations of the storage type may be either vented or unvented. Figures 3.10 and 3.11 illustrate the main features of each.

   System choice will be made in conjunction with choice of cold water system, and will depend on the need, or otherwise, to maintain a reserve of cold water to keep the system in operation at all times, or a preference to avoid any disadvantages of high-level cold water storage.

Vented hot water storage system

This is fed from a high-level cistern which provides the necessary pressure at outlets, accommodates expansion due to heated water, and is fitted with an open safety vent pipe to permit the escape of air or steam, and to prevent explosion without the need for any mechanical device.

The vented hot water storage system provides a reserve supply of hot water at a constant low pressure.

The feed cistern needs protection against the entry of contaminants. A protected drinking water cistern that meets the requirements of Water Regulations will guard against this.

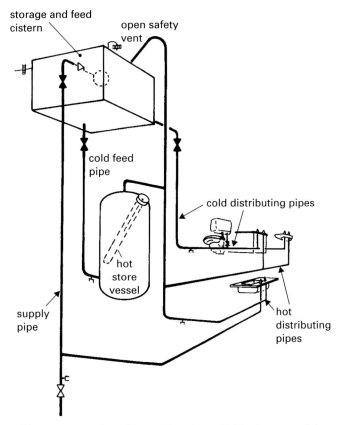

The open vent pipe will provide safety relief in the event of the system overheating.

Heated water will expand to the cold feed pipe.

The system shown is heated by immersion heater but could alternatively be heated by boiler.

Figure 3.10 Vented hot water system

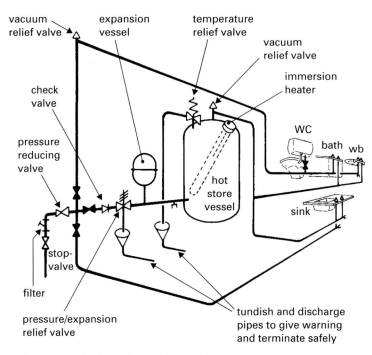

Pressure reducing valve reduces mains pressure to pressure suitable for operating system.

Pressure relief valve guards against excess pressure.

Expansion vessel accommodates expansion of water when heated.

Thermostat (not shown) controls temperature at normal level.

High temperature cut-out (not shown) protects against over-heating of water.

Temperature relief valve allows boiling water and steam to escape if thermostat and thermal cut-out should fail.

Tundish and discharge pipes take relief water to safe place.

(a) **Basic outline of system and components**

Figure 3.11 Unvented hot water storage system

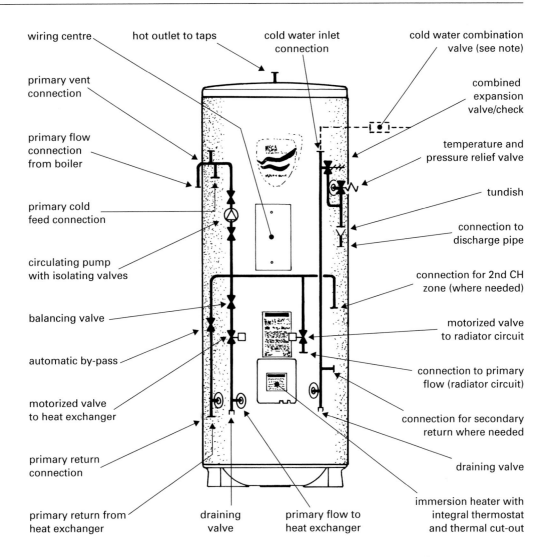

wiring centre

hot outlet to taps

cold water inlet connection

cold water combination valve (see note)

primary vent connection

combined expansion valve/check

primary flow connection from boiler

temperature and pressure relief valve

primary cold feed connection

tundish

connection to discharge pipe

circulating pump with isolating valves

connection for 2nd CH zone (where needed)

balancing valve

motorized valve to radiator circuit

automatic by-pass

connection to primary flow (radiator circuit)

motorized valve to heat exchanger

connection for secondary return where needed

primary return connection

draining valve

immersion heater with integral thermostat and thermal cut-out

primary return from heat exchanger

draining valve

primary flow to heat exchanger

Note In this system the cold water combination valve containing pressure reducer, isolating valve, line strainer and check valve is supplied separately for connection to cold water inlet.

(b) Unit type unvented hot water storage system pre-plumbed for site installation

Figure 3.11 continued

Unvented hot water storage system

The unvented hot water storage system is usually fed directly from the supply pipe under mains pressure. It does not require the use of a feed cistern, or an open vent pipe to atmosphere. Instead, it relies on mechanical devices for the safe control of heat energy along with other devices for the control of hot water expansion (see figure 3.11).

The main features of the unvented hot water storage system are as follows:

- needs continuity of mains supply pressure and flow, otherwise hot water cannot be guaranteed;
- eliminates the need for cold water storage and risk of frost damage;
- may require a larger supply pipe but eliminates some duplication of pipework;
- contains no reserve water supply in case of supply failure;
- eliminates cistern refill noise;
- relies on mechanical safety devices for protection from explosion which need regular inspection and maintenance;
- gives higher pressure at outlets, particularly at showers;
- allows quicker installation than vented system but involves more costly components.

Building Regulations Approved Document G3 states that a hot water storage system that has a storage vessel with no vent pipe to atmosphere (unvented system) is required to:

a)   be installed by a competent person;
b)   have safety devices that prevent the temperature of the stored water at any time exceeding 100°C;
c)   have pipework that safely conveys the discharge of hot water from safety devices to where it is visible but will cause no danger to persons, in or about the building.

However, the above requirement does not apply to:

a)   a hot store vessel that has a capacity of 15 l or less; or
b)   a system of space heating only; or
c)   a system used for an industrial process.

Unvented hot water storage systems should be of the 'unit' or 'package' type and both the system and its safety devices should be approved by:

a) a member body of the European Organisation for Technical Approvals (EOTA) operating a technical approvals scheme, e.g. the British Board of Agrément (BBA). Approval under this scheme will show that the system meets the relevant requirement of regulation G3; or

b) a certification body accredited under the National Accreditation Council for Certification Bodies (NACCB), e.g. accreditation and testing to BS 7206, will, under this scheme, ensure that the requirement of Regulation G3 is met; or

c) independent assessment which clearly demonstrates that the system meets an equivalent level of verification and performance to that shown in (a) or (b) above.

A **unit system** (see figure 3.11(b)) has all safety devices and other operating devices fitted by the manufacturer at the factory, ready for site installation.

A **packaged system** (see figure 4.17) has all safety devices factory fitted, with all other operating devices supplied by the manufacturer in 'kit' form for site assembly.

An unvented system is required to be installed by a competent person. Under Building Regulations G3 a competent person is described as a person who holds a current Registered Operative Identity Card for the installation of unvented domestic hot water storage systems issued by:

a) the Construction Industry Training Board (CITB); or

b) the Institute of Plumbing and Heating Engineering; or

c) the Association of Installers of Unvented Hot Water Systems (Scotland and Northern Ireland); or

d) individuals who are designated Registered Operatives and employed by companies included on the list of Approved Installers published by the BBA up to 31 December 1991; or

e) an equivalent body.

Safety devices

Directly heated systems are required to have at least two temperature activated safety devices operating in sequence to prevent the water from reaching 100°C:

1. a non self-resetting thermal cut-out to BS EN 70730-2-9 for electrically heated systems, or to BS 4201 for thermostats for gas burning appliances; and

2. one or more temperature relief valves to BS 6283 part 2 or BS EN 1490, suited to system pressures from 1 bar to 10 bar.

These devices are in addition to any thermostatic control used to control the day-to-day temperature of the stored water.

In an indirect system where the water is heated by a boiler the thermal cut-out may be fitted directly on the boiler. Some systems are fitted with more than one heat source, e.g. a boiler and an immersion heater for emergency use. In these situations a separate and independent thermal cut-out device is needed to protect against the danger of overheating from each heat source.

The temperature relief valve should be located directly on the storage vessel, near the top where the water is hottest, to ensure the stored water does not exceed 100°C.

- The valve should be sized to give a measured discharge rating that is at least equal to the power input to the water.
- The temperature relief valve should not be disconnected except for replacement. It should not be relocated or replaced by any other device or fitting.
- The valve should discharge through a metal discharge pipe to an air gap over a tundish arrangement located close to the valve.

Further information on temperature relief valves and discharge pipes can be seen in chapter 4, 'Prevention of bursting'.

Larger systems over 500 l capacity or more than 45 kW input

Unvented hot water storage systems with a storage vessel of more than 500 l capacity or with a power input of more than 45 kW are likely to be individually designed to suit a specific project and may not be suited to approval under EOTA or NACCB standards. However, systems should be designed to ensure that safety standards are at least equal to those specified previously for smaller systems. Systems should be designed by a qualified engineer and installed by a competent person.

Inspection and approval of installations

The installation of an unvented hot water storage system is designated under Building Regulations as 'building work' and any person who intends to carry out building work should give notice to a local building control body before the work is commenced. Following notification, it may be considered necessary for the Building Control Officer or Approved Inspector to inspect the installation.

Site inspection of an individual installation is not likely to be required where:

- an unvented system is approved by a member of EOTA (e.g. BBA) or an NACCB accredited body or where an equivalent level of safety protection can be proven; and
- the unvented system is installed by a competent person who holds a current Registered Operative Identity Card issued by an approved certification body in compliance with part G3 of Schedule 1 of the Building Regulations (see above).

When a building notice or full plans are submitted to a building control body and they include an unvented system, the notice must specify:

- the name, make and model of the system being installed;
- the name of the body that has approved or certificated that the system conforms to G3 of the Regulations;

- the name of the body that has issued the current Registered Operator Identity Card to the installer.

Similar notification should be made to the local water supplier in respect of the Water Supply (Water Fittings) Regulations 1999 which also apply to unvented hot water installations. See chapter 1.

Non-pressure or inlet-controlled water heaters

These are generally seen as single point heaters fitted either above the appliance with a swivel outlet spout, or under the appliance using special taps to control the cold water inlet to the heater inlet. Where special taps and mixer taps are used, it is important to the safety of the user that the tap outlet remains unobstructed. No hose or other connection should be made to the outlet of a non-pressure or inlet-controlled storage-type water heater in which the tap mechanism controls the cold water inlet to the heater.

These heaters may be heated by either gas or electricity (see figure 3.12).

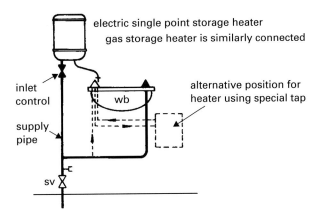

Expansion of heated water overflows through outlet spout.

Outlet must not be obstructed, nor must any connection be made to it.

Can be connected to wash basin or sink if special taps are used to control the inlet and leave the outlet unobstructed.

Figure 3.12   Non-pressure or inlet-controlled water heater

Pressure or outlet-controlled water heaters

These may be heated by either gas or electricity. Although these are called pressure type heaters, they are generally designed for supply from a feed cistern and are not usually suitable for operation under direct mains pressure. See figures 3.13 and 3.14. Care should be taken to ensure that the heater will withstand the pressures to which it is to be subjected. In smaller dwellings a capacity of 100 l to 150 l is considered sufficient. Where off-peak electricity is used the cylinder should have a capacity of at least 200 l.

Cold feed pipe provides for expansion of heated water.

Open safety vent terminates over feed cistern to provide safe route for boiling water and steam, should the system overheat.

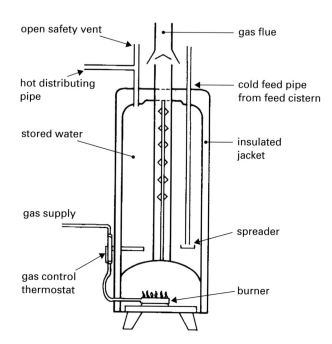

Figure 3.13   Outlet-controlled gas water heater

For dwellings, cistern must be 'protected'.

Heater capacity (minimum):
  o for small dwellings, 100 l to 150 l,
  o for off-peak electricity systems, 200 l.

Electricity systems are usually in the form of factory lagged and cased hot water cylinder with connections similar to those shown here for a gas installation.

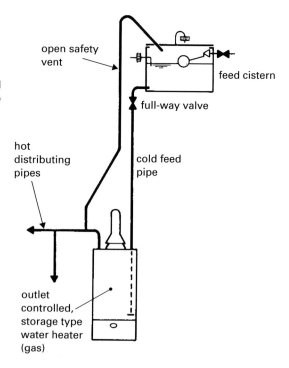

Figure 3.14   System using outlet-controlled heater

Direct type will often be used with an electric immersion heater as the sole source of heat.

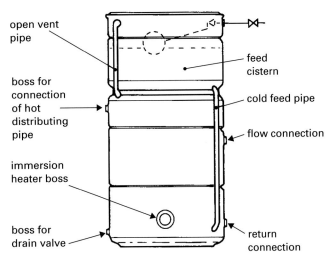

(a) Direct type

Flow and return from heat exchanger may be connected to centralized boiler for main heat source, with immersion heater used as supplementary heat source.

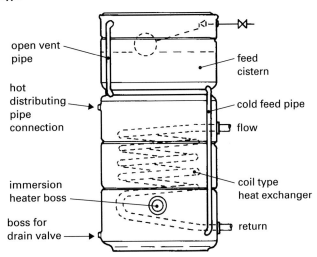

Also available:
- o factory lagged or purpose-made unit with lagging and metal outer casing;
- o single feed indirect type.

(b) Double feed indirect type

Figure 3.15 Combination storage heaters

Cistern or combination type storage heaters

A combination type storage water heater incorporates a cold water cistern which should be located so that at the very least its base is not lower than the level of the highest connected hot water outlet, and is high enough to give adequate flows at outlets; see figures 3.15 and 3.16.

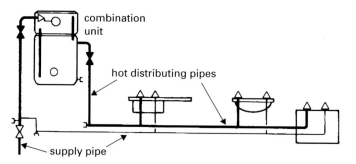

combination
unit

hot distributing pipes

supply pipe

Unit can be directly heated by immersion heater or indirectly heated from boiler.

Advantages:
- low installation costs;
- useful for flats when space is limited.

Disadvantages
- cold storage space limited;
- low pressure at hot taps;
- cannot be used for showers.

Figure 3.16   Typical combination unit installation

Electric immersion heater type storage heaters

Immersion heaters can be used as an independent heat source, or to provide supplementary heat to other centralized boiler systems. Figure 3.17 gives an example of a vented system.

Systems using immersion heaters may be vented or unvented. Safety devices must be used as appropriate to the system.

Immersion heaters should comply with BS EN 60335-2-21 whilst electrical controls should comply with BS 60730-2.

Immersion heaters and controls should be located so they are readily accessible for insertion, removal and adjustment. Figure 3.18 gives alternative immersion heater positions. Figure 3.19 shows information on a bottom entry immersion heater relative to the cold feed connection.

When installing an immersion heater, a minimum gap of 15 mm should be maintained between the end of the element and the inside surface of the vessel. 11″ (280 mm) and 14″ (355 mm) heaters should be installed horizontally and longer ones should be installed vertically.

The *Domestic Heating Compliance Guide* recommends that cylinders heated primarily by electricity should be provided with two electrical heating elements or immersion heaters, both of which should have thermostatic control. The lower element should be capable of heating up at least 85% of the cylinder content and be heated through the 'off-peak' electricity tariff. The upper element should be used to boost the system from the normal 'daytime' tariff and should be capable of heating at least 60 l of water. After heating the cylinder to 60°C from the off-peak element, it should be possible to draw off at least 80% of the cylinder's contents at 45°C or above at a flow rate of 0.25 l/s.

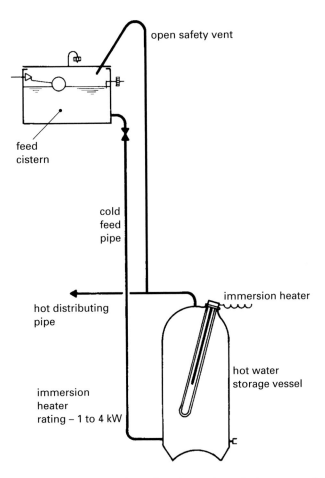

The hot store vessel may be a hot water cylinder or a combination unit, or a specially made, encased and lagged cylinder.

Figure 3.17 Vented storage system using immersion heater

(a) Single element – top entry

Most common and cheapest arrangement. May not heat cylinder to bottom.

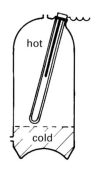

Thermostats and elements in hotter part of cylinder

(b) Double element – top entry

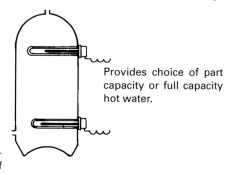

Dual provision from twin element to give full heat, or short boost from smaller element.

Switches and thermostat to comply with requirements of BS EN 60730-2.

(c) Single element – bottom entry

Better arrangement for supply of hot water with good temperature control.

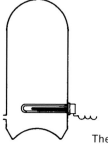

(d) Double element – twin side entry

Provides choice of part capacity or full capacity hot water.

Thermostats and elements in cooler part of cylinder.

Figure 3.18 Positioning immersion heaters

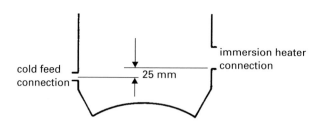

cold feed connection

immersion heater connection

25 mm

Immersion heater connection to be at least 25 mm above the centre line of the cold water inlet.

Adequate clearance needed around immersion heater for removal and replacement.

Figure 3.19 Immersion heater position and cold feed connection

All immersion heaters are required to incorporate a safety cut-out device that will prevent water in the cylinder from boiling. In the event that the thermostat fails, the non-self resettable overheat cut-out will activate to shut off electrical power to the heater before the water reaches a temperature of 98°C. After checking for electrical faults, the device may be manually reset by the press of a button. A diagram of an immersion heater with safety cut-out is shown in figure 3.20.

Electric immersion heaters used in hot water cylinders should conform to BS EN 60335-2-73 and controls to BS EN 60730-2. Wiring should comply with BS 7671.

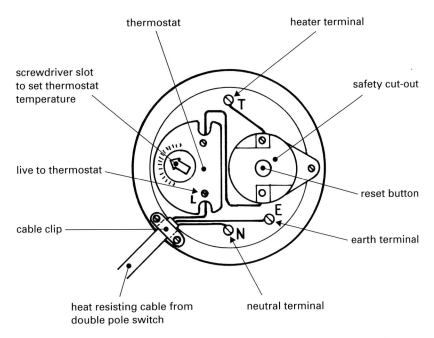

Range of thermostat settings from 50°C to 82°C. Recommended setting 60°C.

Safety cut-out set at 90° to prevent water reaching 100°C.

Immersion heaters to conform to BS EN 60335-2-73.

Figure 3.20 Top of immersion heater with cover removed

Gas-fired circulators

These are essentially small gas-fired boilers. They may be independent or used in conjunction with a system using some other fuel; see figure 3.21.

Boiler-heated hot water systems

BS 6700 deals with independent water heating appliances (boilers) using gas, oil, solid fuel and electricity, and includes direct, indirect, vented and unvented systems.

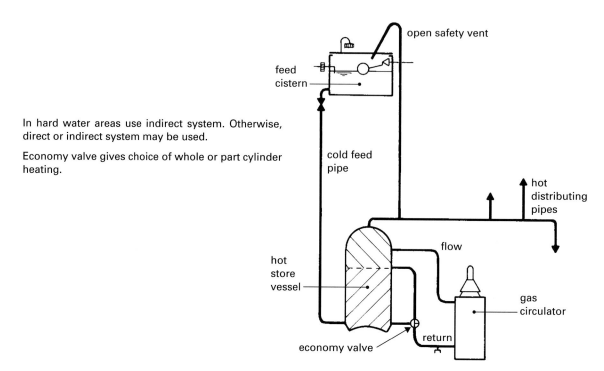

In hard water areas use indirect system. Otherwise, direct or indirect system may be used.

Economy valve gives choice of whole or part cylinder heating.

Figure 3.21 System using gas-fired circulator

Direct systems are those in which the boiler or heat source 'directly' heats the water that is drawn off. The principles of this system are shown in figure 3.22. The electric immersion heater system shown in figure 3.17 is also a type of direct system.

In **indirect systems** the primary water (heated by the boiler) is physically separated from the secondary water (water drawn off from the taps) by a heat exchanger in the form of a coil, or inner annular cylinder; see figure 3.23.

Where systems combine both hot water and central heating functions, indirect systems are essential to prevent excessive furring in hard water areas, to maintain a cleaner secondary hot water supply and to permit the use of corrosion inhibitors, thus increasing the life of components.

Gravity flow hot water circulation

Circulation in hot water systems may be attained by using natural convection currents (gravity circulation) or, alternatively, by the use of a circulation pump to 'force' the water around the system. Hence the terms 'forced circulation' and 'pumped circulation'. Gravity circuits should not be used in modern systems because they are inefficient and slow to react compared with fully pumped circuits.

In gravity circuits, good circulation depends on system height and the difference in densities between the hot flow and cooler return to provide the motive force to overcome frictional resistances in pipes and fittings and to provide effective hot water circulation.

Figure 3.24 shows examples of good and poor circulation in gravity primary circuits and table 3.1 relates the density of water to various temperatures.

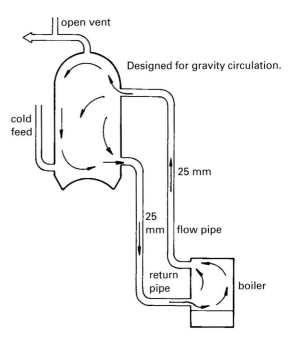

Designed for gravity circulation.

open vent

cold feed

25 mm

25 mm

flow pipe

return pipe

boiler

System directly heated by a boiler will only be seen in old systems.

Not for use in modern systems because:
- they are inefficient
- they cannot be protected from corrosion
- they readily fur up in hard water areas
- water circulated from boiler to cylinder will be drawn off from taps

Figure 3.22   Direct hot water system (vented)

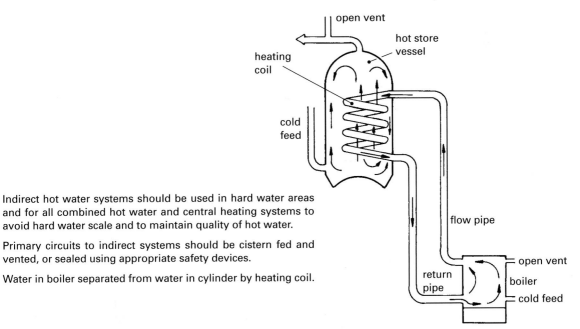

open vent

hot store vessel

heating coil

cold feed

flow pipe

return pipe

open vent

boiler

cold feed

Indirect hot water systems should be used in hard water areas and for all combined hot water and central heating systems to avoid hard water scale and to maintain quality of hot water.

Primary circuits to indirect systems should be cistern fed and vented, or sealed using appropriate safety devices.

Water in boiler separated from water in cylinder by heating coil.

Figure 3.23   Indirect hot water system (vented)

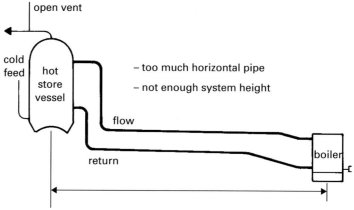

open vent

cold feed

hot store vessel

– too much horizontal pipe
– not enough system height

flow

return

boiler

(a) Poor circulation

open vent

flow

hot store vessel

cold feed

3 m minimum system height

return

boiler

In gravity circuits good circulation depends on system height and pressure differentials between hot flow and cooler return to overcome resistances in pipes and bends.

(b) Good circulation

Figure 3.24   Circulation in primary circuits (gravity)

Table 3.1   Examples of water temperature and density

Temperature °C	Density kg/m²
4	1000
40	992
50	988
60	983
70	978
82	974

Example of calculation of circulating head in gravity circuits

The circulating head (CP) can be calculated by the formula:

$$CP = (Dr - Df)\,9.81 \times \text{circulating height (m)}$$

where

Dr = density of return pipe (kg/m³);
Df = density of flow pipe (kg/m³);
9.81 = value of gravitational force in Newtons (N)

So taking figure 3.24(b) as an example and assuming flow and return temperatures of 82°C and 60°C, respectively:

$$\begin{aligned} CP &= (Dr - Df)\,9.81 \times \text{circulating height (3 m)} \\ &= (983 \times 974) \times 9.81 \times 3 \\ &= 264.87 \text{ N/m}^2 \text{ or } 264.87 \text{ Pa} \end{aligned}$$

3.5   Primary circuits

Primary circuits are used for circulation of hot water between the boiler and the hot store vessel and include any radiator circuits (see figures 3.25 and 3.26). They may be vented or unvented. Circulation may be gravity or pumped, direct or indirect although the use of gravity circulation is discouraged under Building Regulations.

Direct systems should not be used in modern systems because of the problems of hard water scale and corrosion in primary circuits due to the constant changing of the water. Corrosion may also lead to discoloration of water drawn off from taps.

Vented primary circuits are required to have two important pipe connections in addition to those for the primary flows and return circulating pipes.

(1)   The first is a permanent 'safety' vent route from the hottest, topmost part of the boiler to a point terminating above and into the feed and expansion cistern. This will provide for the immediate release of boiling water and steam should the boiler overheat.
(2)   The second connection is the cold feed pipe which provides a route for filling and replacing primary system water, and also provides a route for the expansion of water when heated, the excess being accommodated in the feed and expansion cistern.

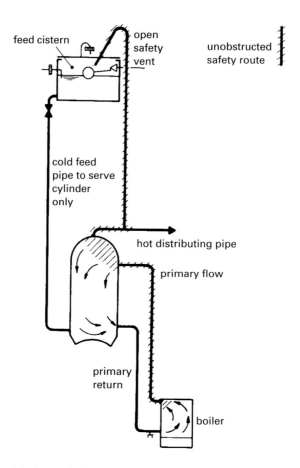

All pipes to be laid to falls to avoid air locks and facilitate draining.

Vent route from top of boiler through cylinder to open vent not to be valved or otherwise closed off.

Minimum sizes for primary circuits to hot store vessels:
- ○ 25 mm for solid fuel gravity systems
- ○ 19 mm for small bore systems

Corrosion inhibitors not to be used.

Figure 3.25   Direct system of hot water (vented)

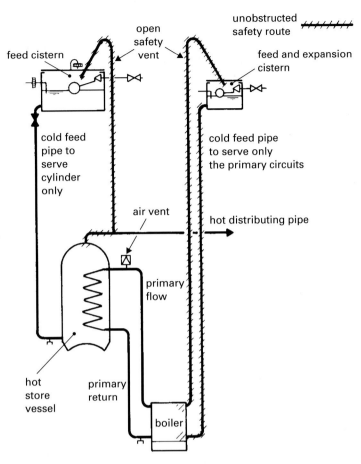

feed cistern

open
safety
vent

unobstructed
safety route

feed and expansion
cistern

cold feed
pipe to
serve
cylinder
only

cold feed pipe
to serve only
the primary circuits

air vent

hot distributing pipe

primary
flow

hot
store
vessel

primary
return

boiler

System shown is of the double feed indirect type.

Open vent and cold feed pipes to primary circuit should *not* have valve, or be otherwise closed off.

The open vent and cold feed pipes may be connected to the primary flow and return pipes respectively.

Where vent pipe is not connected to highest point in primary circuit, an air release valve should be installed.

Figure 3.26   Indirect system of hot water (vented)

Hot water systems using the single feed indirect cylinder

The single feed indirect cylinder has little application today, although many systems still remain in use. It is not recommended in modern pumped systems because its air seal is likely to be lost. However, manufacturers can still supply them, if asked, for replacement purposes.

The single feed indirect cylinder permits cost savings on materials and installation because there is no need for a separate open vent and cold feed for the primary circuits; see figure 3.27.

They are, however, limited in use as the volume of the cylinder is directly related to the volume of the water in the primary circuit (including radiators). If the system is oversized or becomes overheated the cylinder might lose its water seal (air lock) and revert to direct circulation.

Single feed indirect cylinders must be vented to atmosphere (over the feed cistern).

BS 6700 gives the following recommendations for the use of single feed indirect cylinders:

(a) Cylinders should conform to BS 1566: Part 2 and be installed as per manufacturers' instructions.
(b) Insulation should conform to the requirements of BS 1566-1.
(c) Where the primary circuit is pumped, the static head must not exceed the pump head.
(d) Corrosion inhibitors or other additives must not be added to primary circuits.
(e) Boiler and radiator manufacturers' recommendations must be followed.

Classes and grades

Single feed indirect cylinders are pressure graded and should be used within the pressure ranges shown later in table 3.5. Their primary heat exchangers may be one of two grades based on the maximum permissible capacity of the primary circuit including pipes, boiler and any radiators installed. More information on single feed indirect cylinders is given in section 3.7.

Combined systems of hot water and central heating

It is usual practice in domestic and many non-domestic premises to combine both hot water and central heating functions in one system using a centralized heat source, e.g. boiler, with in some cases a supplementary heat source, such as a gas circulator or electric immersion heater for summer use.

The primary circuits to these systems will invariably incorporate the use of a circulating pump to distribute some or all of the heat produced by the boiler. Circulation methods for these systems may include:

• gravity hot water and pumped central heating;
• fully pumped hot water and central heating.

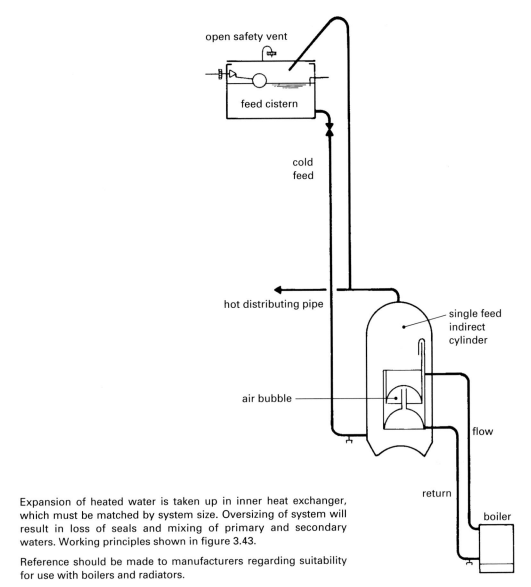

open safety vent

feed cistern

cold feed

hot distributing pipe

single feed indirect cylinder

air bubble

flow

return

boiler

Expansion of heated water is taken up in inner heat exchanger, which must be matched by system size. Oversizing of system will result in loss of seals and mixing of primary and secondary waters. Working principles shown in figure 3.43.

Reference should be made to manufacturers regarding suitability for use with boilers and radiators.

Installation similar to direct system.
When system is pumped, static head should exceed pump head.
No corrosion inhibitor to be added.

Figure 3.27 System using single feed indirect cylinder

These systems are illustrated in figures 3.28–3.31.

There is an increasing need to use the fully pumped system which has the advantage of smaller circulating pipework, leading to lower costs and quicker installation times. It is quick to respond and makes good use of more efficient thermostatic and timed controls.

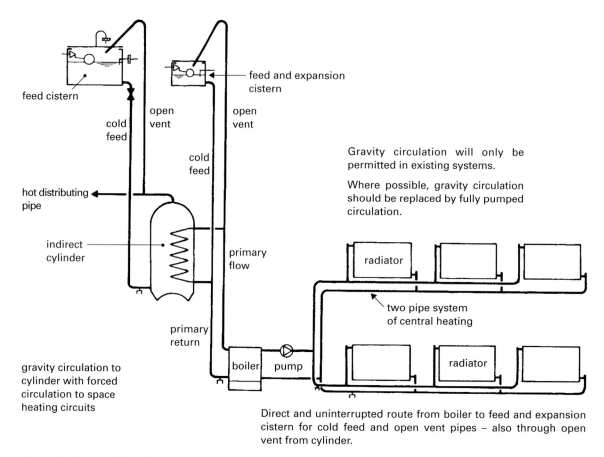

feed cistern

open vent

cold feed

hot distributing pipe

indirect cylinder

gravity circulation to cylinder with forced circulation to space heating circuits

primary return

feed and expansion cistern

open vent

cold feed

primary flow

boiler   pump

radiator

radiator

two pipe system of central heating

Gravity circulation will only be permitted in existing systems.

Where possible, gravity circulation should be replaced by fully pumped circulation.

Direct and uninterrupted route from boiler to feed and expansion cistern for cold feed and open vent pipes – also through open vent from cylinder.

Figure 3.28   Combined hot water and central heating system with vented primary circuits

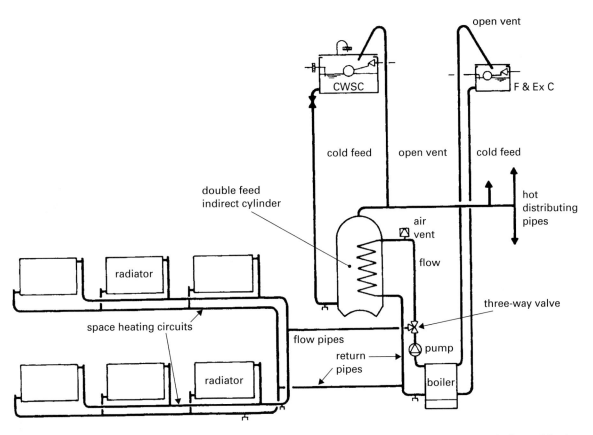

CWSC

F & Ex C

open vent

cold feed   open vent   cold feed

hot distributing pipes

double feed indirect cylinder

air vent

flow

radiator

three-way valve

space heating circuits

flow pipes

return pipes

pump

radiator

boiler

Cold feed and vent pipes uninterrupted between boiler and feed and expansion cistern.

Figure 3.29   Fully pumped system of hot water and central heating (vented)

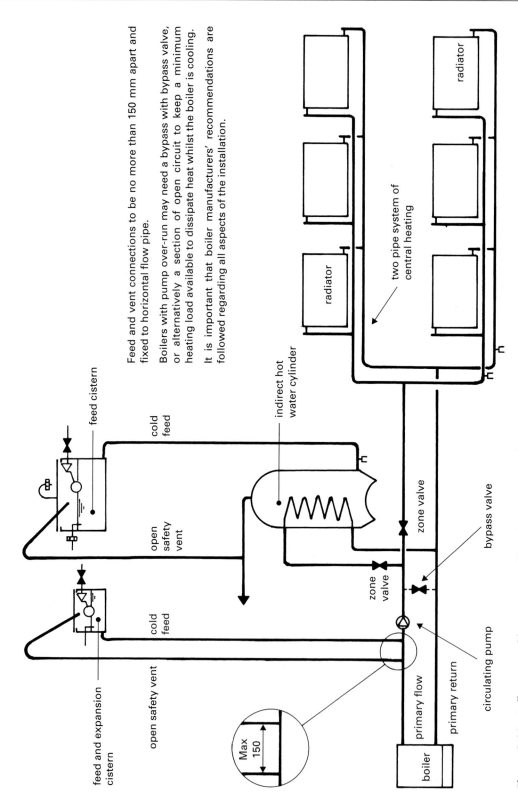

Feed and vent connections to be no more than 150 mm apart and fixed to horizontal flow pipe.

Boilers with pump over-run may need a bypass with bypass valve, or alternatively a section of open circuit to keep a minimum heating load available to dissipate heat whilst the boiler is cooling.

It is important that boiler manufacturers' recommendations are followed regarding all aspects of the installation.

Figure 3.30 Fully pumped hot water and heating system with close coupled feed and vent

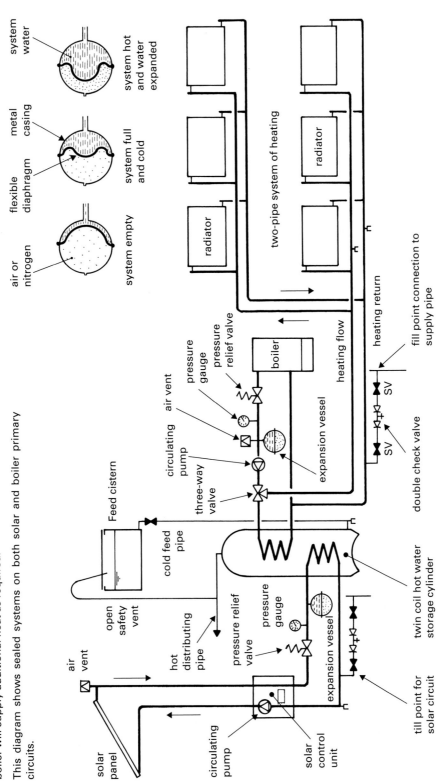

Detail to figure 3.31 showing operation of expansion vessel

Expansion vessel must accommodate expansion of the total volume of water in the system including space heating circuits (approximately 4%).

Sealed system requires the use of mechanical devices for safety in the control of temperature and expansion of heated water.

Solar panel will supply up to 60% of annual hot water requirement and boiler will supply additional heat as required.

This diagram shows sealed systems on both solar and boiler primary circuits.

Figure 3.31   Sealed central heating system with solar power to vented hot water

Sealed central heating systems with vented hot water

Sealed (unvented) primary circuits are often associated with the combination boiler with which they are commonly used. They can, however, be used with both vented and unvented secondary hot water systems as an alternative to the vented primary circuits previously discussed.

Sealed systems require the use of an expansion vessel to accommodate the increased volume of water when the system is heated through temperatures between 10°C and 100°C.

Sealed systems also require the use of a number of mechanical safety devices which are indicated in figure 3.31.

Primary heaters (coil or annular cylinder) should be suitable for operation at pressures of 0.35 bar above the pressure relief valve setting.

Supplementary water heating and independent summer water heating

It is common practice for supplementary water heating and independent summer water heating to be provided by use of electric immersion heaters in the storage vessel or by gas circulators. Supplementary water heating provided from solar energy or heat pumps is growing in popularity.

Where supplementary electric immersion heating is to be used in conjunction with a boiler, the storage vessel should be positioned at least 1 m above the boiler to prevent circulation of hot water from the storage vessel to the boiler when only the immersion heater is in use.

Solar heating

As global temperatures continue to rise, the need to reduce CO_2 emissions to the atmosphere is given increasing priority. This need is reflected by changes in legislation to encourage the use of 'alternative' energy sources in houses and other buildings. By using solar energy to heat hot water the rate of CO_2 emission could be significantly improved and the cost of heating water diminished.

Of the total 'heat load' in domestic premises, some 10% to 15% goes on hot water. Solar systems, correctly designed and properly installed, can provide up to 90% of domestic hot water needs in summer, although in winter, with shorter daylight hours, this is reduced to around 25% to 30%. Over the year, solar energy can account for 50% to 60% of the annual hot water energy requirements in a typical home.

Many different designs of solar system are possible. Systems may be individually designed for a particular location but manufacturers' packaged systems are proving to be very popular. Solar energy can be used with both vented and unvented secondary systems and designers can choose between a feeder system, or the more economically viable twin-coil cylinders. The two types of solar system are illustrated in figure 3.32.

The 'feeder' system uses a solar heated cylinder, which in turn feeds preheated warm/hot water to the main hot store vessel, to be topped up as necessary with heat from other energy sources, e.g. gas or electricity. The twin coil cylinder performs both feeder and hot water storage functions and is more efficient. A sealed primary circuit is preferred and this is illustrated in figure 3.31 where a twin coil solar system is used in conjunction with a sealed central heating system.

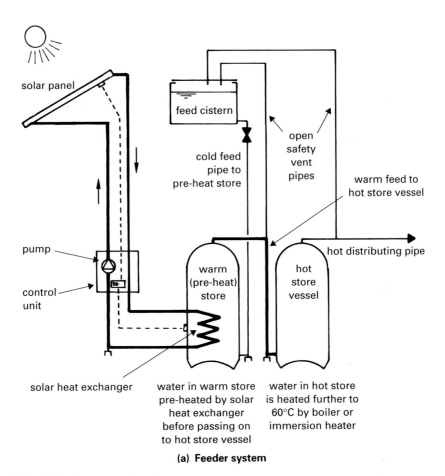

solar panel

feed cistern

pump

control unit

cold feed pipe to pre-heat store

open safety vent pipes

warm feed to hot store vessel

hot distributing pipe

warm (pre-heat) store

hot store vessel

solar heat exchanger

water in warm store pre-heated by solar heat exchanger before passing on to hot store vessel

water in hot store is heated further to 60°C by boiler or immersion heater

(a) Feeder system

Figure 3.32 Solar hot water heating systems

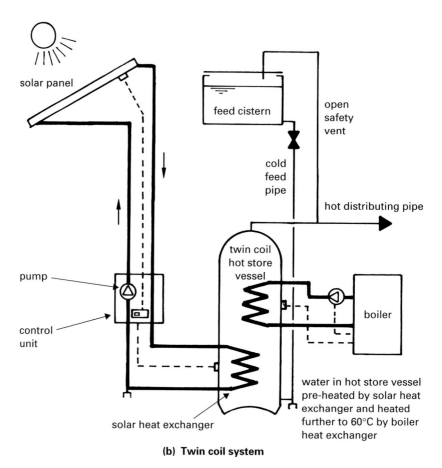

(b) Twin coil system

Figure 3.32 continued

Designers/installers should check if planning permission is required before commencing the work and installations must comply with Water and Building Regulations. Account should be taken of relevant codes or standards and manufacturers' recommendations should be followed. Components should be durable and have a useful life that compares favourably with that of other building services equipment. When fixing collector panels, care should be taken not to compromise roof loading, weather tightness, fire resistance or insulation of the building.

System design should ensure that flow rates through each collector are equal and air-venting arrangements are included so that all air can be expelled. Flow velocities should not exceed 2 m/s (preferably 1 m/s) to reduce risk of noise and erosion.

The electrical power input to the circulation pump in the solar primary circuit should be less than 50 W, or 2% of the peak thermal power of the collector, whichever is the higher.

System components

Solar systems consist of one or more solar collectors linked to a hot store cylinder, a primary pipework system and a pump station/control unit with electrical controls.

Solar collectors are available in two main types, flat plate panel and vacuum tube. Flat plate types cost less but vacuum tubes are said to collect energy a little more efficiently. Collectors should be independently certified as complying with all tests relating to safety, thermal performance and identification according to BS EN 12975. Copies of the full test report should be made available upon request.

Collectors should ideally be located in full sun on a south to south-west facing roof although any position between south-east and south-west will give good results (see figure 3.33). Any roof pitch between 5° and 60° will provide good heat absorption, but a 27° roof pitch will give the best results (see figure 3.33). Where a suitable roof is not available, a vertical wall position will give good results albeit with some loss of performance. On flat roofs a roof stand will be required.

If shaded, collectors can suffer some loss of performance although modern collectors, designed for northern European weather, still give reasonable results in diffused light and cloudy conditions.

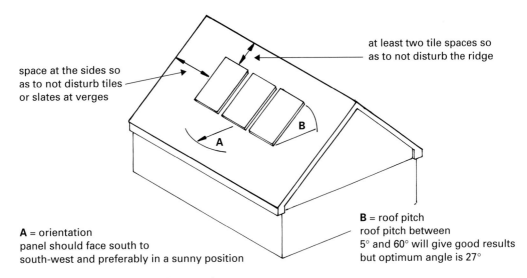

Figure 3.33 Positioning solar panels

Solar water storage. The *Domestic Heating Compliance Guide* advises that solar heated water storage vessels should conform with the following minimum provisions which apply to both new and replacement installations:

a) Vented copper hot water storage vessels should comply with the heat loss and auxiliary heating heat exchanger requirements of BS 1566-1.
b) Unvented hot water storage system products should:
   o   meet the insulation requirements of BS 7206; or
   o   be certified by the British Board of Agrément, the Water Research Council; or
   o   be certified by another accredited body as complying with Building Regulations.
c) Primary storage systems should meet the insulation requirements of Sections 4.3.1 or 4.3.2 of the Water Heater Manufacturers Association *Performance specification for thermal stores*. Primary store temperatures are much higher than those of normal hot water storage so it is important that these are well insulated.

Author's note   Readers should be aware that, at the time of writing, the advice given in the *Domestic Heating Compliance Guide* for certification of unvented systems differs from the guidance given in Building Regulations, Approved Document G3. A revision of Part G should hopefully remove this anomaly. If in doubt the local building control body should be consulted.

The ratio of solar preheated water storage volume to collector area should be specified as follows. The dedicated solar storage volume should be:

• at least 25 l (or equivalent heat capacity) per net m^2 of the solar collector absorber area; or
• a volume (or equivalent heat capacity) that is equivalent to 80% or more of the daily hot water demand (as defined by SAP 2005).

Solar primary circuit fluid. The primary circulation loop from collector to hot water storage vessel heat exchanger should be filled with a suitable liquid that will resist the formation of lime scale, inhibit corrosion and prevent frost damage, any of which might otherwise restrict circulation or impair heat transfer into the collector and from collector to hot water cylinder. Manufacturers will recommend a suitable liquid for their systems.

The primary circuit liquid can become extremely hot, up to 200°C, so measures should be taken to prevent or reduce the formation of lime scale in the secondary system so that performance is not significantly affected.

Solar circulation pipes should be insulated to reduce heat loss to a minimum to retain heat energy for transfer into the cylinder heat exchanger. The distance between the preheat vessel and the hot water storage vessel it serves should be kept to a minimum to reduce heat energy loss.

Solar heat exchangers should be sized to ensure a low return temperature to the solar collector. Criteria for the sizing heat exchangers can be seen in table 3.2.

BS 5918 (withdrawn) suggests that the size of the preheat vessel should be related to the amount of hot water used. In a house with three to five occupants, the daily consumption is about 30 l per person at 55°C, but for one or two person households this figure rises to 35 l to 40 l per person. The volume of preheat water should equal the expected daily hot water consumption. Cylinders should be chosen to encourage stratification as this will help increase the amount of solar energy obtained.

Where a domestic solar hot water system is used in conjunction with an auxiliary heated thermal store, designers/installers should consider the use of a separate preheat storage vessel. This is because of the expected higher temperatures and lack of stratification, particularly in thermal stores with open pumped circuits.

Pump station and controller unit. This is the central part of the control system and is important to the safety and efficiency of the solar circulation system. System controls should make sure that:

• best use is made of energy gained from collectors within the system's hot store vessel(s);
• heat energy stored in the domestic hot water system is not lost other than by normal use of hot water;
• auxiliary heat sources do not switch on when there is adequate preheated solar water available at a suitable temperature for domestic use; and
• the solar system is secure against the adverse affects of excessive primary temperatures and pressures.

Table 3.2   Criteria for sizing solar heat exchangers

System flow rates	Heat exchanger area required
Up to 0.5 l per minute per m² of solar collector	At least 0.1 m² of heat exchanger area per m² of solar collector net absorber area
At or above 0.5 l per minute per m² of collector	At least 0.2 m² of heat exchanger area per m² of solar collector net absorber area

Additionally, controls are needed to:

- ensure that fluid is pumped around the system only when needed;
- sense the temperature difference at collector and preheat vessel;
- limit water storage to a safe temperature; and
- prevent excessive pump cycling.

Insulation

The temperature difference between collector and preheat vessel will affect the performance of the system. Good insulation will help to maintain hot temperatures from collector to preheat vessel, thus giving improved heat transfer. Extremes of temperature are a particular problem as collectors and circulation pipes may be subject to freezing weather in winter and extreme heat in summer. Correct use of insulation and antifreeze will cope with the cold. High temperatures within the pipes and collectors could lead to furring if precautions are not taken.

At times when there is no fluid flow, collector and pipe temperatures in summer could reach up to 200°C. Where copper pipes are fitted, soft soldered joints should not be used. Instead joints should be brazed or other types of heat resistant joints used.

There is also the need to insulate against condensation damage that might occur where collector fixings penetrate roofs.

Installation

Relevant safety procedures should be followed and safety legislation complied with. Suitable access equipment should be used when fixing panels to roof structures and special attention should be paid to any lifting requirements recommended by the manufacturer.

Fixings should be such that no maintenance is required during the life of the collector. Because solar panels are exposed to the weather, e.g. wind, snow and rain, they should be adequately supported to resist external loads in addition to the obvious weight of the collector. The additional weight of the collector should not impair the stability of the roof and/or its weather tightness. Fixing methods should ensure that no holes are left to permit entry by small mammals, birds or large insects, particularly where pipes and fixings pass through the surface of the roof.

When fixing to sloping roofs, space should be allowed for the following:

- at least two tile spaces above so as not to disturb the ridge;
- space at the sides so as not to disturb tiles/slates at verges;
- room within the roof for access to make top and bottom connections to the panel.

Collector temperatures well in excess of the boiling point of water may be achieved during installation before the circulation liquid is added. Care should be taken to avoid burns, the shock of which could cause the installer to fall. To prevent this, it is recommended that collectors are shaded during installation with a removable cover. Shading will also reduce risk of damage to the collector.

Systems should be inspected and hydraulically tested as part of the commissioning procedures.

Heat pumps

Heat pumps extracting energy from the ground, water or air at ambient temperatures can be used to preheat conventional hot water systems, to augment existing systems, or to supply full hot water and central heating requirements (figure 3. 34).

Under Building Regulations, heating systems using a heat pump as the heat generator are required to meet the conditions set out in Section 8 of the *Domestic Heating Compliance Guide*. These conditions apply equally to new and existing dwellings. The Guide looks at three types of heat pump system. The systems are described in table 3.3. The principle of heat pump operation is shown in figure 3.34 and ground source heat pumps are illustrated in figure 3.35.

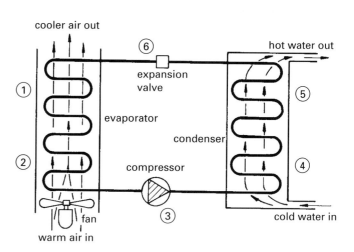

The heat pump works on the 'refrigeration cycle' principle (air to water)

(1) Refrigerant liquid forced under pressure into evaporator.
(2) Heat energy from air absorbed by refrigerant as it is vaporized from liquid to gas.
(3) Compressor 'squeezes' gas which becomes hotter and more compact.
(4) The hotter compacted gas is pumped under pressure into the condenser.
(5) In the condenser the gas becomes liquid and throws off heat which is absorbed into the hot water.
(6) Condensed refrigerant liquid passes through expansion valve, to lower its pressure, before flowing through to evaporator to begin the cycle again.

Figure 3.34   The heat pump principle (air to water)

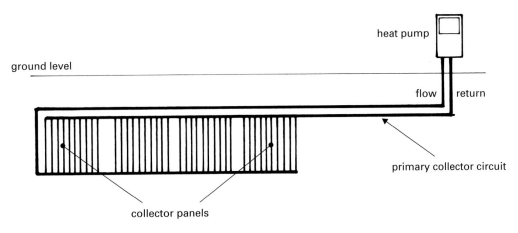

collector panels

(a) Compact type collector

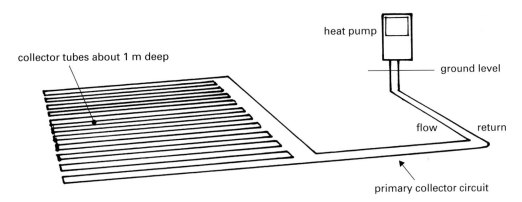

(b) Horizontal type collector

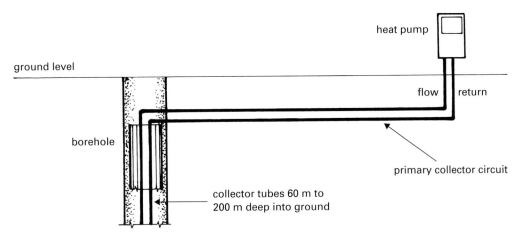

(c) Vertical type collector

Figure 3.35   Collectors for ground source heat

Table 3.3   Heat pump systems

Heat pump types	Warm (hot) water systems	Warm air Systems
Ground source systems extract their heat from below the ground, either from horizontal pipes laid in trenches or from vertical pipes bored deep into the ground. Circulation fluid should be added as recommended by the heat pump manufacturer	Ground to water	Ground to air
Water source systems extract their heat indirectly from a stable water source such as a river, lake or pond. The circulation fluid may be water but a water/propylene glycol or other acceptable equivalent antifreeze mixture may be used, depending on operating temperatures	Water to water	Water to air
Air source systems take their heat energy directly from the atmosphere at ambient temperatures. The heat may be transferred to the building via a water-based heating/hot water system	Air to water	Air to air

Water temperatures and efficiency

Heat pumps can be used in dwellings to provide the entire hot water and space heating needs of the building or may be used in conjunction with a supplementary heat source.

Where the heat pump serves hot water, independently or with a space heating requirement, the heat pump should be capable of supplying water temperatures in the range 60°C to 65°C. However, where only radiator circuits are supplied, circulation water temperatures of 40°C to 55°C are preferred using radiators of the high-efficiency type with high water volume. If the heat pump is unable to deliver these temperatures, supplementary heating should be provided. For optimum efficiency, the source temperature should be as high as possible and the heat distribution temperature as low as possible. The domestic hot water system should be fitted with a thermostat and a time clock to regulate the water temperature and the time taken to heat the water.

Electrically driven heat pumps should be designed for a *performance coefficient* of 2.0 or more at the heating system design condition. Typically a heat pump can be expected to provide between 3 and 5 kW of heat energy output for every kW of energy put in. For example, if a heat pump uses 3 kW of electrical energy to provide 9 kW of heat, the coefficient of performance (COP) would be 3.

Installation

It is recommended that a sealed (pressurized) system is used for primary hot water and heating circuits and a constant water flow should be maintained through the heat pump. Manufacturer's recommendations for pipe sizes should be followed. To maximize efficiency and ease of commissioning and maintenance, heating circuits should be arranged for reverse return operation.

Insulation is important to the efficiency of the system and the *Domestic Heating and Compliance Guide* suggests that the recommendations set out in *TIMSA HVCA guidance for achieving compliance with part L of the Building Regulations* should be followed. It is important that insulation be applied to all pipework between the dwelling and the ground heat exchanger, and to any pipes within the building that do not contribute to space heating needs of the building. See also chapters 7 and 8 for further information on insulation and efficiency in hot water.

The ground loop water circuit should be protected with a suitable antifreeze solution and primary heating circuits within the building with a corrosion inhibitor. In this, the heat pump manufacturer's recommendations should be followed. Both the ground loop circuit and primary heating circuits should be thoroughly flushed before the protective liquids are added.

Installation of heat pumps and associated equipment should be carried out by an approved installer, which, according to the *Domestic Heating Compliance Guide*, means a person who is approved by the manufacturer. Manufacturers provide courses designed to raise the competence levels of the installer.

If the refrigeration circuit should need attention, a competent refrigeration and air conditioning engineer should carry out the work. In this case a competent person is one who has a valid refrigerant handling certificate and/or an appropriate engineering services skillcard.

Installation of hot water and heating systems should be undertaken by a person who is competent in that area of work.

3.6 Secondary hot water distributing systems

Secondary hot water distributing systems include any hot store vessel, pipes and components used to store the hot water and convey it to the point of use.

A secondary hot water distribution system may be one of the following types:

(1) gravity fed (vented) from a cold water feed cistern;
(2) gravity fed (unvented) from a water storage cistern;
(3) directly supplied under pressure from mains, through an instantaneous water heater or a water jacketed tube heater;
(4) unvented storage type, directly supplied under pressure from mains.

Connections to hot water storage vessels should be arranged so that:

(a) the cold feed pipe is connected near the bottom, below any primary return connection or central heating element, thus ensuring that only the hot water is drawn off;
(b) the hot distributing pipe is connected to the top of the cylinder above any primary flow connection or heating element. This will minimize mixing of hot and cold water during draw-off and refilling, and reduce cold shock to the heating element.

Dead legs and secondary circulation

To promote maximum economy of energy and water the hot water distributing system should be designed so that hot water appears quickly at draw-off taps when they are opened. The Water Regulations Guide states that water should reach the tap at a temperature of 50°C within 30 seconds of fully opening the tap. The length of pipe measured from the tap to the water heater or hot water storage vessel should be as short as possible and should not exceed the lengths shown in figure 3.36. Where these lengths exceed the values given in table 3.4 the pipe should be insulated to at least the standards given in table 8.5.

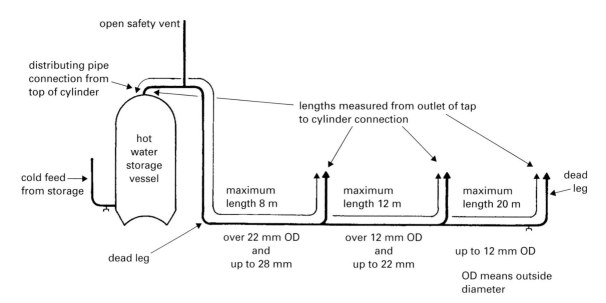

Figure 3.36   Maximum length of dead legs without insulation

Table 3.4   Maximum lengths of uninsulated distributing pipes

Outside diameter of distributing pipe mm	Maximum length m
Not exceeding 12	20
Exceeding 12 but not exceeding 22	12
Exceeding 22 but not exceeding 28	8
Over 28	3

Building Regulations require that pipes should be insulated using an insulation material that has a thermal conductivity value of 0.045 W/m.K or less and a thickness equal to the pipe diameter, up to a maximum of 40 mm.

Secondary circulation minimizes delays in obtaining hot water from taps and reduces waste of water during any delay but can lead to energy losses. Secondary circulation should be considered only when short dead legs are impractical and the circuit should be well insulated to reduce the inevitable heat losses from pipe runs. Secondary circuits should be avoided where possible. A diagrammatic arrangement of secondary circulation is shown in figure 3.37.

In secondary systems where it is not possible to attain gravity circulation, a non-corroding circulating pump should be installed to ensure that water within the secondary circuit remains hot. The pump should be located on the return pipe close to the cylinder.

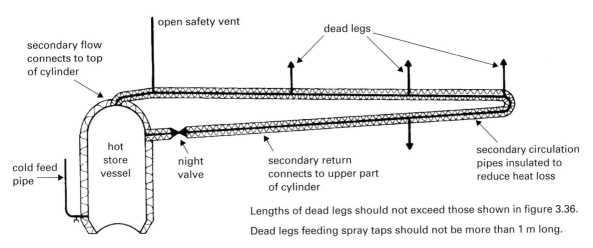

Figure 3.37 **Distributing system with secondary circulation**

Secondary returns should connect to the top quarter of the hot store vessel to take best advantage of the hottest water.

To economize on energy and prevent circulation during off-peak periods, a night valve should be fitted to the return near to the cylinder connection. This may be manually operated, but preferably an electrically operated time control valve will be used.

3.7 Components for hot water systems

Cold feed pipes

A cold feed pipe supplies hot water apparatus with cold water from storage (see figure 3.38). In vented systems of secondary hot water, the cold feed water is supplied from a feed cistern at high level and should be connected near the bottom of the hot store vessel. A servicing valve that will not inhibit or restrict the flow should be fitted to the cold feed pipe as near as is practical to the feed cistern.

The cold feed pipe should not be used to supply any fitting other than the hot store vessel. In a direct system, the cold feed pipe should connect directly to the boiler and not to the return pipe.

In a vented indirect system, other than one using a single feed indirect hot water cylinder, the primary circuit must be supplied through a separate cold feed pipe from a separate feed and expansion cistern. The cold feed pipe should be connected to the lowest point in the primary circuit near to the boiler (see figure 3.29) and preferably directly into the boiler.

Alternatively, in systems having gravity circulation to the hot store vessel, it is permissible to connect to the gravity return as shown in figure 3.28. Where feed and expansion cisterns have a capacity of 18 l or more a servicing valve should be fitted on the cold feed pipe near to its connection to the feed and expansion cistern. On smaller systems this valve is not necessary.

Cold feed pipes should be sized as part of the secondary distribution system. See chapter 6.

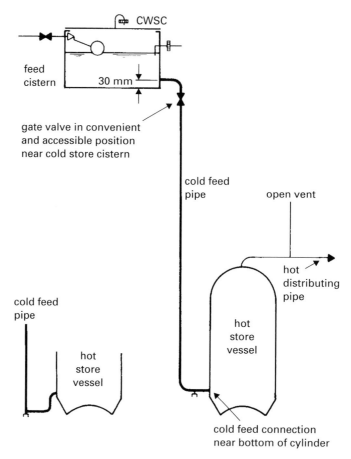

The dipped entry connection reduces the risk of hot water circulation in the cold feed pipe.

Cold feed must serve hot water apparatus only and no other appliance.

The cold feed pipe must *not* be connected to the boiler on direct systems.

No valve to be fitted to cold feed pipe on primary circuits.

Figure 3.38   Cold feed pipes

The open vent pipe

The open safety vent, as it is often called, is connected to the top of the hot storage vessel and rises to terminate above the cold feed cistern. Recommendations for the installation of the open safety vent are given in figure 3.39.

Vented primary circuits are similarly treated. The vent must terminate over a feed and expansion cistern. Due allowance should be made for any head produced by a circulation pump (if fitted) in order to prevent either continuous discharge from the vent, or air entrainment into the system.

The lower connection of the vent may be made to the highest point in the primary flow pipe, or preferably to the top of the boiler.

Vent connection offset to reduce parasitic (one pipe) circulation and loss of heat in vent pipe. The minimum offset recommended is 450 mm.

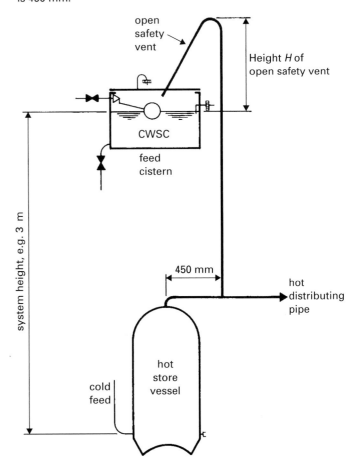

Rules for installation of the open vent:
- No valve to be fitted to vent pipe.
- Pipe to rise continuously from connection at hot store vessel to highest point over cistern.
- Vent pipe not less than 19 mm bore.
- Vent pipe *not* to be connected to cold feed.
- Formula for height *H* of open safety vent over cistern to prevent expanded water from overflowing:
  $$H = 150 \text{ mm} + 40 \text{ mm per metre of system height}$$
  For a system height of 3 m
  $$H = 150 + (3 \times 40)$$
  $$= 270 \text{ mm}$$

Note The formula shown above is for gravity systems. The height of the vent over the cistern in pumped systems should be related to the pump head.

Figure 3.39   The open safety vent

It should be noted that the open vent and cold feed pipes serve separate functions and should *not* be connected together unless the energy supply to each heater:

- is under thermostatic control;
- is fitted with a temperature-operated manually reset energy cut-out independent of the thermostatic control; and
- has a temperature relief valve in accordance with BS 6283-2, or a combined temperature and pressure relief valve in accordance with BS EN 1490, e.g. as required by BS 7206 and BS EN 60335-2-21.

Note   BS 7206 is withdrawn and replaced by BS EN 12897.

Cylinders supplied from common feed cisterns

Common cold feed pipes are permissible, provided the installation is as shown in figure 3.40 and where individual separate systems are not practicable. This is an arrangement commonly used in flats under single ownership, i.e. local authority or housing association developments. It can lead to problems where flats become privately owned and where pipes and cisterns are a shared responsibility. It is therefore not generally recommended.

Hot water storage vessels

Building Regulations (Approved Document L) require that hot store vessels be efficiently heated, well insulated and provide water at a suitable temperature for the user. Boiler-heated vessels should be fitted with an effective heat exchanger for the transfer of heat from the primary circuit to the secondary hot water. They should be fitted with thermostatic control to shut off the heat when the recommended cylinder temperature is reached, and a time switch to shut off the heat supply at times of no demand.

Hot water storage capacities should be related to likely consumption and the recovery rate of the hot water storage vessel.

For domestic use the temperature of stored hot water should not exceed 65°C and must not reach 100°C at any time. A temperature of 60°C should be adequate to meet all normal domestic requirements. A 60°C temperature will minimize scale formation and provide a suitable temperature for domestic use.

In other cases, such as in larger kitchens and in laundries, water temperatures may need to be higher. However, high temperatures should be avoided wherever possible because of furring in hard water areas and the danger of scalding users. See section 3.8.

In dwellings a hot water storage capacity of 35 l to 45 l per person per day is recommended subject to a minimum storage capacity of 100 l for solid fuel boiler systems, and 200 l for systems using off-peak electricity. However, for speculative housing where occupancy is unknown, a storage capacity of 115 l is usual. Domestic hot water requirements are given in table 3.5. Calculation of hot water storage capacities for larger buildings is dealt with later in section 5.4.

To reduce the risk of corrosion to the cylinder or tank, a protector rod (sacrificial anode) may be fitted by the manufacturer. The need for this will depend on the type of water supplied and usually applies to deep well waters. The water supplier's advice should be sought before deciding whether a protector rod is to be used. However, for vented copper

Table 3.5   Hot water supply requirements

Requirement (dwellings)	Quantity l	Temperature °C	Flow rate l/s
Stored water per dwelling	135	60	n/a
Hot water per person per day	35–45	40–60	n/a
Bath per use	100	40 mixed	0.2–0.3
Shower per use	25	32–40	0.05–0.1
Power shower per use	50	32–40	0.2–0.3
Wash basin per use	4.5	40–60	0.1–0.15
Kitchen sink per use	18	60	0.1–0.2

Separate vents are required to prevent hot water circulation from one cylinder to another.

Anti-siphon pipe is shown as an alternative backflow prevention device and is not needed if mechanical devices are used at each floor level.

Pipes to be aligned so as to avoid air locks and facilitate filling and draining.

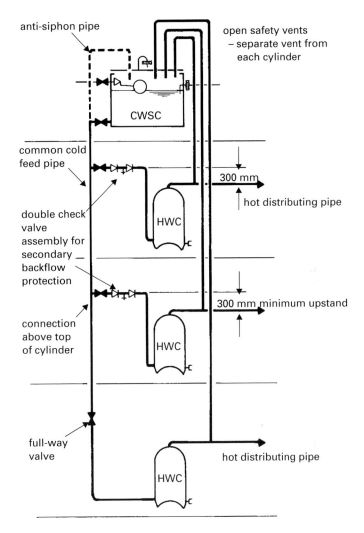

Figure 3.40 Common cold feed arrangement

cylinders to BS 1566-1 aluminium protector rods have been replaced by an increase in the thickness of the cylinder base and corrosion resistance testing. An aluminium protector rod may still be fitted when specified by the purchaser, but the cylinder might not then be designated as corrosion resistant in accordance with the standard.

Cylinders and tanks should be installed with the long side vertical to assist effective stratification or 'layering' of hot and cold water (see figure 3.41). The ratio of height to width or diameter should not be less than 2:1. An inlet baffle should be fitted, preferably near the cold inflow pipe, to spread the incoming cold water.

All hot water storage vessels should be labelled to show the following information:

- type of vessel;
- nominal size in litres;
- standing heat loss in kWh/day;

- heat exchanger performance in kW;
- BS number and Kitemark or other recognized quality control scheme reference; and
- insulation to BS 1566-1 or equivalent standard.

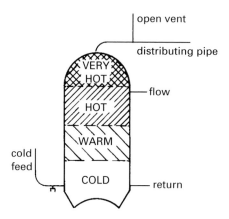

(a) Vertical cylinder

Vertical cylinder preferred to give better stratification and ensure that the hottest water is drawn off.

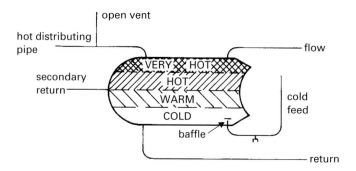

(b) Horizontal cylinder

Baffle will help prevent water from cold feed mixing with hot water at top of cylinder.

Hot water layer is very shallow and when drawn off may allow water layers to mix.

Figure 3.41   Cylinders showing stratification

Types of hot water cylinder

Hot water storage vessels are manufactured in a number of materials and in a variety of types and grades, and only cylinders and tanks to a relevant British Standard should be considered. Except in special cases, they should conform to BS 853-1, BS 1566-1, BS 1566-2, BS 3198 or BS 7206, as appropriate. Special cases would include cylinders outside the range of sizes covered by the above standards and the more energy efficient 'high performance cylinders'.

It is important that the correct grade of cylinder or tank is fitted, having adequate wall thickness to suit the internal pressure to which the vessel will be subjected (see table 3.6). Most cylinders for domestic use are made of copper but some are made from galvanized steel and stainless steel.

Connection bosses to cylinders can be fitted to suit individual requirements, but 'standard' connections may include those shown in the various diagrams in this book. For standard connections and dimensional details for cylinders to BS 1566-1, see figure 3.42 and table 3.7.

Table 3.6 Grades and pressure rating for hot store vessels

Type and BS number	Grade	Max. head	Grade	Max. head	Grade	Max. head	Grade	Max. head
Galvanized steel cylinders and tanks:								
tank, direct to BS 417: Part 2	A	4.5 m	3.0 m	18 m	n/a	n/a	n/a	n/a
cylinder, direct to BS 417: Part 2	A	30 m	B	18 m	C	9 m	n/a	n/a
cylinder, indirect to BS 1565 (withdrawn)	1	25 m	2	15 m	3	10 m	n/a	n/a
Copper cylinders:								
direct to BS 1566-1	1	25 m	2	15 m	3	10 m	4	6 m
double feed indirect to BS 1566-1	1	25 m	2	15 m	3	10 m	4	6 m
single feed indirect to BS 1566-2	n/a	n/a	2	15 m	3	10 m	4	6 m

n/a = not applicable.

Vented copper cylinders to BS 1566-1

BS 1566-1 includes specifications for both direct and double feed indirect cylinders. They are available in a range of capacities from 96 l to 1800 l as shown in table 3.7(a). Cylinders to BS 1566-1 are required to conform to one of three grades (1, 2 or 3) and to one of three types of construction (G, D or P).

Grades. Before ordering a cylinder the likely maximum working head (pressure) within the system should be established and the appropriate grade of cylinder chosen. The thickness of metal used to manufacture cylinders is calculated to suit the maximum working head or internal pressure that it is to be subjected to:

- **Grade 1** 25 m maximum working head
- **Grade 2** 15 m maximum working head
- **Grade 3** 10 m maximum working head

Cylinder types. The type of cylinder chosen is dependent on the type of system in which it is to be used. Indirect cylinders may be manufactured with annular or coil type primary heaters. For higher primary water pressures or higher primary water circulation rates, coil types are preferred.

- **Type G** Double feed indirect; for use in either gravity or pumped primary systems
- **Type P** Double feed indirect; for use in pumped primary systems only
- **Type D** Directly heated; for heating by internal immersion heater

Note Type G and type P cylinders may also include provision for direct heating.

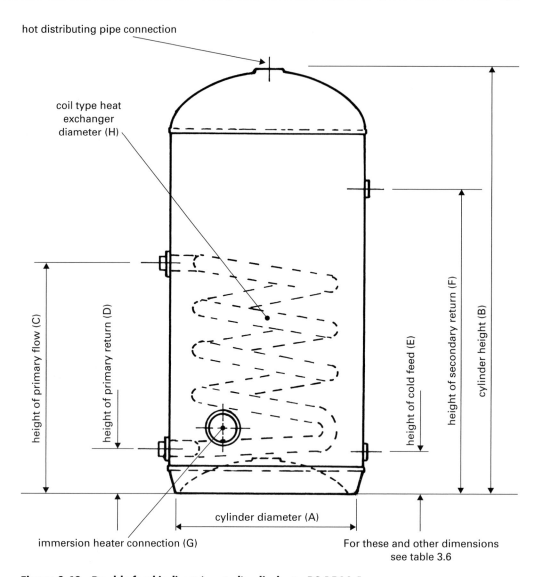

hot distributing pipe connection

coil type heat
exchanger
diameter (H)

height of primary flow (C)

height of primary return (D)

height of cold feed (E)

height of secondary return (F)

cylinder height (B)

cylinder diameter (A)

immersion heater connection (G)

For these and other dimensions
see table 3.6

Figure 3.42   Double feed indirect (vented) cylinder to BS 1566-1

Table 3.7   Vented copper cylinders to BS 1566-1

(a) Capacities and secondary connection details

| BS 1566-1 type reference | Cylinder diameter (mm) | Cylinder height (mm) | Nominal storage capacity (l) | Height of connection above base of cylinder | | | Secondary connection designations |
				Secondary return (mm)	Secondary feed (mm)	Immersion heater (mm)	
	A	B		L	H		
0	300	1 600	96	1 250	100	150	G1
1	350	900	72	700	100	150	G1
2	400	900	96	700	100	150	G1
3	400	1 050	114	800	100	150	G1
4	450	675	84	450	100	150	G1
5	450	750	95	550	100	150	G1
6	450	825	106	625	100	150	G1
7	450	900	117	700	100	150	G1
8	450	1 050	140	800	100	150	G1
9	450	1 200	162	950	100	150	G1
9E	450	1 500	206	1 200	100	150	G1
10	500	1 200	190	950	150	200	$G1\frac{1}{2}$
11	500	1 500	245	1 200	150	200	$G1\frac{1}{2}$
12	600	1 200	280	950	150	200	G2
13	600	1 500	360	1 200	150	200	G2
14	600	1 800	440	1 350	150	200	G2

(b) Primary connection details

| BS 1566-1 type reference | Minimum heating surface area (mm) | Diameter of primary heater tube (mm) | Height of connection above base of cylinder | | Primary flow/return connection designation |
			Primary return (mm)	Primary flow (mm)	
			J	M	
0	0.42	28	100	540	G1 B
1	0.31	28	100	400	G1 B
2	0.42	28	100	400	G1 B
3	0.50	28	100	470	G1 B
4	0.37	28	100	300	G1 B
5	0.48	28	100	340	G1 B
6	0.53	28	100	370	G1 B
7	0.61	28	100	400	G1 B
8	0.70	28	100	470	G1 B
9	0.79	28	100	540	G1 B
9E	0.96	28	100	620	G1 B
10	0.88	35	150	540	$G1\frac{1}{4}$ B
11	1.10	35	150	670	$G1\frac{1}{4}$ B
12	1.18	42	150	540	$G1\frac{1}{2}$ B
13	1.57	42	150	670	$G1\frac{1}{2}$ B
14	1.97	42	150	800	$G1\frac{1}{2}$ B

The reheat performance requirements for indirectly heated cylinders are specified in the *Domestic Heating Compliance Guide*. Primary heat exchangers should be capable of withstanding an internal working pressure of at least 3.5 bar and should have a minimum heating surface area and tube diameter as shown in table 3.7(b). When reheated, the cylinder should achieve a temperature of 60°C in 25 minutes or less. Type P cylinders (and type G cylinders, where required) must be able to deliver at least 85% of the nominal storage capacity at a temperature above 50°C.

Single feed indirect copper cylinders to BS 1566-2

Whereas double feed indirect cylinders require separate feed cisterns for both primary and secondary circuits, the single feed indirect cylinder requires only one feed cistern, the feed water to the primary circuit being obtained from within the cylinder through the primary heat exchanger. Expansion of primary water is accommodated within the primary heat exchanger where an air gap (bubble) is used to prevent primary water from coming into contact with secondary water. The working principles of the single feed indirect type hot water cylinder are shown in figure 3.43.

Single feed indirect cylinders have a limited application in direct pump circulation and are *not* suitable for use in sealed systems. This type of cylinder is *not* recommended for use in modern pumped systems because its air bubble is likely to be lost. The single feed cylinder is rarely used today but manufacturers can supply them, if asked, for replacement purposes.

Classes of cylinder. The cylinder class should be chosen to suit the capacity of the primary system including any radiators that might be installed. Cylinders should *not* be used with primary water capacities greater than that shown below, otherwise the air gap in the primary heater may be lost through expansion and the cylinder may revert to direct operation.

- Class 110   suitable for systems having a primary capacity of not more than 110 l
- Class 180   suitable for systems having a primary capacity of not more than 180 l

Cylinder grades. Before ordering a cylinder the likely maximum working head (pressure) within the system should be established and the appropriate grade (2, 3 or 4) of cylinder chosen. The metal thickness from which the cylinder is made will vary according to the grade:

- **Grade 2**   15 m maximum working head   2.20 bar test pressure
- **Grade 3**   10 m maximum working head   1.45 bar test pressure
- **Grade 4**   6 m maximum working head   1.0 bar test pressure

A specification for aluminium protector rods is included in BS 1566-2 with the recommendation that they be fitted to copper cylinders (during manufacture) for use in areas where the water is likely to cause pinhole corrosion. This, however, is in contradiction to BS 1566-1 where cylinders have extra metal thickness applied to the cylinder base, which is the preferred option for protection against pinhole corrosion failure.

Hot water storage combination units to BS 3198

There are two types, cylindrical (see figure 3.15) or non-cylindrical. The hot store vessel is supplied from a cold water feed cistern incorporated in the unit adjacent to the hot water

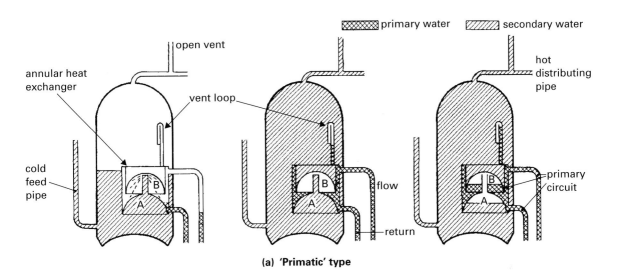

(a) 'Primatic' type

System filling Water from cold feed enters cylinder and from there into inner tank A. It then passes through inner tank B before finally filling the annular heat exchanger and the primary circuits. Air escapes from the vent loop.

System full Inner tank A is full. Inner tank B retains an air pocket. Annular heat exchanger is full and vent loop retains an air pocket. The air pockets prevent mixing of primary and secondary waters.

System heating up Hot water expands in primary circuit and annular heat exchanger. The expanded water pushes the air pocket from tank B to be retained in tank A and the air pocket in the vent loop is moved but also retained.

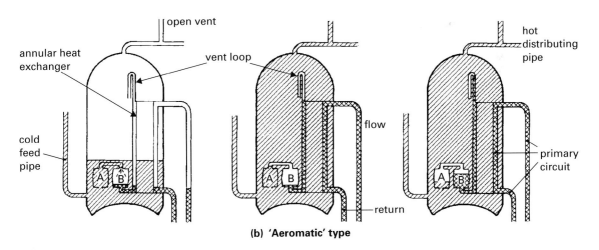

(b) 'Aeromatic' type

Figure 3.43   Single feed indirect cylinders – working principles

vessel. Direct, double feed indirect or single feed indirect types are available having hot water storage capacities ranging between 65 l and 210 l. These are illustrated in figures 3.15 and 3.16.

Preferred hot water storage capacities (litres)
85   115   130   140   150   180   210

Cylinders heated by electricity

Cylinders heated primarily by electricity should be fitted with two independent heating elements or immersion heaters, which may be factory fitted or inserted into bosses on site. The lower element should be connected to use off-peak electricity and should be able to heat at least 80% of the water contained in the cylinder. The upper element should be connected to use normal daytime tariff electricity. It should be able to heat at least 60 l of water but only to boost the off-peak heated water at times of heavy use. When the water has been fully heated to 60°C, the design of the vessel should permit at least 80% of the contents to be drawn off at a temperature of at least 45°C at a flow rate of 0.25 l/s.

High performance cylinders

The use of high efficiency cylinders should be considered. These give much improved performances compared with standard hot water vessels and are considered to be 'best practice' (see figure 3.44). Features of these cylinders include:

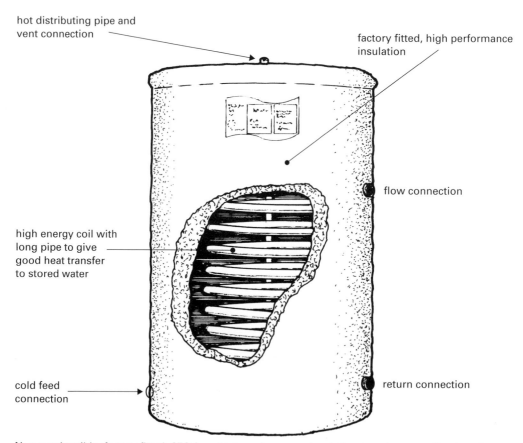

hot distributing pipe and vent connection

factory fitted, high performance insulation

flow connection

high energy coil with long pipe to give good heat transfer to stored water

cold feed connection

return connection

Non-combustible, factory fitted, CFC-free, high performance insulation to reduce standing heat loss.

High efficiency coil increases efficiency, improves recovery rate and permits smaller water storage capacity.

Figure 3.44   Typical high performance cylinder

- high efficiency coil to help save energy by reducing boiler cycling;
- faster recovery time, in some cases down to 15 minutes depending on the boiler used;
- better insulation (factory fitted) to reduce heat loss from the cylinder;
- standards exceeding those required under BS schemes; and
- highly suited to use with high efficiency condensing boilers.

Insulation of hot water cylinders and tanks

In order to minimize energy consumption, hot water cylinders and tanks are required under Building Regulations to be insulated. For compliance with the insulation requirements of Parts L1A and L1B, insulation in dwellings should follow the guidance given in the *Domestic Heating Compliance Guide*. Briefly, the guide says that vented copper hot water cylinders should comply with the insulation requirements of BS 1566-1.

Vented cylinders not within the scope of BS 1566-1 such as single feed indirect cylinders to BS 1566-2 and combination units to BS 3198 should also be insulated to a standard equivalent to that given in BS 1566-1. Unvented cylinders are required to be insulated to the standards set in BS 7206. Primary storage systems, because they work at higher temperatures, up to 82°C, should be insulated to the requirements of Sections 4.3.1 and 4.3.2 of the Water Heater Manufacturers' Association *Performance specification for thermal stores*.

For newly installed cylinders, the best insulating medium is that applied at the factory as part of the manufacturing process. This will also provide some protection to the cylinder during transit to site and during installation. For existing cylinders an insulating jacket to BS 5615 may be used, and fitted so that there are no parts of the cylinder wall exposed. Fixing bands should not be over-tightened, as this will reduce insulation efficiency.

Cisterns for supplying cold water to hot water apparatus

Feed cisterns used to supply the hot water system should conform to the requirements for cold water storage cisterns outlined in section 2.3. The feed cistern should be fitted at a height that will ensure satisfactory flow from all hot taps, and have a capacity at least equal to that of the hot store vessel that it feeds. A larger capacity will be required for combined storage and feed cisterns that also supply water to cold taps. Clause 5.3.9.4 of BS 6700 recommends at least 230 l minimum capacity for storage and feed cisterns, but this is inconsistent with other parts of the Standard.

Feed and expansion cisterns to primary circuits (see figure 3.45) should comply with the requirements of BS 417-2 and BS 4213 and with the requirements of Water Regulations for cold water storage cisterns. The cistern should be situated so that its water level is below that of the feed cistern at all times (see figure 3.45b).

Overflow and warning pipes from feed cisterns and feed and expansion cisterns should run separately and should terminate outside the building in a prominent position.

Boilers

All boilers, firing equipment, and components of a combined boiler and cylinder unit should comply with appropriate British Standards wherever possible.

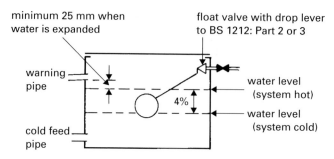

(a) Space for expansion water

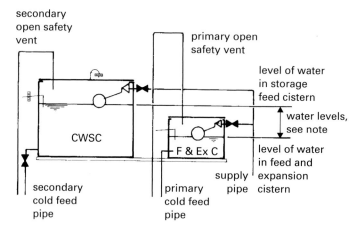

(b) Water level in feed and expansion cistern

Space must be allowed to accommodate expansion equal to 4% of water in circuit.

Cistern and float valve to resist temperature of 100°C.

Figure 3.45   Feed and expansion cistern

General provisions relating to boilers are as follows:

(1)   Domestic boilers should provide rapid heat-up of stored hot water, including any towel rails, airing coils and secondary circulating pipes. Modern systems will also provide for the heating of rooms.

(2)   Boilers and rooms containing boilers must be adequately ventilated to provide:
  (a)   sufficient air for combustion;
  (b)   enough air to remove the products of combustion and to discharge them safely and properly without danger to occupants of the building (see Building Regulations and the Gas Safety (Installation and Use) Regulations).

(3)   Boilers and flues must be installed so they will not cause excess heat or fire in the building (see Building Regulations).

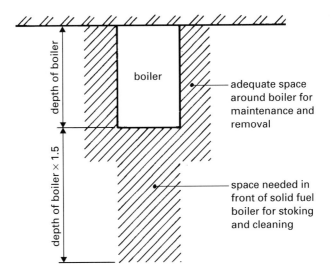

Figure 3.46   Space requirements for access to boilers

(4)   Adequate space for access (see figure 3.46) is needed for:
   (a)   boiler maintenance, removal of burners, pumps and pipe connections, and its eventual replacement;
   (b)   stoking and cleaning, i.e. 1.25 times the back to front dimension of the boiler, subject to a minimum of 1 m.

(5)   Precautions should be taken to ensure that combustible construction materials are not placed near boilers (see Building Regulations).

(6)   Fuel stores should be at a safe distance from the boiler (refer to fuel suppliers' recommendations).

(7)   Precautions should be taken to prevent explosion should water temperatures exceed 100°C. An explanation of these precautions follows in chapter 4.

(8)   In all cases, boiler manufacturers' recommendations should be followed for correct installation, commissioning and maintenance.

(9)   Solid fuel boilers produce heat energy for a long time after being stoked and cannot readily be shut down. Sufficient allowance should be made for the safe dissipation of that heat within the hot water or heating system.

(10)   Appliances should be tested and inspected to ensure they are suited to their purpose and labelled to indicate their performance rating.

(11)   Boilers for use with sealed primaries and unvented systems require specific safety controls and must therefore be selected from those that are specifically suitable for these purposes.

(12)   Boilers installed in dwellings are required under Building Regulations to meet SEDBUK energy efficiency ratings (see chapter 8).

Gas fired boilers should be appropriate for the gases with which they are used. Only high efficiency boilers may be used.

Under the Gas (Installation and Use) Regulations, gas installers are required to be registered with CORGI (The Council for Registered Gas Installers) and operatives must be able to prove they are competent in the gas work they do. Operatives are required to carry an identity card to show the customer that they are listed on the CORGI Data Base and are competent to carry out the work. The card will show proof that their competencies are up to date and will indicate the range of work that the operative is permitted to carry out.

CORGI has a duty to promote and enhance gas safety, standards and quality. As part of its duty CORGI maintains an up-to-date register of competent and qualified gas installation businesses. Registration is a legal requirement for businesses and self-employed people working on gas fittings or appliances. A person registered with CORGI is allowed to self-certify that their gas installations comply with Building Regulations.

Solid fuel boilers. Installers of solid fuel appliances in the United Kingdom should be registered with The Equipment Testing and Approval Scheme (HETAS). HETAS is an independent organization, recognized by government for testing and approval of domestic solid fuels, solid fuel burning appliances and associated equipment and services. To become registered under the scheme installers are obliged to prove their competence by successful completion of an appropriate HETAS course of training and assessment. A person registered under the HETAS scheme is allowed to self-certify that their installation of solid fuel burning combustion appliances meets the requirements of Building Regulations.

Oil fired boilers should be selected from the Domestic Oil Burning Equipment Testing Association (DOBETA) list of tested and approved domestic oil-burning appliances and installations.

Installers of oil fired boilers, storage tanks and associated equipment should be registered with The Oil-Fired Technical Association for the Petroleum Industry Ltd (OFTEC). OFTEC also runs the competent persons scheme covering oil-fired combustion appliances, and installers who are registered with them are considered to be qualified in oil installation work and are permitted to self-certify that their oil installation complies with Building Regulations.

Electrical work to boilers (or water heaters) in dwellings must comply with Building Regulations Part P 'Electrical Safety'. Compliance can be achieved if the installation follows the technical rules of BS 7671 *Requirements for electrical installations – IEE Wiring Regulations*, and is installed by a person who is either:

- an installer listed on an approved Part P competent person scheme as competent to carry out the work and qualified to complete a BS 7671 certificate; or
- a certificate holder of the National Inspection Council for Electrical Installation Contracting (NICEIC) and qualified to complete a BS 7671 certificate.

A person who is registered with a Part P competent persons scheme will be permitted to self-certify that his or her work is in compliance with building regulations.

The Benchmark Scheme

The Benchmark Scheme provides a code of practice for the installation, commissioning and servicing of central heating systems, the heart of which is the boiler. Central to the scheme is a 'Benchmark Commissioning Check List', supplied with each new boiler, to be used by the installer/maintenance engineer to record details of installation, commissioning and maintenance of the boiler. Within the log book, the installer should also complete information about themselves and their qualifications to do the work and, where appropriate, include their CORGI registration number. It is an important part of the installation procedure that the installer completes this book and signs it off to confirm that the work has been carried out in accordance with manufacturer's recommendations and that it complies with the energy requirements of Building Regulations Part L.

The Commissioning Check List should be left with the customer and every time the installer carries out a service the 'Service Record Sheet' section of the log book should be completed. This will then provide a complete and ongoing record to show the customer that the boiler (and heating system) continues to be kept in good working order.

Flues for boilers and water heaters

BS 6700 does not give details of requirements for flues and reference should be made to various other codes and in particular to the Gas Safety (Installation and Use) Regulations and to the Building Regulations. In general terms the following three points should be considered:

(1)   An adequate flue shall be provided wherever necessary.
(2)   All materials and components shall comply with the requirements of the appropriate standard.
(3)   The use of an inadequate or badly maintained flue can have fatal consequences.

Circulating pumps

Pump circulation is needed in all cases where natural circulating pressure is insufficient (see figure 3.47).

Immersed rotor (glandless) type circulating pumps must be used on primary circuits only. Pumps used for secondary circulation must be resistant to corrosion.

Inlet and outlet connections should be fitted with full-way valves, and space allowed for renewal or repair.

Circulating pumps should be quiet in operation and suitably suppressed to prevent radio or television interference.

An electrical isolating switch must be fixed adjacent to and within sight of the circulating pump. The whole of the wiring, earthing, etc. must be carried out by a competent electrician in accordance with BS 7671 *Requirements for electrical installations – IEE Wiring Regulations.*

Circulating pumps should comply with the requirements of BS EN 1151-1 and BS EN 60335-2-51, and should be installed in accordance with manufacturers' recommendations.

Valves and taps

Draw-off taps and combination taps should comply with the requirements of BS 5412 or BS EN 200. Most taps are manufactured from brass. Taps made of acetal materials are suited to both hot and cold water applications.

Valves used for isolating a section of the water service should provide a positive seal when closed. See also servicing valves, in section 2.4.

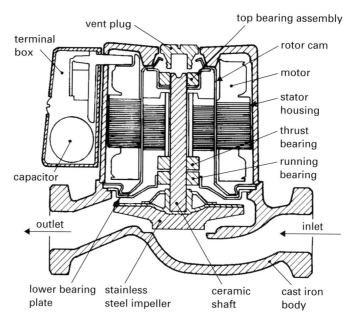

terminal box

vent plug

top bearing assembly

rotor cam

motor

stator housing

thrust bearing

running bearing

capacitor

outlet

inlet

lower bearing plate

stainless steel impeller

ceramic shaft

cast iron body

(a) Section through circulating pump

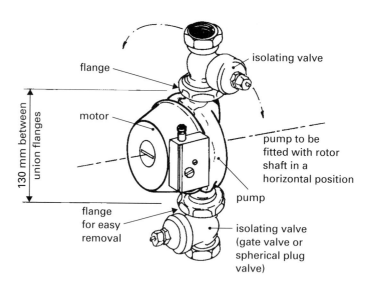

flange

isolating valve

motor

130 mm between union flanges

pump to be fitted with rotor shaft in a horizontal position

pump

flange for easy removal

isolating valve (gate valve or spherical plug valve)

(b) Typical pump arrangement

Figure 3.47   Circulating pump

Pressure operated, temperature operated and combined relief valves, check valves, pressure reducing valves, anti-vacuum valves and pipe interrupters should be fitted in accordance with Building Regulations and Water Regulations (see chapters 4 and 5).

Draining taps should comply with the requirements of BS 1010 or BS 2879 (see figure 7.8), and be suitable for hosepipe connection.

A sufficient number of draining taps should be fitted so that the entire water system can be drained down when not in use, thus avoiding frost damage.

A further advantage is that systems can be readily drained before repair work is carried out, so reducing the risk of mess and damage from pipes and fittings that are disconnected (see also chapter 7).

Mixing valves

Valves used for the mixing of hot and cold water (see figures 3.48–3.50) must be connected so that protection is provided against the possible contamination of water supplies. Where a mixer is used to control the water to more than one outlet, e.g. showers, it should be of the thermostatic type. Non-thermostatic mixers should not be used to control more than one outlet. Reference should also be made to the Health and Safety Executive Note (G) 104 'Safe' hot water and surface temperatures (see also section 3.8).

Tap mixers or combination taps

There are two types of tap mixers:

(1)   Those that mix water within the valve body. These should be supplied from the same source, i.e. both hot and cold from a common storage cistern, or both under direct mains pressure. This ensures that no crossflow can occur between water sources, and gives better and safer temperature and flow control.

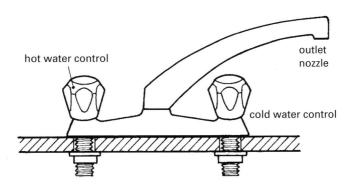

Figure 3.48   Sink mixer tap

(2)   Twin outlet types that mix water outside the valve outlet. These are of two designs. First, the divided outlet which creates no crossflow risk and can be connected to differing sources, and second, the type with a tube within a tube which is suitable for supply from separate sources, provided care is taken to ensure that the cold is on the outside, and the hot on the inside (see figure 3.49). This also reduces the risk of scalding.

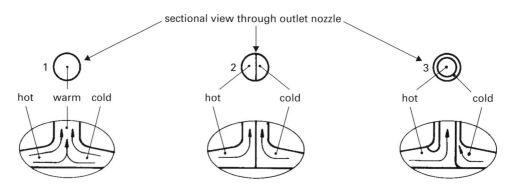

Internal mixing type

Mixing occurs in body of valve.

Precautions required against crossflow.

Difficult to maintain temperature where hot and cold inlet pressures vary.

Divided outlet with external mixing

Prevents crossflow between hot and cold supplies.

Some risk of scalding from hot outflow.

Double tube arrangement with external mixing

Prevents crossflow provided cold is on the outside and hot on the inside.

Reduced risk of scalding from outflow.

Figure 3.49   Types of mixer tap

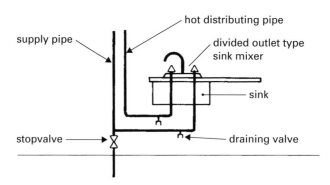

Hot and cold supplies from mixed sources may be connected as shown if the mixer is of the divided outlet type.

Where water is blended within the mixer, supplies should be from a common source. However, the above arrangement may be permitted if the hot water is supplied from a cistern complying with Water Regulations.

Figure 3.50   Sink mixer installation

Shower mixers

There are two types: thermostatic and non-thermostatic. Thermostatic types (see figure 3.51) are preferred because they provide automatic temperature control. They sense changes in temperature and adjust hot and cold flows accordingly. They also provide the user with a degree of protection against the effects of irregular flow and pressure variations caused when other taps are used. Pipework arrangements to showers are shown in figure 3.52.

High usage shower mixers supplying multiple shower outlets may need to be notified to the water undertaker before installation.

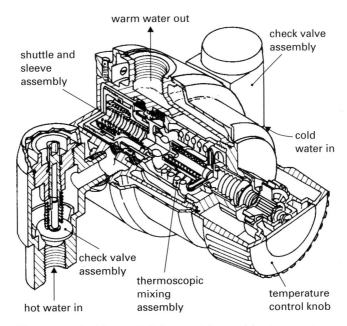

Thermoscopic (thermostatic) assembly provides temperature balance.

Shuttle and sleeve assembly will shut off hot water in the event of cold water failure.

Figure 3.51   Thermostatic shower mixer

(a) Storage-fed shower

Separate supplies to shower:
  o decrease risk of flow and pressure variations;
  o eliminate backflow risk.

Shower mixer supplied from one source, e.g. hot and cold from storage or both hot and cold direct from mains.

With a fixed shower head which has an air gap maintained at all times, there is no backflow risk.

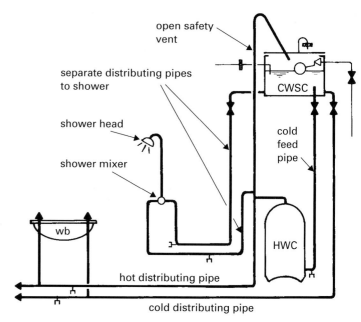

(b) Mains-fed shower

Where shower hose is constrained and type AA air gap is maintained at nozzle at all times, backflow prevention devices such as the double check valve assembly are not needed.

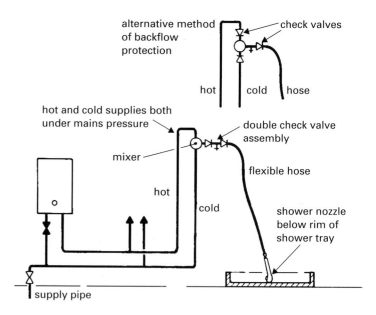

Figure 3.52   Shower installation

3.8   Hot water provision for the less able

Under Part M of the Building Regulations, provision must be made within buildings for people with various forms of disability. This will include the provision of accessible sanitary conveniences and safe and easy to use taps and valves.

To comply with Part M, taps on baths and wash basins in non-residential premises should either:

- be controlled automatically, e.g. by remote control; or
- be capable of being operated using a closed fist, e.g. by lever action.

Additionally, terminal fittings should conform to G18.5 of the DETR Guidance Note to Schedule 2 of the Water Supply (Water Fittings) Regulations 1999.

Under G18.5 taps installed in schools, public buildings, and in other facilities used by the public should be supplied with water through thermostatic mixing valves so that water is discharged from outlets at a safe temperature.

Shower fittings should also comply with G18.5 and shower controls should be clearly and logically marked.

These requirements are reinforced by information from the Deputy Prime Minister's Office highlighting the dangers of bath-time scalding injuries.

Shower and WC facilities

Within the provision of showers and WC facilities in buildings, whether for use by the public or by staff, Part M requires a proportion of these to be made accessible to the less able.

Individual self-contained bathing and showering arrangements should follow the recommendations shown in figures 3.53 and 3.54. Details of WC compartments for wheelchair users and ambulant disabled are shown in figures 3.55 and 3.56 respectively. Further information can be found in Approved Document M, *Access and use of buildings*.

Taps for use by disabled people

Taps for use by disabled people should be easy to use by people whose strength or manual dexterity is limited. Under Building Regulations (Section 5.4 of Approved Document M) any bath or basin in sanitary accommodation, other than dwellings, must be capable of being controlled automatically, or operated by a closed fist which means lever operated (see figure 3.57). Good practice would suggest that there is a case for these provisions to be extended to taps at kitchen sinks and to all taps in dwellings where the disabled have access.

Safe hot water temperatures

Information from the Office of the Deputy Prime Minister suggests that almost 600 serious scalding incidents each year are caused by hot bath water. Of these about 20 people die and around 570 suffer severe scald injuries. Statistics show that almost three quarters of the fatalities are aged 65 years or above and over three quarters of the severe scalds are suffered by children under the age of 5 years. (*continued on page 153*)

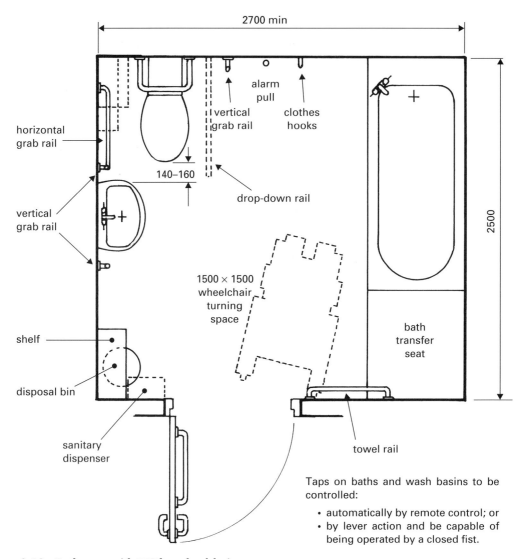

Figure 3.53   Bathroom with WC for wheelchair users

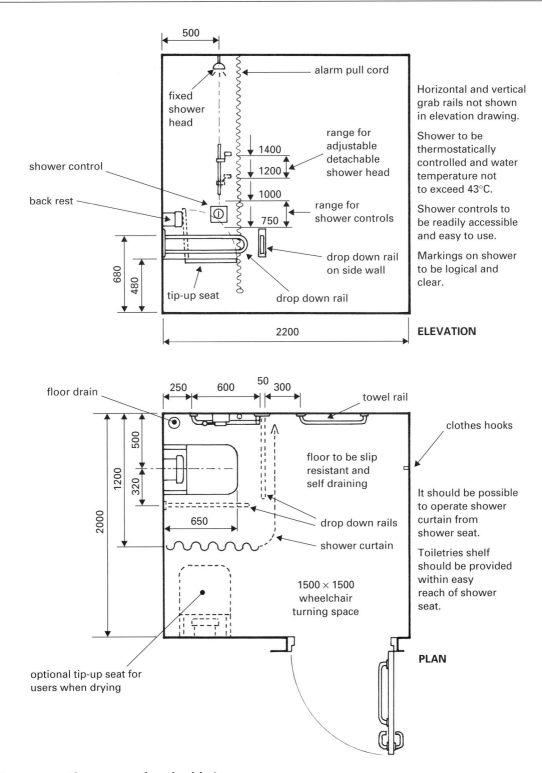

Figure 3.54   Shower room for wheelchair users

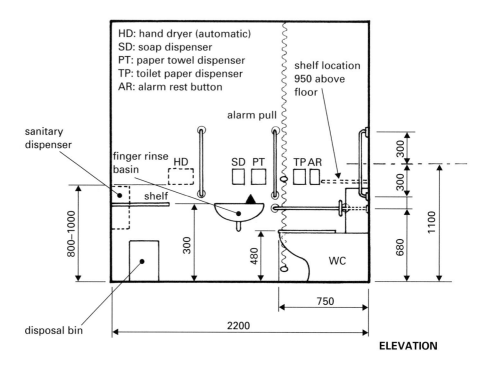

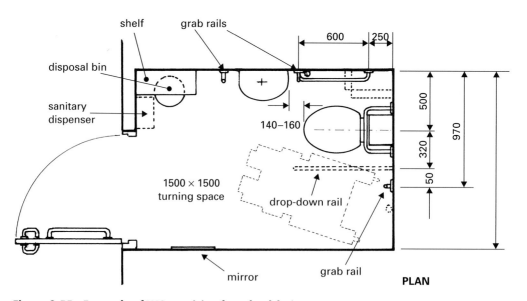

Figure 3.55   Example of WC provision for wheelchair users

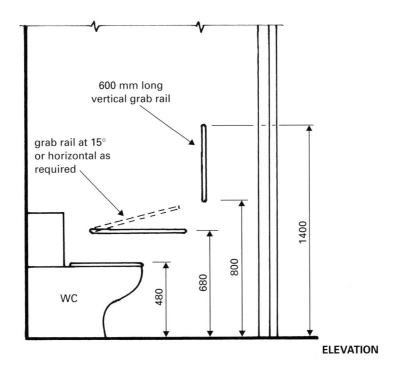

ELEVATION

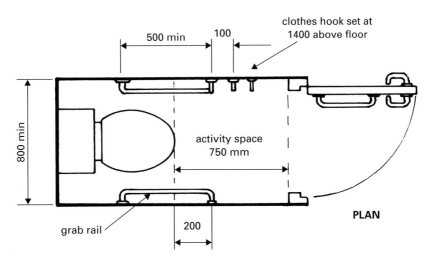

PLAN

Figure 3.56 Example of WC cubicle for ambulant disabled users

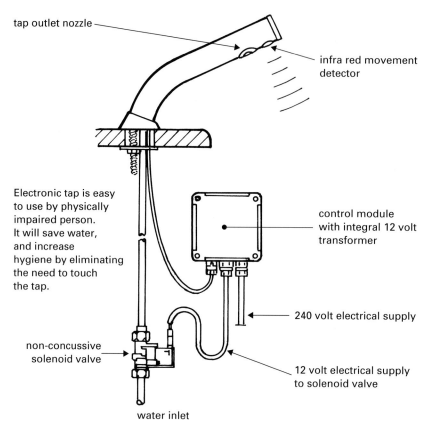

tap outlet nozzle

infra red movement detector

Electronic tap is easy
to use by physically
impaired person.
It will save water,
and increase
hygiene by eliminating
the need to touch
the tap.

control module
with integral 12 volt
transformer

240 volt electrical supply

non-concussive
solenoid valve

12 volt electrical supply
to solenoid valve

water inlet

(a) Tap with electronic flow control

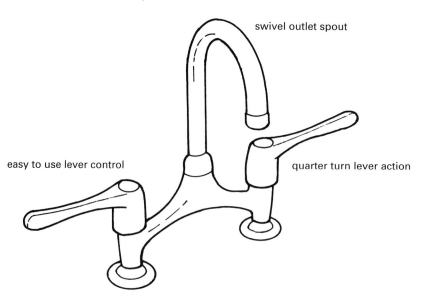

swivel outlet spout

easy to use lever control

quarter turn lever action

(b) Mixer tap, lever operated for ease of use

Figure 3.57   Taps for use by the less able

The degree of scalding depends the temperature of the water and the sensitivity of the skin. Research has shown that healthy adult skin is likely to suffer third-degree burns after 30 seconds' exposure to a temperature of 54–55°C. Third-degree burns are likely after 5 seconds' exposure at 60°C and only 1 second at 70°C.

The skin of young children and older people is more sensitive than that of the average healthy adult so third-degree burns are likely for them at lower temperatures than stated above.

Disabled people may be susceptible to scalds because of a variety of ailments. People with learning difficulties, sight or hearing difficulties, or problems with movement of limbs, and particularly people with low tactile sensitivity, may easily be injured by contact with hot surfaces or hot water.

To reduce the incidence of scalding, recommended water temperatures from draw-offs should be limited to the following:

- 46°C for bath fill;
- 41°C for showers and wash basins; and
- 38°C for bidets.

It should be borne in mind that although the maximum recommended safe temperature for hot water from a bath tap is 46°C, this is not considered to be a safe bathing temperature. The British Burns Association recommends 37°C as a comfortable bathing temperature for children. In premises covered by the Care Standards Act 2000, the maximum permitted outlet temperature is 43°C. In NHS premises, the maximum temperature from bath taps is generally given as 44°C.

It is expected that new provisions for the control of hot water temperatures will be introduced as and when a review of Part G of the Building Regulations (Hygiene) is completed.

Temperature control

Risk of proliferation of *Legionella* bacteria in water systems has led to recommendations that hot water be stored above 60°C and delivered above 50°C within 1 minute of opening a tap. This of course conflicts with the above recommendations for the prevention of scalding.

This problem can be overcome by the use of thermostatic mixing valves fitted at or near to the point of use. These can be very effective in accurately reducing the temperature of hot water issued from taps and maintaining it at a safe preset temperature even when water temperatures vary due to other draw-off points being used simultaneously. These valves should be positioned on the hot and cold distributing pipes near to the connection with the bath or basin (figure 3. 58).

Use of these valves will allow water to be stored and delivered at temperatures high enough to limit the growth of *Legionella* bacteria within the hot water supply system without compromising hot draw-off temperatures from the tap.

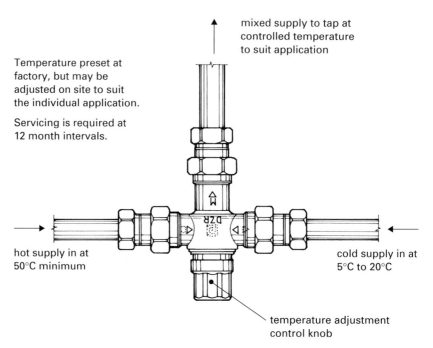

Temperature preset at factory, but may be adjusted on site to suit the individual application.

Servicing is required at 12 month intervals.

mixed supply to tap at controlled temperature to suit application

hot supply in at 50°C minimum

cold supply in at 5°C to 20°C

temperature adjustment control knob

(a) Typical in-line thermostatic mixing valve

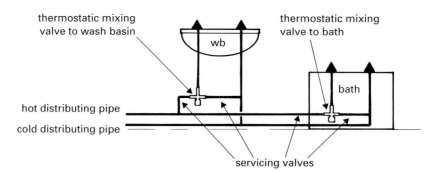

thermostatic mixing valve to wash basin

thermostatic mixing valve to bath

wb

bath

hot distributing pipe

cold distributing pipe

servicing valves

Water temperatures
 • Storage temperature 60°C to 65°C
 • Delivered water temperature 50°C minimum
 • Cold water temperature between 5°C and 20°C
 • Mixed water temperature as required:
   – 41°C for showers and wash basins
   – 46°C for bath, and
   – 38°C for bidets
   – not more than 43°C in premises covered by the Care Standards Act

(b) Mixing valves positioned near taps

Figure 3.58   Mixing valves for temperature control

Chapter 4
Prevention of bursting

It is extremely dangerous to heat water in a filled enclosed vessel to a temperature above 100°C. Water boils at 100°C at normal atmospheric pressure, but at 3 bar pressure, for example, the boiling point rises to 143°C and the heated water will have expanded and/or its pressure will have increased. If the water expands to such an extent that a small split develops in the cylinder, there will be an immediate drop in pressure and the water will flash to steam, increasing in volume by about 1600 times. This will result in the total rupture or explosion of the vessel.

Building Regulations require that the temperature of stored hot water shall not exceed 100°C.

Successful and safe operation of a hot water system depends on:

- the right equipment, properly installed in a system that is well designed and maintained, and not exposed to misguided interference;
- reliability and durability of safety devices and equipment;
- the use of approved appliances having all necessary safety devices fitted during manufacture to ensure correct assembly and calibration;
- good maintenance, for continued safety, with reasonable expectation that this will continue;
- safe and efficient control and use of energy, temperature, expansion and pressure.

Materials for hot water systems should conform to relevant British or European Standards, or be approved under a recognized fittings approval scheme, e.g. the Water Research Centre Water Fittings Scheme. Systems and components should, where appropriate, be approved under a recognized scheme of approval and certification, e.g. EOTA (European Organisation for Technical Approvals).

4.1   Energy control and safety devices

In hot water systems and appliances, control devices are needed to conserve energy and for the safety of the occupants of buildings, except:

- in systems where the heat source is incapable of raising the temperature above 100°C; or
- for instantaneous electric water heaters with a capacity of 15 l or less that are fitted with a CE mark or conform to BS EN 26 as appropriate.

Three independent forms of energy control are used: thermostatic control, temperature relief and heat dissipation. Table 4.1 gives an overview of hot water energy and safety controls.

Table 4.1   Energy and safety controls for hot water systems – an overview

Hot water systems	Method of energy and safety control		
	Level 1	Level 2	Level 3
Vented systems heated by solid fuel	Thermostatic temperature control	n/a	An open safety vent discharging to a feed cistern/feed and expansion cistern; and a temperature relief valve to BS 6283-2, or a combined temperature and pressure relief valve to BS EN 1490
Vented systems heated by gas or electricity	Thermostatic temperature control	Temperature-operated manually reset energy cut-out, independent of the thermostatic control	Open safety vent discharging to a feed cistern/feed and expansion cistern
Unvented systems up to 15 l capacity	Thermostatic temperature control	Temperature-operated manually reset energy cut-out, independent of the thermostatic control	(i) For an electric storage water heater to BS EN 60335-2-21, a temperature relief valve to BS 6283-2, or a combined temperature and pressure relief valve to BS EN 1490; (ii) For a boiler conforming to BS EN 483, BS 7977-2 or BS EN 625, or fitted with a CE mark, as appropriate, a pressure relief valve
Unvented systems over 15 l capacity	Thermostatic temperature control, e.g. cylinder thermostat or immersion heater thermostat	Temperature-operated manually reset energy cut-out, independent of the thermostatic control	A temperature relief valve complying with BS 6283-2 or a combined temperature and pressure relief valve complying with BS EN 1490, e.g. as required by BS 7206 and BS EN 60335-2-21
Water jacketed tube heaters	Thermostatic temperature control	Temperature-operated manually reset energy cut-out, independent of the thermostatic control	A temperature relief valve complying with BS 6283-2, or a combined temperature and pressure relief valve complying with BS EN 1490 or a second temperature-operated non-self-resetting cut-out with diversity of operation and different from that shown in column 2

Notes:
- Feed cisterns to vented systems to be capable of withstanding temperatures of 100°C.
- Thermostats, temperature-operated energy cut-outs and temperature relief valves or combined temperature and pressure relief valves must be set so that they operate in sequence (1, 2, 3) as temperature rises.
- Controls for unvented systems of more than 15 l capacity are mandatory under Building Regulations (G3).
- For further information consult BS 6700 or system/appliance manufacturers.

Thermostatic control

Effective thermostatic control is needed to prevent the temperature of stored water from rising above the design temperature, which for domestic hot water is 60°C to 65°C. This need applies to both vented and unvented systems. In boilers, water jacketed tube heaters and some industrial or commercial applications, a higher temperature may be required. This day-to-day temperature control of stored water can be achieved by using either a cylinder thermostat (figure 4.1) or an immersion heater thermostat (figure 4.2) depending on the methods used to heat the water. This is in addition to any thermostatic control needed on boilers and other types of water heating appliance (figures 4.3 and 4.4).

Thermostatic control should normally have the effect of shutting off the energy source (heat) immediately the design temperature is reached, and will, of course, allow the heat to be switched back on when the stored water temperature is lowered through use. There are, however, two exceptions:

(1) In fully pumped primary circuits where a low water content boiler is used and the manufacturer recommends the use of a pump over-run to dissipate any excess heat generated by the boiler after the boiler is switched off.
(2) In solid fuel boiler installations where the temperature of the stored water is controlled directly by the temperature of the water in the boiler. As the residual heat in

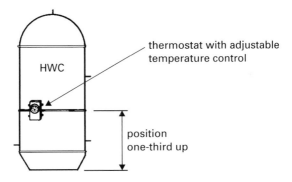

Figure 4.1   Cylinder thermostat

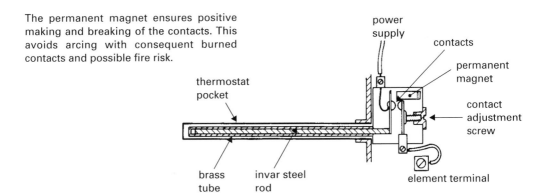

Figure 4.2   Section through immersion heater thermostat

Heat cannot be shut off immediately
as fuel takes time to cool down.

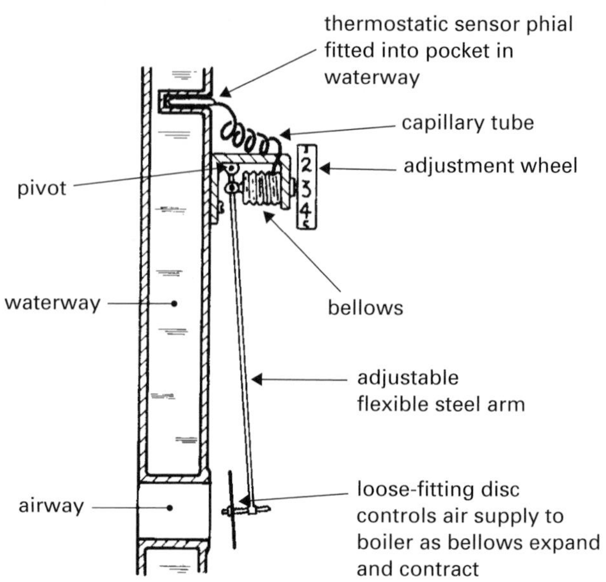

thermostatic sensor phial
fitted into pocket in
waterway

capillary tube

adjustment wheel

pivot

waterway

bellows

adjustable
flexible steel arm

airway

loose-fitting disc
controls air supply to
boiler as bellows expand
and contract

Figure 4.3   Thermostatic control for small solid fuel boiler

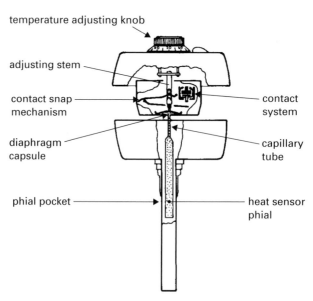

temperature adjusting knob

adjusting stem

contact snap
mechanism

contact
system

diaphragm
capsule

capillary
tube

phial pocket

heat sensor
phial

Used in gas or oil burners to give instant control.

Figure 4.4   Boiler thermostat, liquid filled

the boiler is slow to die down, the system should have adequate capacity to absorb any excess heat generated during the cooling period. For this reason systems heated by solid fuel boilers are required to have a minimum capacity of 100 l hot water storage.

Temperature-operated energy cut-out device

This device, which may also be known as a 'thermal cut-out' device, is used in unvented systems and will only operate if the normal thermostat fails and the hot store vessel over-heats. It must be of the non-self-resetting type, be independent of the thermostat and be designed to prevent the water temperature reaching 100°C. It can be expected to cut off the heat source at a predetermined temperature of about 90°C. There must be a separate energy cut-out device for each heat source.

In directly heated systems, with an immersion heater to provide the heat source, the energy cut-out will be situated within the thermostat. Heaters having two or more immersion heaters should have independent cut-out devices within each immersion heater. Immersion heaters used as supplementary heaters to indirectly heated systems should have their own cut-out device independent of any device fitted to control the heat from the boiler.

The indirectly heated system will usually have its energy cut-out device arranged to shut off a motorized valve on the flow to the cylinder, or directly to shut off the boiler, or alternatively there may be an energy cut-out device located on the boiler itself. Whichever method is employed, it is important that the water in the cylinder is not allowed to reach boiling point, which is the purpose of the energy cut-out device.

Temperature-operated energy cut-out devices are illustrated in figure 4.5.

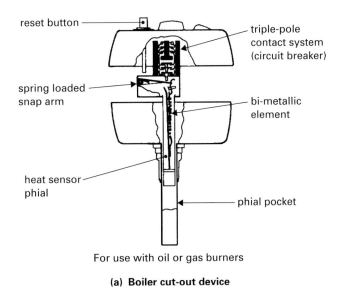

reset button

triple-pole
contact system
(circuit breaker)

spring loaded
snap arm

bi-metallic
element

heat sensor
phial

phial pocket

For use with oil or gas burners

(a) Boiler cut-out device

Figure 4.5   High energy cut-out devices

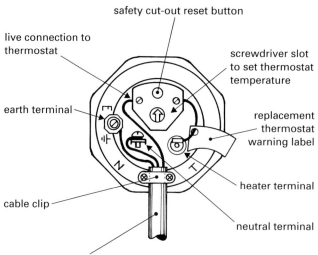

safety cut-out reset button

live connection to thermostat

screwdriver slot to set thermostat temperature

earth terminal

replacement thermostat warning label

heater terminal

cable clip

neutral terminal

heat resisting cable from double pole switch minimum cable size, 1.5 mm²

Recommended thermostat setting 60°C.

Safety cut-out set at 90° to prevent water reaching 100°C.

For safety, any replacement thermostat should be of the same type.

The thermostat is used to maintain day-to-day temperatures. If water temperatures rise excessively above the thermostat setting (60°C) the thermal cut-out will shut off the energy supply and prevent boiling. Once the thermal cut-out has operated, the reason for this should be investigated and any fault rectified before the cut-out is re-set.

The stored water must be allowed to cool before the cut-out can be re-set by lightly pushing the re-set pin into the cover until it reaches a stop.

(b) Thermal cut-out in immersion heater

Figure 4.5   continued

Temperature relief and heat dissipation

Adequate means of dissipating the heat input are needed in case both the temperature thermostat and the energy cut-out fail. Three methods of temperature relief/dissipation are given below:

1.   A **vent pipe** to atmosphere, often known as the open safety vent, is used in the traditional vented hot water system, and shown in figures 3.25 and 3.26. Its purpose is to provide immediate release of boiling water should the thermostatic controls fail and cause the hot water to overheat.
2.   A **temperature relief valve** to BS 6283-2 (figure 4.6) is used in unvented systems and must be fitted in the top of the storage vessel, within the top 20% volume of the water.

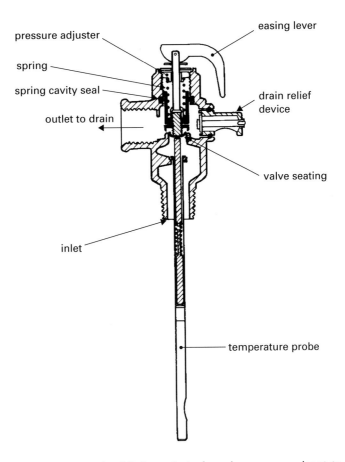

pressure adjuster

easing lever

spring

spring cavity seal

drain relief device

outlet to drain

valve seating

inlet

temperature probe

Discharge capacity 1.5 times that of maximum energy input to heater.

No valve to be fitted between temperature relief valve and heater.

Figure 4.6 Temperature relief valve to BS 6283: Part 2

This device is used as a safety back-up in case both the temperature thermostat and the thermal cut-out device fail. There is also an application for this valve in vented systems heated by solid fuel.

3. A **combined temperature and pressure relief valve** to BS EN 1490 may be used in unvented systems as a safety back-up or as an alternative to a temperature relief valve.

The water discharged from a temperature relief valve or combined temperature and pressure relief valve must be removed from the point of discharge to a safe place. In operation the valve will open at a pre-set temperature to permit the overheated water to escape safely from the hot water storage vessel before it boils. It will usually operate at about 95°C.

Maintenance and periodic easing of temperature relief valves are particularly important for continued efficiency. A notice drawing attention to this should be provided in a prominent position for the user.

Sealed primary circuits

When sealed primaries are used, it is permissible for a second temperature-operated energy cut-out to be employed in place of the temperature relief valve. Sealed primary circuits must not be heated by solid fuel. Both a vent and a temperature relief valve must be fitted in addition to a thermostat if water in a primary circuit or in a direct system is heated by solid fuel. This is because complete thermostatic control and an effective temperature-operated energy cut-out are not available.

Sequence of operation

Thermostats, temperature-operated cut-outs and temperature relief valves must be set to operate in this sequence as the temperature increases. These three devices are not essential where water is only heated indirectly by a primary circuit which is already protected, or from a source of heat that is incapable of raising the temperature above 90°C.

Discharges from temperature relief valves and expansion/pressure relief valves

Building Regulations, in Approved Document G3, state 'there shall be precautions to ensure that the hot water discharged from safety devices is safely conveyed to where it is visible but will cause no danger to persons in or about the building'.
   Two important points are stated here.

(1)   Any discharge from a temperature relief valve or expansion relief valve must be readily visible. This will:
   (a)   show there is a fault on the system that requires maintenance;
   (b)   reduce wastage of water because faults will be seen and rectified.
(2)   Any discharge must be to a safe place. This will apply more importantly to the temperature relief valve, which will discharge very hot water from the top of the hot storage vessel, whereas the discharge from the pressure relief valve situated on the cold supply pipe will be relatively cold. See figures 4.7 and 4.8.

The use of a tundish (see figure 4.9) will permit greater flexibility in the positioning of appliances and discharge outlet and will also provide visibility where discharges are more than 9 m from the appliance.
   Discharges from expansion or pressure relief valves can be treated a little more leniently than those from temperature relief valves as the danger from the discharge is considerably less.

Additional considerations for multiple installations, e.g. flats

The problems here are ensuring that visibility is maintained and also knowing which appliance is at fault should a discharge occur (see figure 4.10).

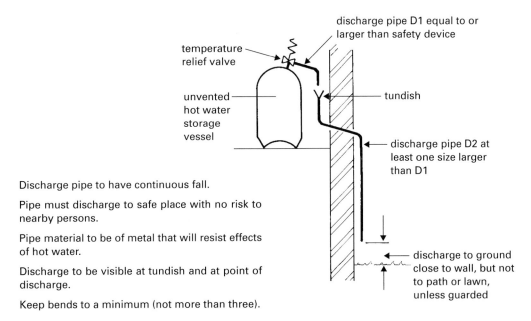

discharge pipe D1 equal to or larger than safety device

temperature relief valve

unvented hot water storage vessel

tundish

discharge pipe D2 at least one size larger than D1

Discharge pipe to have continuous fall.

Pipe must discharge to safe place with no risk to nearby persons.

Pipe material to be of metal that will resist effects of hot water.

Discharge to be visible at tundish and at point of discharge.

Keep bends to a minimum (not more than three).

discharge to ground close to wall, but not to path or lawn, unless guarded

Figure 4.7   Discharge from temperature relief valve

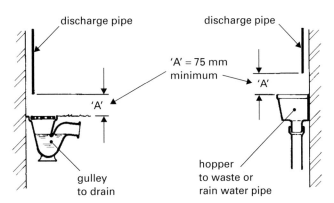

discharge pipe

discharge pipe

'A' = 75 mm minimum

'A'

'A'

gulley to drain

hopper to waste or rain water pipe

Consider suitability of drain or down pipe material for hot water.

Guidance document G3 to the Building Regulations shows the discharge pipe terminating below the gulley grating but above the trap water level. In this case the discharge will not be readily visible.

(a) To gulley   **(b) To RWP hopper**

Figure 4.8   Alternative methods of discharge

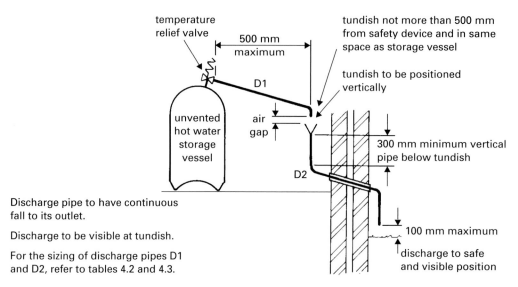

temperature
relief valve

|← 500 mm →|
maximum

D1

tundish not more than 500 mm
from safety device and in same
space as storage vessel

tundish to be positioned
vertically

unvented
hot water
storage
vessel

air
gap

300 mm minimum vertical
pipe below tundish

D2

100 mm maximum

Discharge pipe to have continuous
fall to its outlet.

Discharge to be visible at tundish.

For the sizing of discharge pipes D1
and D2, refer to tables 4.2 and 4.3.

discharge to safe
and visible position

Figure 4.9   Positioning and design of tundish

Not more than six appliances to
discharge to any one common
discharge pipe

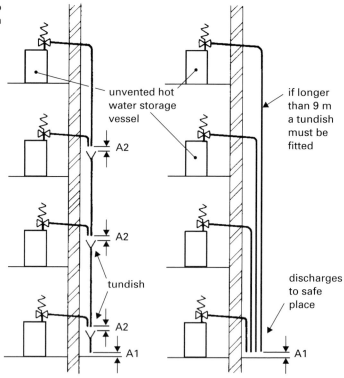

unvented hot
water storage
vessel

A2

if longer
than 9 m
a tundish
must be
fitted

A2

tundish

discharges
to safe
place

A2

A1

A1

(a) Combined discharges

Not recommended unless a
responsible person is on site
at all times

(b) Separate discharges

Best method for visibility of discharge.
A1 = 100 mm maximum
A2 = 75 mm suggested

Figure 4.10   Discharges from multiple installations

Pipework may be within the building, provided that tundishes and discharge terminals are readily visible and adequate provision is made to remove any discharges from the building safely, e.g. by trapped connections to drains.

Discharges from a temperature relief valve and an expansion relief valve may be combined in a common discharge pipe; see figure 4.11.

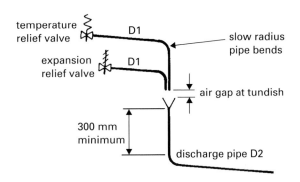

Refer to table 4.2 for sizes of pipes D1 and D2.   Discharge to safe and visible position.

Figure 4.11   Combined discharge from temperature relief valve and expansion relief valve

Sizing discharge pipes

Discharge pipes must be capable of removing all water discharged from temperature relief valves and expansion relief valves and should be sized in accordance with table 4.2. This will ensure that frictional resistances are minimized and any discharges are able to flow unhindered to their discharge point.

Pipe D1 (see figure 4.9) should be at least as large as the outlet of the temperature relief valve it serves.

Pipe D2 should be at least one size larger than the safety device it serves and due account taken of the pipe length and any bends or elbows used, either of which may require the pipe

Table 4.2   Sizing of discharge pipes D2

Temperature relief valve outlet size	Minimum size of pipe D1 (mm)	Minimum size of pipe D2 (mm)	Maximum length of pipe (including equivalent length due to resistance in bends or elbows)
$G\frac{1}{2}$	15	22 28 35	Up to 9 m Up to 18 m Up to 27 m
$G\frac{3}{4}$	22	28 35 42	Up to 9 m Up to 18 m Up to 27 m
G1	28	35 42 54	Up to 9 m Up to 18 m Up to 27 m

Note: pipe diameters shown are for copper tube.

Table 4.3   Equivalent pipe lengths for bends and elbows

Nominal size of pipe (mm)	22	28	35	42	54
Equivalent pipe lengths	0.8	1.0	1.4	1.7	2.3

diameter to be increased. Equivalent pipe lengths for bends and elbows can be added to the measured pipe length to give an effective pipe length for sizing purposes. Equivalent pipe lengths for bends and elbows are given in table 4.3.

Example calculation

A discharge pipe D1 is connected to a $G\frac{1}{2}$ temperature relief valve outlet. By reference to table 4.2, it can be seen that pipe D1 is required to have a diameter at least 15 mm.

For the purposes of this calculation, pipe D2 has a measured pipe length of 8 m from the tundish to its discharge point and is fitted with three elbows. Referring again to table 4.2, we can see that pipe D2 must be at least one size larger than D1, in this case 22 mm. Also, by reference to table 4.3 we find that its effective length must be adjusted to account for frictional resistance in the three elbows.

Assume pipe D2 to be size 22 mm
Measured pipe length + equivalent length (3 elbows) = effective pipe length
Effective pipe length = 8 m + (3 × 0.8) = **10.4 m**

Referring now to table 4. 2, it can be seen that the calculated length is longer than 9 m, so pipe D2 must be resized.

Now assume pipe D2 to be 28 mm
Measured pipe length + equivalent length (3 elbows) = effective pipe length
Effective pipe length = 8 m + (3 × 1.0) = **11 m**

By reference to table 4.2 we can see that the calculated pipe length is well within the permitted length of 18 m. Therefore the size of the discharge pipe D2 will be 28 mm.

Note   This calculation is repeated in tabular form in figure 4.12.

1. Assumed pipe size (mm)	2. Measured pipe length (m)	3. Equivalent length for bend or elbow (m)	4. Effective pipe length (4 + 3) (m)	5. Permitted pipe length (m)	6. Comments
22	8	2.4	10.4	9	Pipe too long, increase size
28	8	3	11	18	28 mm pipe is suitable

Figure 4.12   Calculation sheet for sizing discharge pipe D2

4.2   Pressure and expansion control

Working pressures

In any hot or cold water system the working pressure must not exceed the safe working pressure of the component parts (see table 4.4).

Where necessary the supply pressure should be controlled by using a break cistern or pressure reducing valve (see figures 4.13 and 4.14).

Table 4.4   Safe working pressures

Circuit	Maximum working pressure bar	Test pressure bar
Sealed primary	3	5
Unvented secondary	6	10

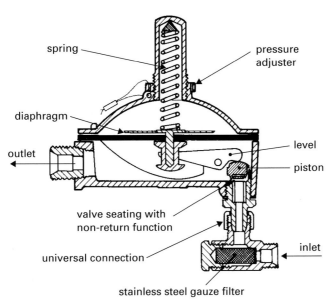

Gives finite control, and constant outlet pressure allowing accurate sizing of system.

Essential where copper hot water cylinder is used.

Figure 4.13   Diaphragm-actuated pressure reducing valve, to BS 6283: Part 4

Provides cruder control than the diaphragm-actuated reducing valve, to maintain supply line at preset maximum pressure.

Use only with glass-lined steel (or similar) cylinder on higher pressure system (up to 6 bar).

Not for use with copper cylinder on low pressure system (up to 3 bar).

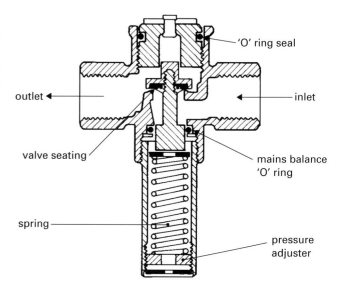

Figure 4.14   Pressure limiting valve

Control of water expansion

Within any hot water system the expansion of the heated water and consequent possible pressure rise must be limited without any discharge of water to waste. That is, expansion water must be accommodated within the hot water system. There are a number of ways of doing this depending upon the system (vented or unvented) and the size of the water heater and/or any storage vessel.

(1)   In the traditional vented system expanded water is permitted to move back into the cold feed pipe towards the feed cistern (see figure 4.15) but only as long as the cold feed pipe is unobstructed so that there will be no resistance to the flow of expanded water. Consequently, there will be no increase in pressure in the supply due to expansion. Any valve installed on the cold feed pipe must be of a type that will readily allow the backflow of water, e.g. a gate valve to BS 5154. A stopvalve with a loose washer plate must not be used.

(2)   In the unvented hot water storage system expansion control is normally achieved by using an expansion vessel or a dedicated air space within the hot store vessel, which should be rated to accommodate a volume of expanded water at least equal to 4% of the total volume of water likely to be heated in the system. (See figures 4.16, 4.17 and 4.18.) The expansion vessel to BS 6144 should be connected on the cold water inlet to the storage vessel. A check valve should be fitted to prevent backflow of expansion water into the cold feed pipe.

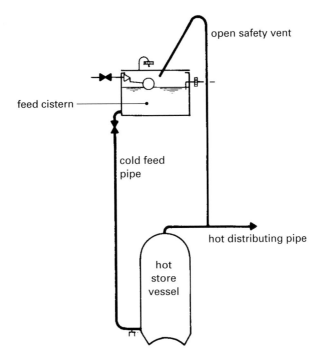

open safety vent

feed cistern

cold feed
pipe

hot distributing pipe

hot
store
vessel

Feed cistern also serves as break cistern to limit pressure in system.

Valve not to have loose jumper.

Cold feed not to be fitted with check valve or other obstruction.

Open safety vent to permit escape of steam in the event of system overheating.

Vent to be open at all times.

Heated water expands safely into cold feed pipe.

Figure 4.15 Expansion in a traditional vented system

(3) Reversed flow into supply pipe. In any mains supplied unvented hot water supply system, whether instantaneous, water-jacketed tube or storage type, reverse flow along the supply pipe may be permitted (see figure 4.19), provided that there is no restriction on the supply pipe such as a check valve, pressure reducing valve or stopvalve which might prevent the reversal of flow as and when water expands. The supply pipe should be large enough to accommodate a volume of water at least equal to 4% of the total volume of water to be heated. In these circumstances, no heated or warm water should reach either the communication pipe or any branch pipe feeding a cold water outlet.

This method, although permitted by Water Regulations, is not widely practised, particularly in systems installed in older properties where the water undertaker's stopvalve might incorporate a loose jumper.

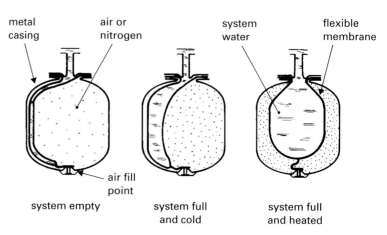

Expansion vessel to conform to BS 6144.

Flexible membrane prevents contact between water and steel casing to minimize corrosion.

Figure 4.16   Expansion vessel

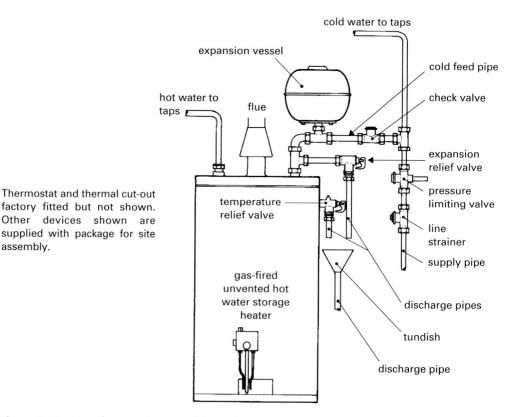

Thermostat and thermal cut-out factory fitted but not shown. Other devices shown are supplied with package for site assembly.

Figure 4.17   Use of expansion vessel in packaged unvented hot water system

(4)   An expansion relief valve (see figure 4.20) may be used as a fail-safe device. However, it is not permitted to be used as the sole means of control of expansion water. This valve should be arranged to open automatically and to discharge water only when the pressure in the system reaches a predetermined level under failure conditions.

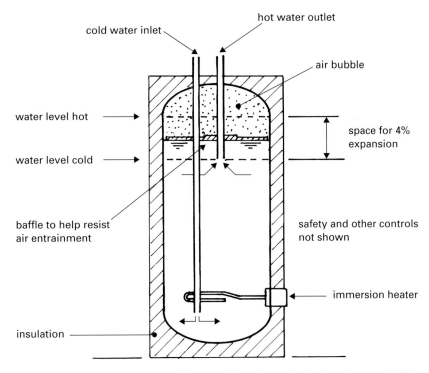

Direct system shown. Indirect systems available.

Figure 4.18   System using air bubble for expansion

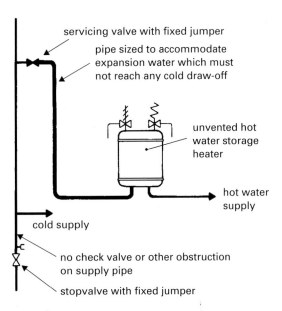

Not recommended because of the possibility that an obstruction to supply pipe could be fitted at a later date.

servicing valve with fixed jumper

pipe sized to accommodate expansion water which must not reach any cold draw-off

unvented hot water storage heater

hot water supply

cold supply

no check valve or other obstruction on supply pipe

stopvalve with fixed jumper

Figure 4.19   System using supply pipe for expansion

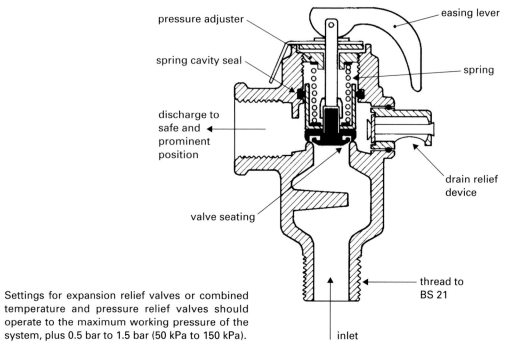

pressure adjuster

easing lever

spring cavity seal

spring

discharge to
safe and
prominent
position

valve seating

drain relief
device

thread to
BS 21

inlet

Settings for expansion relief valves or combined
temperature and pressure relief valves should
operate to the maximum working pressure of the
system, plus 0.5 bar to 1.5 bar (50 kPa to 150 kPa).

Figure 4.20   Expansion relief valve

Any water discharged from an expansion valve must be discharged safely to a conspicuous position in a similar fashion to that of the temperature relief valve (see section 4.1).

4.3   Control of water level

Connections should be made so that water cannot be drawn from any primary or closed circuit, and to ensure that the hottest water can be drawn only from the top of a hot water storage vessel and above any primary flow connection or heating element. (See figure 4.21.) This includes the position of connections from secondary returns.

Unintentional draining down of any system (particularly storage types) is dangerous and should be avoided, as it may:

• expose temperature sensing controls, thus impairing their operation;
• expose the heating element, which could then become overheated;
• result in steam being produced with possibly disastrous consequences.

Vented primary circuits should be fitted with an adequate supply of make-up water, i.e. cold feed.

Sealed primary circuits, having no permanent supply of make-up water, should have a notice displayed drawing attention to the need for regular inspection and maintenance to keep the system and its pressure at the design level. Sealed systems must be under complete thermostatic control.

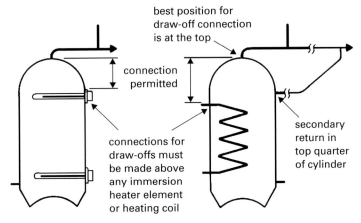

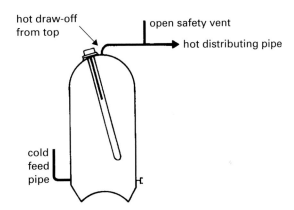

(a) Draw-off connections to hot water cylinder

(b) Cylinder with top entry heater element

Where immersion heater or primary coil is inserted at top of cylinder any draw-off connection must be above immersion heater connection. The only exception is the use of a draining valve which must have a removable key for operation.

Figure 4.21   Control of water level

Chapter 5
Pipe sizing

Pipes and fittings should be sized so that the flow rates for individual draw-offs are equal to the design flow rates shown in table 5.1. During simultaneous discharges, flows from taps should not be less than the minimum flow rates shown in table 5.1.

BS 6700 recommends that flow velocities should not exceed 3 m/s. BS EN 805 recommends 0.5 m/s to 2 m/s with a maximum of 3.5 m/s in exceptional circumstances. Filling times for cisterns may range from 1 to 4 hours depending on their capacity and the flow rate available from the local water supply. In dwellings the filling time should not exceed 1 hour.

Design flow rates may be calculated by dividing the cistern capacity by the required filling time.

Table 5.1   Design flow rates and loading units

Outlet fitting	Design flow rate l/s	Minimum flow rate l/s	Loading units
WC flushing cistern single or dual flush – to fill in 2 minutes	0.13	0.05	2
WC trough cistern	0.15 per WC	0.10	2
Wash basin tap size $\frac{1}{2}$ – DN 15	0.15 per tap	0.10	1.5 to 3
Spray tap or spray mixer	0.05 per tap	0.03	–
Bidet	0.20 per tap	0.10	1
Bath tap, nominal size $\frac{3}{4}$ – DN 20	0.30	0.20	10
Bath tap, nominal size 1 – DN 25	0.60	0.40	22
Shower head (will vary with type of head)	0.20 hot or cold	0.10	3
Sink tap, nominal size $\frac{1}{2}$ – DN 15	0.20	0.10	3
Sink tap, nominal size $\frac{3}{4}$ – DN 20	0.30	0.20	5
Sink tap, nominal size 1 – DN 20	0.60	0.40	–
Washing machine size – DN 15	0.20 hot or cold	0.15	–
Dishwasher size – DN 15	0.15	0.10	3
Urinal flushing cistern	0.004 per position served	0.002	–
Pressure flushing valve for WC or urinal	1.5	1.2	–

Notes:
(1)   Flushing troughs are advisable where likely use of WCs is more than once per minute.
(2)   Mixer fittings use less water than separate taps, but this can be disregarded in sizing.
(3)   Flow rates to shower mixers vary according to type fitted. Manufacturers should be consulted.
(4)   Manufacturers should be consulted for flow rates to washing machines and dishwashers for other than a single dwelling.
(5)   For cistern fed urinals demand is very low and can usually be ignored. Alternatively, use the continuous flow.
(6)   Loading units should not be used for outlet fittings having high peak demands, e.g. those in industrial installations. In these cases use the continuous flow.
(7)   BS 6700 does not give loading units for sink tap DN 20 or pressure flushing valve for WCs or urinals.

Correct pipe sizes will ensure adequate flow rates at appliances and avoid problems caused by oversizing and undersizing; see figure 5.1.

Oversizing will mean:

- additional and unnecessary installation costs;
- delays in obtaining hot water at outlets;
- increased heat losses from hot water distributing pipes.

Undersizing may lead to:

- inadequate delivery from outlets and possibly no delivery at some outlets during simultaneous use;
- some variation in temperature and pressure at outlets, especially showers and other mixers;
- some increase in noise levels.

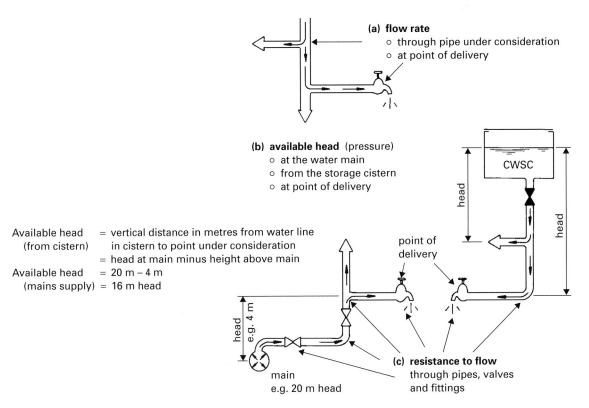

Figure 5.1   Pipe sizing considerations

In smaller, straightforward installations such as single dwellings, pipes are often sized on the basis of experience and convention.

In larger and more complex buildings, or with supply pipes that are very long, it is necessary to use a recognized method of calculation such as that shown in sections 5.1 and 5.2.

BS EN 806-3 gives an alternative 'simplified method' of pipe sizing that can be used for 'standard installations'.

5.1   Sizing procedure for supply pipes

The procedure below is followed by an explanation of each step with appropriate examples.

(1)   Assume a pipe diameter.
(2)   Determine the flow rate:
    (a)   by using loading units;
    (b)   for continuous flows;
    (c)   obtain the design flow rate by adding (a) and (b).
(3)   Determine the effective pipe length:
    (d)   work out the measured pipe length;
    (e)   work out the equivalent pipe length for fittings;
    (f)   work out the equivalent pipe length for draw-offs;
    (g)   obtain the effective pipe length by adding (d), (e) and (f).
(4)   Calculate the permissible loss of head:
    (h)   determine the available head:
    (i)   determine the head loss per metre run through pipes;
    (j)   determine the head loss through fittings;
    (k)   calculate the permissible head loss.
(5)   Determine the pipe diameter:
    (l)   decide whether the assumed pipe size will give the design flow rate in (c) without exceeding the permissible head loss in (k).

Explanation of the procedure

Assume a pipe diameter (1)

In pipe sizing it is usual to make an assumption of the expected pipe size and then prove whether or not the assumed size will carry the required flow.

Determine the flow rate (2)

In most buildings it is unlikely that all the appliances installed will be used simultaneously. As the number of outlets increases the likelihood of them all being used at the same time decreases. Therefore it is economic sense to design the system for likely peak flows based on probability theory using loading units, rather than using the possible maximum flow rate.

(a)   *Loading units.* A loading unit is a factor or number given to an appliance which relates the flow rate at its terminal fitting to the length of time in use and the frequency of use for a particular type and use of building (probable usage). Loading units for various appliances are given in table 5.1.

    By multiplying the number of each type of appliance by its loading unit and adding the results, a figure for the total loading units can be obtained. This is converted to a design flow rate using figure 5.2.

    An example using loading units is given in figure 5.3.

(b) *Continuous flows.* For some appliances, such as automatic flushing cisterns, the flow rate must be considered as a continuous flow instead of applying probability theory and using loading units. For such appliances the full design flow rate for the outlet fitting must be used, as given in table 5.1.

However, in the example shown in figure 5.3, the continuous flow for the two urinals of 0.008 l/s (from table 5.1) is negligible and can be ignored for design purposes.

(c) *Design flow rate.* The design flow rate for a pipe is the sum of the flow rate determined from loading units (a) and the continuous flows (b).

Determine the effective pipe length (3)

(d) *Find the measured pipe length.* Figure 5.4 is an example showing how the measured pipe length is found.

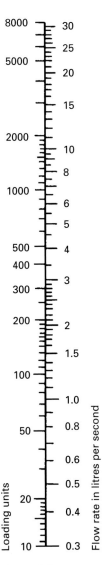

**Figure 5.2
Conversion
chart – loading
units to flow rate**

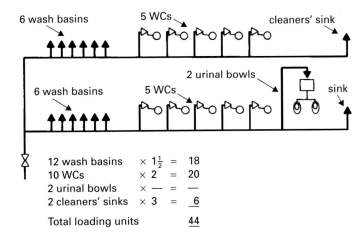

12 wash basins	× 1½	=	18
10 WCs	× 2	=	20
2 urinal bowls	× —	=	—
2 cleaners' sinks	× 3	=	6
Total loading units			44

Therefore, from figure 5.2, the required flow rate for the system is 0.7 l/s.

Figure 5.3   Example of use of loading units

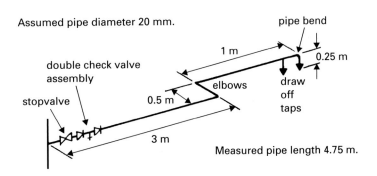

Note   There is no need to consider both branch pipes to taps.

Figure 5.4   Example of measured pipe length

Table 5.2   Equivalent pipe lengths (copper, stainless steel and plastics)

Bore of pipe	Equivalent pipe length			
	Elbow	Tee	Stopvalve	Check valve
mm	m	m	m	m
12	0.5	0.6	4.0	2.5
20	0.8	1.0	7.0	4.3
25	1.0	1.5	10.0	5.6
32	1.4	2.0	13.0	6.0
40	1.7	2.5	16.0	7.9
50	2.3	3.5	22.0	11.5
65	3.0	4.5	–	–
73	3.4	5.8	34.0	–

Notes:
(1)   For tees consider change of direction only. For gate valves losses are insignificant.
(2)   For fittings not shown, consult manufacturers if significant head losses are expected.
(3)   For galvanized steel pipes in a small installation, pipe sizing calculations may be based on the data in this table for equivalent nominal sizes of smooth bore pipes. For larger installations, data relating specifically to galvanized steel should be used. BS 6700 refers to suitable data in the *Plumbing Engineering Services Design Guide* published by the Institute of Plumbing.

(e, f)   *Find the equivalent pipe lengths for fittings and draw-offs.* For convenience the frictional resistances to flow through fittings are expressed in terms of pipe lengths having the same resistance to flow as the fitting. Hence the term 'equivalent pipe length' (see table 5.2).

For example, a 20 mm elbow offers the same resistance to flow as a 20 mm pipe 0.8 m long.

Figure 5.5 shows the equivalent pipe lengths for the fittings in the example in figure 5.4.

(g)   *Effective pipe length.* The effective pipe length is the sum of the measured pipe length (d) and the equivalent pipe lengths for fittings (e) and draw-offs (f).

Therefore, for the example shown in figure 5.4 the effective pipe length would be:

Measured pipe length   4.75 m
Equivalent pipe lengths
    elbows 2×0.8 =   1.6 m
    tee 1×1.0 =   1.0 m
    stopvalve 1×7.0 =   7.0 m
    taps 2×3.7 =   7.4 m
    check valves 2×4.3 =   8.6 m

Effective pipe length   =   30.35 m

Permissible loss of head (pressure) (4)

Pressure can be expressed in the following ways.

(i)   In pascals, the pascal (Pa) being the SI unit for pressure.
(ii)   As force per unit area, N/m^2.
        $1 \ N/m^2 = 1$ pascal (Pa).

Using the example from figure 5.4:

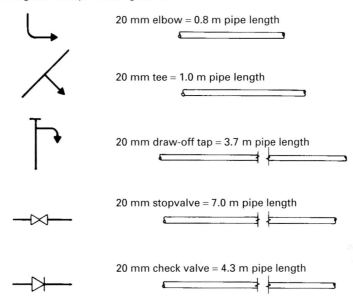

20 mm elbow = 0.8 m pipe length

20 mm tee = 1.0 m pipe length

20 mm draw-off tap = 3.7 m pipe length

20 mm stopvalve = 7.0 m pipe length

20 mm check valve = 4.3 m pipe length

Figure 5.5   Examples of equivalent pipe lengths

(iii)   As a multiple of atmospheric pressure (bar).
Atmospheric pressure = 100 kN/m^2 = 100 kPa = 1 bar.

(iv)   As metres head, that is, the height of the water column from the water level to the draw-off point.
1 m head = 9.81 kN/m^2 = 9.81 kPa = 98.1 mb.

In the sizing of pipes, any of these units can be used. BS 6700 favours the pascal. However, this book retains the use of metres head, giving a more visual indication of pressure that compares readily to the height and position of fittings and storage vessels in the building.

(h)   *Available head*. This is the static head or pressure at the pipe or fitting under consideration, measured in metres head (see figure 5.1).

(i)   *Head loss through pipes*. The loss of head (pressure) through pipes due to frictional resistance to water flow is directly related to the length of the pipe run and the diameter of the pipe. Pipes of different materials will have different head losses, depending on the roughness of the bore of the pipe and on the water temperature. Copper, stainless steel and plastics pipes have smooth bores and only pipes of these materials are considered in this section.

(j)   *Head loss through fittings*. In some cases it is preferable to subtract the likely resistances in fittings (particularly draw-offs) from the available head, rather than using equivalent pipe lengths.

Table 5.3 gives typical head losses in taps for average flows compared with equivalent pipe lengths. Figures 5.6 and 5.7 provide a method for determining head losses through stopvalves and float-operated valves respectively.

Note   Where meters are installed in a pipeline the loss of head through the meter should be deducted from the available head.

Table 5.3   Typical head losses and equivalent pipe lengths for taps

Nominal size of tap	Flow rate	Head loss	Equivalent pipe length
	l/s	m	m
G$\frac{1}{2}$–DN 15	0.15	0.5	3.7
G$\frac{1}{2}$–DN 15	0.20	0.8	3.7
G$\frac{3}{4}$–DN 20	0.30	0.8	11.8
G 1 –DN 25	0.60	1.5	22.0

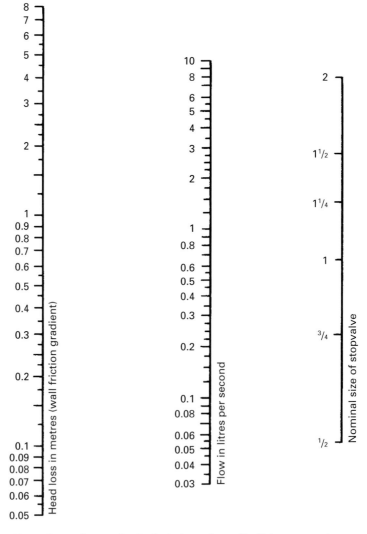

Note   Gate valves and spherical plug valves offer little or no resistance to flow provided they are fully open.

Figure 5.6   Head loss through stopvalves

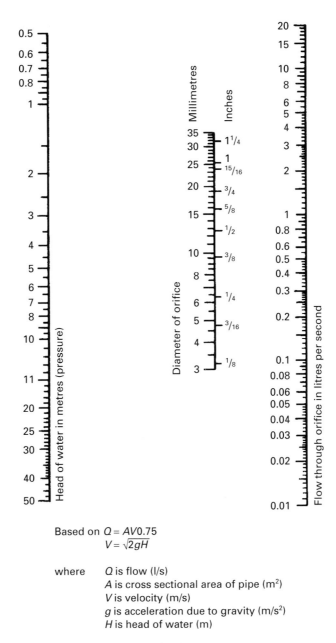

Based on $Q = AV0.75$
$V = \sqrt{2gH}$

where Q is flow (l/s)
           A is cross sectional area of pipe (m^2)
           V is velocity (m/s)
           g is acceleration due to gravity (m/s^2)
           H is head of water (m)

Figure 5.7   Head loss through float-operated valves

(k)   *Permissible head loss.* This relates the available head to the frictional resistances in the pipeline. The relationship is given by the formula:

$$\text{Permissible head loss (m/m run)} = \frac{\text{Available head (m)}}{\text{Effective pipe length (m)}}$$

This formula is used to determine whether the frictional resistance in a pipe will permit the required flow rate without too much loss of head or pressure. Figure 5.8 illustrates the permissible head loss for the example in figure 5.4.

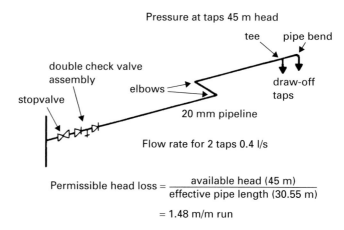

$$\text{Permissible head loss} = \frac{\text{available head (45 m)}}{\text{effective pipe length (30.55 m)}}$$

$$= 1.48 \text{ m/m run}$$

Figure 5.8   Example of permissible head loss

Determine the pipe diameter (5)

In the example in figure 5.4 a pipe size of 20 mm has been assumed. This pipe size must give the design flow rate without the permissible head loss being exceeded. If it does not, a fresh pipe size must be assumed and the procedure worked through again.

Figure 5.9 relates pipe size to flow rate, flow velocity and head loss. Knowing the assumed pipe size and the calculated design flow rate, the flow velocity and the head loss can be found from the figure as follows.

(1)   Draw a line joining the assumed pipe size (20 mm) and the design flow rate (0.4 l/s).
(2)   Continue this line across the velocity and head loss scales.
(3)   Check that the loss of head (0.12 m/m run) does not exceed the calculated permissible head loss of 1.48 m/m run.
(4)   Check that the flow velocity (1.4 m/s) is not too high by referring to table 5.4.

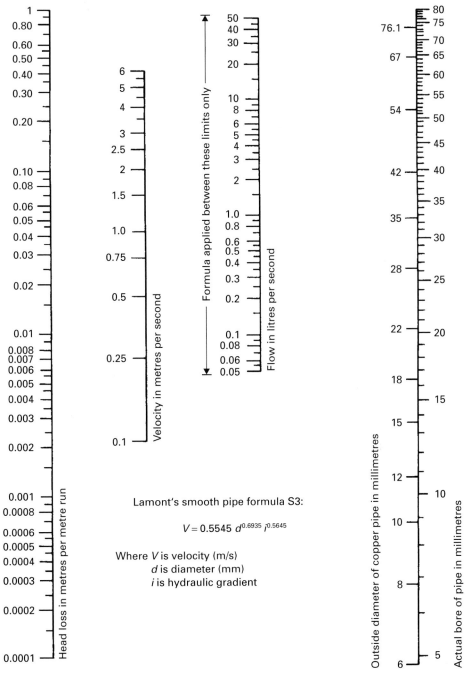

Lamont's smooth pipe formula S3:

$$V = 0.5545\ d^{0.6935}\ i^{0.5645}$$

Where V is velocity (m/s)
d is diameter (mm)
i is hydraulic gradient

Notes   Figures shown are for cold water at 12°C.
Hot water will show slightly more favourable head loss results.

BS 6700 gives head loss in kPa.

1 m head = 9.81 kPa.

Figure 5.9   Determination of pipe diameter

Table 5.4   Maximum recommended flow velocities

Water temperature	Flow velocity	
	Pipes readily accessible	Pipes not readily accessible
°C	m/s	m/s
10	3.0	2.0
50	3.0	1.5
70	2.5	1.3
90	2.0	1.0

Note   Flow velocities should be limited to reduce system noise.

5.2   Tabular method of pipe sizing

Pipe sizing in larger and more complicated buildings is perhaps best done by using a simplified tabular procedure. BS 6700 gives examples of this but for more detailed data readers should refer to the Institute of Plumbing's *Plumbing Engineering Services Design Guide*.

The data used in the tabular method that follows are taken from BS 6700 but the author has simplified the method compared with that given in the standard.

The tabular method uses a work sheet which can be completed as each of the steps is followed in the pipe sizing procedure. An example of the method follows with some explanation of each step.

Explanation of the tabular method

Pipework diagram

(1)   Make a diagram of the pipeline or system to be considered (see figure 5.10).
(2)   Number the pipes beginning at the point of least head, numbering the main pipe run first, then the branch pipes.
(3)   Make a table to show the loading units and flow rates for each stage of the main run. Calculate and enter loading units and flow rates; see figure 5.10.

Calculate flow demand

(1)   Calculate maximum demand (see figure 5.10):
   • add up loading units for each stage (each floor level);
   • convert loading units to flow rates;
   • add up flow rates for each stage.
(2)   Calculate probable demand (see figure 5.10):
   • add up loading units for all stages;
   • convert total loading units to flow rate.
(3)   Calculate percentage demand (number of stages for which frictional resistances need be allowed). See figure 5.12.

Work through the calculation sheet

See figure 5.11, using the data shown in figures 5.10 and 5.12.

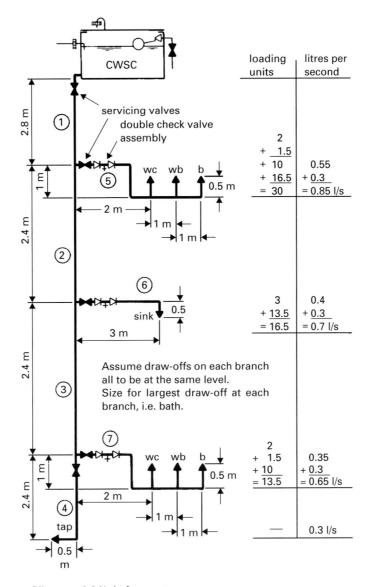

Bib tap at 0.3 l/s in frequent use.

Note Figure is not to scale for convenience, water level in cistern taken to be at base of cistern. Servicing valves assumed to be full-flow gate valves having no head losses.

Refer also to figure 5.12.

Figure 5.10   Pipe sizing diagram

This is an example of a suitable calculation sheet with explanatory notes.

Calculation sheet

(1) Pipe reference	(2) Loading units	(3) Flow rate (l/s)	(4) Pipe size (mm diameter)	(5) Loss of head (m/m run)	(6) Flow velocity (m/s)	(7) Measured pipe run (m)	(8) Equivalent pipe length (m)	(9) Effective pipe length (m)	(10) Head consumed (m)	(11) Progressive head (m)	(12) Available head (m)	(13) Final pipe size (mm)	(14) Remarks
Enter pipe reference on calculation sheet	Determine loading unit (table 5.1)	Convert loading units to flow rate (figure 5.2)	Make assumption as to pipe size (inside diameter)	Work out frictional resistance per metre (figure 5.9)	Determine velocity of flow (figure 5.9)	Measure length of pipe under consideration	Consider frictional resistances in fittings (table 5.2 and figures 5.6 and 5.7)	Add totals in columns 7 and 8	Head consumed – multiply column 5 by column 9	Add head consumed in column 10 to progressive head in previous row of column 11	Record available head at point of delivery	Compare progressive head with available head to confirm pipe diameter or not	Notes

Note If, for any pipe or series of pipes, it is found that the assumed pipe size gives a progressive head that is in excess of the available head, or is noticeably low, it will be necessary to repeat the sizing operation using a revised assumed pipe diameter.

Figure 5.11   Calculation sheet – explanation of use

Estimated maximum demand = 1.4 l/s

Probable demand = 0.85 l/s

$$\text{Percentage demand} = \frac{\text{probable demand}}{\text{estimated maximum demand}} \times \frac{100}{1}$$

$$= \frac{0.85}{1.4} \times \frac{100}{1} = 60\%$$

Therefore only 60% of the installation need be considered.

For example, if we were designing for a multi-storey building 20 storeys high, only the first 12 storeys need to be calculated.

However, in the example followed here, the whole system has been sized because the last fitting on the run has a high flow rate in continuous use.

For branches only the pipes to the largest draw-off, i.e. the bath tap, need be sized.

Calculation sheet

(1) Pipe reference	(2) Loading units	(3) Flow rate (l/s)	(4) Pipe size (mm diameter)	(5) Loss of head (m/m run)	(6) Flow velocity (m/s)	(7) Measured pipe run (m)	(8) Equivalent pipe length (m)	(9) Effective pipe length (m)	(10) Head consumed (m)	(11) Progressive head (m)	(12) Available head (m)	(13) Final pipe size (mm)	(14) Remarks
1	30	0.85	32	0.05	1.2	2.8	1.4	4.2	0.21	0.21	2.8	32	
5	13.5	0.35	20	0.095	1.25	5.5	12.0	17.5	1.66	1.87	3.3	20	
2	16.5	0.7	25	0.12	1.5	2.4	–	2.4	0.29	2.16	5.2	25	
6	3	0.3	20	0.07	1.0	3.5	10.4	13.9	0.97	3.13	5.7	20	
3	13.5	0.65	25	0.1	1.4	2.4	–	2.4	0.24	3.37	7.6	25	
7	13.5	0.35	20	0.095	1.25	5.5	12.0	17.5	1.66	5.03	8.1	20	
4	—	0.3	20	0.07	1.0	2.9	1.6	4.5	0.31	5.34	10.0	20	

Refer also to figure 5.10.

Figure 5.12   Calculation sheet – example of use

5.3   Sizing cold water storage

In Britain, cold water has traditionally been stored in both domestic and non-domestic buildings, to provide a reserve of water in case of mains failure. However, in recent years we have seen an increase in the use of 'direct' pressure systems, particularly for hot water services where many combination boilers and unvented hot water storage vessels are now being installed.

BS 6700 no longer gives storage capacities for houses. The following figures are based on the 1997 edition.

Smaller houses	cistern supplying cold water only	– 100 l to 150 l
	cistern supplying hot and cold outlets	– 200 l to 300 l
Larger houses	per person where cistern fills only	– 80 l
	at night per person	– 130 l

However, in clause 5.3.9.4 it recommends a minimum storage capacity of 230 l where the cistern supplies both cold water outlets and hot water apparatus, which was a requirement of byelaws in the past. The author still favours the old byelaw requirements, which are more specific and which still seem to be the normal capacity installed.

Cold water storage cistern	– 115 l minimum
Feed cistern	– No minimum but should be equal to the capacity of the hot store vessel supplied
Combined feed and storage cistern	– 230 l minimum

For larger buildings the capacity of the cold water storage cistern depends on:

- type and use of buildings;
- number of occupants;
- type and number of fittings;
- frequency and pattern of use;
- likelihood and frequency of breakdown of supply.

These factors have been taken into account in table 5.5, which sets out minimum storage capacities in various types of building to provide a 24-hour reserve capacity in case of mains failure.

Calculation of minimum cold water storage capacity

Determine the amount of cold water storage required to cover 24 hours interruption of supply in a combined hotel and restaurant. Number of hotel guests 75, number of restaurant guests 350.

Storage capacity = number of guests × storage per person from table 5.5
Hotel storage capacity $= 75 \times 200 = 15\,000$ l
Restaurant storage capacity $= 350 \times 7 = 2450$ l
Therefore total storage capacity required is $15\,000$ l $+ 2450$ l $= 17\,450$ l

Assuming 12-hour fill time, the design flow rate to the cistern would be:

$$\begin{aligned}
\text{Design flow rate} &= 17\ 450\ \text{l cistern capacity} \div 12\ \text{hours} \\
&= 17\ 540 \div (12 \times 3600\ \text{seconds}) \\
&= 0.4\ \text{l/s}
\end{aligned}$$

Table 5.5   Recommended minimum storage of hot and cold water for domestic purposes

Type of building	Minimum cold water storage litres (l)	Minimum hot water storage litres (l)
Hostel	90 per bed space	32 per bed space
Hotel	200 per bed space	45 per bed space
Office premises:		
with canteen facilities	45 per employee	4.5 per employee
without canteen facilities	40 per employee	4.0 per employee
Restaurant	7 per meal	3.5 per meal
Day school:		
nursery / primary	15 per pupil	4.5 per pupil
secondary / technical	20 per pupil	5.0 per pupil
Boarding school	90 per pupil	23 per pupil
Children's home or residential nursery	135 per bed space	25 per bed space
Nurses' home	120 per bed space	45 per bed space
Nursing or convalescent home	135 per bed space	45 per bed space

Note   Minimum cold water storage shown includes that used to supply hot water outlets.

5.4   Sizing hot water storage

Minimum hot water storage capacities for dwellings, from BS 6700, are:

- 35 l to 45 l per occupant, unless the heat source provides a quick recovery rate;
- 100 l for systems heated by solid fuel boilers;
- 200 l for systems heated by off-peak electricity.

The feed cistern should have a capacity at least equal to that of the hot storage vessel.

Information on storage capacities for larger buildings is given in table 5.5 using data based on the Institute of Plumbing's *Plumbing Engineering Services Design Guide*. The calculations are similar to those in section 5.3 for the minimum cold water storage capacity in larger buildings, and require the number of people to be multiplied by the storage per person, shown in table 5.5.

BS 6700 takes a different approach to the sizing of hot water storage and suggests that when sizing hot water storage for any installation, account must be taken of the following:

- pattern of use;
- rate of heat input to the stored water (see table 5.6);

- recovery period for the hot store vessel;
- any stratification of the stored water.

Table 5.6   Typical heat input values

Appliance	Heat input kW
Electric immersion heater	3
Gas-fired circulator	3
Small boiler and direct cylinder	6
Medium boiler and indirect cylinder	10
Directly gas-fired storage hot water heater (domestic type)	10
Large domestic boiler and indirect cylinder	15

Stratification (see figure 3.41) means that the hot water in the storage vessel floats on a layer of cold feed water. This enables hot water to be drawn from the storage vessel without the incoming cold feed water mixing appreciably with the remaining hot water. In turn, this allows a later draw-off of water at a temperature close to the design storage temperature, with less frequent reheating of the contents of the storage vessel and savings in heating costs and energy.

Stratification is most effective when cylinders and tanks are installed vertically rather than horizontally, with a ratio of height to width or diameter of at least 2:1.

The cold feed inlet should be arranged to minimize agitation and hence mixing, by being of ample size and, if necessary, fitted with a baffle to spread the incoming water.

Stratification is used to good effect in off-peak electric water heaters (see figure 5.13). In this case no heat is normally added to the water during the daytime use and consequently very little mixing of hot and cold water takes place. In other arrangements the heating of the water will induce some mixing.

Calculation of hot water storage capacity

As noted above, the storage capacity required in any situation depends on the rate of heat input to the stored hot water and on the pattern of use. For calculating the required storage capacity BS 6700 provides a formula for the time M (in min) taken to heat a quantity of water through a specified temperature rise:

$$M = VT/(14.3P)$$

where

V is the volume of water heated (in l);
T is the temperature rise (in °C);
P is the rate of heat input to the water (in kW).

This formula can be applied to any pattern of use and whether stratification of the stored water takes place or not. It ignores heat losses from the hot water storage vessel, since over the relatively short times involved in reheating water after a draw-off has taken place, their effect is usually small.

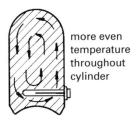

(a) Bottom entry heater

With a bottom entry immersion heater mixing will occur when the water is being heated.

more even temperature throughout cylinder

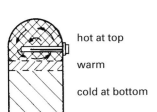

(b) Top entry heater

With top entry immersion heater stratification will prevent mixing.

hot at top

warm

cold at bottom

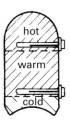

(c) Twin entry immersion heater

With a twin entry immersion heater the top entry element can provide economical energy consumption for normal use with the bottom entry element operating when large quantities of water are needed, for example, for bathing and washing.

With off-peak heaters the top element can be brought into use during on-peak periods when needed to top up hot water, and stratification will ensure that on-peak electricity is not used to excess.

hot

warm

cold

Figure 5.13 Effects of stratification

The application of this formula to the sizing of hot water cylinders is best illustrated by the following examples, in which figures have been rounded.

In these examples a small dwelling with one bath installed has been assumed. Maximum requirement: 1 bath (60 l at 60°C plus 40 l cold water) plus 10 l hot water at 60°C for kitchen use, followed by a second bath fill after 25 min.

Thus a draw-off of 70 l at 60°C is required, followed after 25 min by 100 l at 40°C, which may be achieved by mixing hot at 60°C with cold at 10°C.

Example 1   Assuming good stratification

Good stratification could be obtained, for example, by heating with a top entry immersion heater. With a rate of heat input of 3 kW, the time to heat the 60 l for the second bath from 10°C to 60°C is:

$$M = VT/(14.3P)$$
$$M = (60 \times 50)/(14.3 \times 3)$$
$$M = 70 \text{ min}$$

Since the second bath is required after 25 min, it has to be provided from storage. But in the 25 min the volume of water heated to 60°C is:

$$V = M\,(14.3)/T$$
$$V = (25 \times 14.3 \times 3)/50$$
$$V = 21\ \text{l}$$

Therefore the minimum required storage capacity is:

$$70 + 6 - 21 = 109\ \text{l}$$

Example 2 *Assuming good mixing of the stored water*

Good mixing of the stored water would occur, for example, with heating by a primary coil in an indirect cylinder.

Immediately after drawing off 70 l at 60°C for the first bath and kitchen use, the heat energy in the remaining water plus the heat energy in the 70 l replacement at 10°C equals the heat energy of the water in the full cylinder.

The heat energy of a quantity of water is the product of its volume and temperature. Then, if V is the minimum size of the storage cylinder and T is the water temperature in the cylinder after refilling with 70 l at 10°C:

$$(V - 70) \times 60 + (70 \times 10) = VT$$
$$T = (60V - 4200 + 700)/V$$
$$T = (60V - 3500)/V$$
$$T = 60 - 3500/V$$

The second bath is required after 25 min. Hence, with a rate of heat input of 3 kW:

$$25 = VT/(14.3 \times 3)$$

and the temperature rise $\quad T = (25 \times 14.3 \times 3)/V$

and $\quad\qquad\qquad\qquad\ \ T = 1072.5/V$

A temperature of at least 40°C is required to run the second bath. Therefore the water temperature of the refilled cylinder after the first draw-off of 70 l, plus the temperature rise after 25 min, must be at least 40°C, or:

$$(60 - 3500/V) + (1072.5V) = 40\ \text{(or more)}$$
$$60 - 2427.5/V = 40$$
$$20 = 2427.5/V$$
$$V = 122\ \text{l}$$

These calculations, which may be carried out for any situation, show the value of promoting stratification wherever possible. They also show the savings in storage capacity that can be made, without affecting the quality of service to the user, by increasing the rate of heat input to the water. Results of similar calculations are shown in table 5.7 and are taken from BS 6700.

Table 5.7   Hot water storage vessels – minimum capacities

Heat input to water kW	Dwelling with 1 bath		Dwelling with 2 baths*	
	With stratification litres (l)	With mixing litres (l)	With stratification litres (l)	With mixing litres (l)
3	109	122	165	260
6	88	88	140	200
10	70	70	130	130
15	70	70	120	130

Note   * Maximum requirement of 130 l drawn off at 60°C (2 baths plus 10 l for kitchen use) followed by a further bath (100 l at 40°C) after 30 min.

5.5   *Legionella* – implications in sizing storage

It has been common practice in the past for water suppliers to recommend cold water storage capacities to provide for 24 hours of interruption in supply. Table 5.5 in this book and Table 1 of BS 6700 reflect this practice.

Recent investigations into the cause and prevention of *Legionella* contamination suggest that hot and cold water storage should be sized to cope with peak demand only, and that in the past, storage vessels have often been over-sized.

Reduced storage capacities would mean quicker turnover of water and less opportunity for *Legionella* and other organisms to flourish.

Stratification in hot water can lead to ideal conditions for bacteria to multiply, as the base of some cylinders often remains within the 20°C temperature range. The base of the cylinder may also contain sediment and hard water scale which provide an ideal breeding ground for bacteria.

Where, however, good mixing occurs, higher temperatures can be obtained at the cylinder base during heating up periods. This is helpful because *Legionella* will not thrive for more than 5 minutes at 60°C and are killed instantly at temperatures of 70°C or more.

Chapter 6
Preservation of water quality

Water undertakers in England and Wales have a duty under the Water Industry Act 1991 to provide a supply of wholesome water which is suitable and safe for drinking and culinary purposes. Similar conditions apply in Scotland and Northern Ireland. At the same time, public demand requires that water supplied is of good appearance with minimal colour, taste or odour.

To enable water undertakers to maintain their supplies in wholesome condition and to preserve the quality of water supplied, the Act provides water undertakers with powers to enforce Water Regulations which came into effect on 1 July 1999. Installers and users must ensure that systems and components which are installed and used comply with Water Regulations.

BS 6700 looks at the preservation of water quality in four main areas.

(1) *Materials in contact with water*. There is not much point in having a good quality water supply if it becomes contaminated by unsuitable pipes, fittings and jointing materials.
(2) *Stagnation of water* and the prevention of bacterial growth particularly at temperatures between 20°C and 50°C.
(3) *Cross connections* must be prevented between pipes supplied directly under mains pressure and pipes supplied from other sources, such as:
    (a) water from a private source;
    (b) non-potable water;
    (c) stored water;
    (d) water drawn off for use.
(4) *Prevention of backflow* from fittings or appliances into services or mains, particularly at point of use, i.e. draw-off taps, flushing cisterns, washing machines, dishwashers, storage cisterns and hose connections. For example, pumps should not be connected so as to cause backflow into the supply pipe (see figure 6.1).

    Backflow prevention forms the largest portion of Clause 2.6 of BS 6700 and is prominent in Paragraph 15 of Schedule 2 of the Water Regulations.

6.1 Materials in contact with water

No material or substance that causes or is likely to cause contamination shall be used in a system that conveys water for domestic or food production purposes.

Contamination of water by contact with unsuitable materials will be avoided if careful attention during design and installation is given to:

• the specification and selection of acceptable materials used in the manufacture of pipes, fittings and appliances;

- the method of installation, and in particular to the method and materials used when jointing and connecting pipes, fittings and appliances;
- the environment into which pipes, fittings and appliances are to be installed;
- the design of the various elements of installation, especially where differing materials are to be used.

Where it is intended to use a pump to increase flow or pressure through a supply pipe, the water supplier should first be consulted and permission granted.

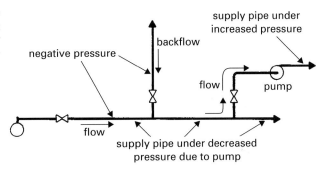

Pump may cause pressure drop sufficient to reverse flow in adjacent pipes.

Figure 6.1   Example of backflow caused by pump installation

Materials selected for use in contact with water intended for domestic purposes should comply with Water Regulations and in particular Paragraph 2 of Schedule 2 'materials and substances in contact with water'. In general this means compliance is assured where any materials used are manufactured to a relevant BS or EN specification, or are listed in the *Water Fittings and Materials Directory* produced under the Water Regulations Advisory Scheme.

Examples of problems caused by unsuitable materials, and situations which should be avoided are shown in figures 6.2 and 6.4.

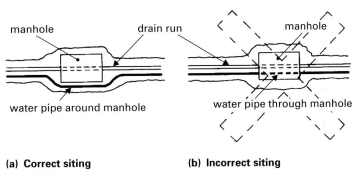

(a) Correct siting   **(b) Incorrect siting**

Pipes not to pass through any foul soil, refuse or refuse chute, ash pit, sewer or drain, cess pool or manhole.

Figure 6.2   Pipes not to pass through manholes, etc.

Pipes made of any material which is susceptible to permeation by gas or to deterioration by contact with substances likely to cause contamination of water should not be laid or installed in a place where permeation or deterioration is likely to occur. Because plastic pipes in particular are liable to a degree of permeation and deterioration by gas and oil, care should be taken when positioning pipelines to avoid contact in the event of leakages occurring from oil or gas lines, for example, at petrol filling stations. Figure 6.3 shows the positioning of pipelines in trenches as suggested by the *National Joint Utilities Group Report No. 6.*

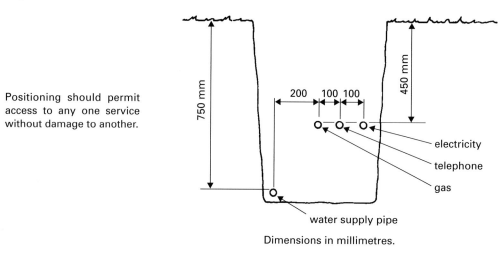

Dimensions in millimetres.

Figure 6.3   Positioning of pipelines in trenches

Substances leached from some materials may adversely affect the quality of water. Although British Standard and other schemes for the testing of pipes, fittings and materials aim to prevent this, much depends on the installer and the material chosen for a particular application, after taking due account of the nature of the water (see figure 6.4).

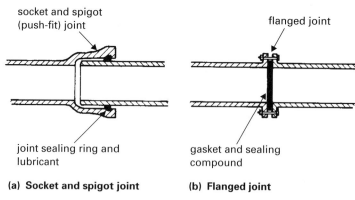

(a) Socket and spigot joint   **(b) Flanged joint**

See table 11.23 for jointing materials and guidance on their use.

The *Water Fittings and Materials Directory* produced under the Water Regulations Advisory Scheme lists materials and fittings which are approved for use.

Figure 6.4   Examples of materials which could cause contamination if not chosen with care

The use of lead for pipes, cisterns or in solders is prohibited because of the obvious danger of plumbo solvency. However, there will be cases when connections have to be made to existing lead pipes (see section 11.15).

For soldered joints to copper pipes, lead-free solders should be specified, e.g. tin/silver alloys.

The use of coal tar for the lining of pipes and cisterns is also prohibited. The direct connection between copper and lead pipes which might lead to electrolytic action and further lead solvency is prohibited in the absence of suitable means to prevent corrosion through galvanic action.

6.2   Stagnation of water and *Legionella*

All waters are likely to contain some bacterial infection. This should not be a problem in hot and cold water supplies as long as systems are properly installed and regularly well maintained. It is important that water in systems is kept clean and used quickly, with it not being allowed to lie in pipes, fittings and storage vessels for long periods.

The Water Regulations Advisory Scheme draws attention to the need to avoid stagnation. In its advisory leaflet '*Commissioning Plumbing System*', the WRAS states '*Water that is left to stagnate in systems can lead to a general deterioration in water quality which for both metal and plastic pipes can give rise to taste and odour problems*'.

Stagnation in new systems may be caused by the presence of flux from soldered joints, swarf, pipe trimmings or builder's debris, all of which can lead to bacterial growth and deterioration of water quality. This problem can be minimized by regular (twice weekly) flushing, or preferably by draining the system down until it is due to be put into use.

To restrict bacterial growth in stored water, temperatures between 20°C and 45°C should be avoided. Water Regulations, in Paragraph 9 of Schedule 2, require that cold water used for domestic purposes is not warmed above 25°C.

This advice is particularly important in the control of legionnaire's disease. Evidence for the occurrence of this disease has generally been found where pipes and components are not regularly maintained, or in parts of systems that cannot be cleaned easily during routine maintenance programmes, e.g. where flushing valves or draining valves are wrongly positioned, or where the accumulation of rust, scale or other debris can settle and build up.

Legionnaire's disease is caused by *Legionella pneumophila*, a pneumonia-like microscopic bacterium that attacks the lungs. The bacteria occur naturally in water and can easily colonize hot and cold water supply systems unless preventative measures are taken. The disease can cause minor illness to healthy people but can prove to be fatal in those who are particularly vulnerable to respiratory diseases or those who are chronically ill. BS 6700 draws attention to the Health and Safety Commission's Approved Code of Practice which in turn draws attention to the following situations of particular concern:

(a)   hot water systems of more than 300 l capacity, and
(b)   hot and cold systems where occupants are particularly susceptible such as health care premises.

Because *Legionella* affects the lungs, and is caused by inhaling very small droplets of infected water, any location or fitting (e.g. showers, spray taps, humidifiers, etc.) that creates a water spray or aerosol mist should be given careful consideration before installation, and water supply systems should be designed, installed and maintained to avoid any risk.

Protective measures that should be taken to minimize colonization by *Legionella* and other harmful bacteria in the water are as follows:

- Cold water should be stored and distributed at as low a temperature as possible and preferably below 20°C.
- Hot water should be stored at 60°C to 65°C and distributed to provide delivery temperatures above 50°C after 1 minute. Note: These temperatures may not be regarded as safe water temperatures. See chapter 3.
- Materials that harbour or promote bacterial growth should be avoided.
- Fittings that tend to create aerosol formation should be avoided.
- Good throughput of water should be maintained.
- Storage capacities should be minimized to give adequate supplies for peak demand but to provide quick and regular replacement of stored water.
- New installations and newly repaired pipework and fittings should be flushed and sterilized before putting into use.
- Measures should be taken to prevent the accumulation of scale, rust and other sediment within the water system, and where possible prevent them from gaining access to the system.
- Regular cleaning and maintenance of cisterns, pipes and fittings are essential.
- The inlet and outlet should be positioned at opposite ends of cold water storage cisterns.
- Drainage connections on larger cisterns should be positioned at the lowest part of the cistern and cisterns arranged with a fall to assist draining down and cleaning.
- Storage vessels and pipelines that will have little use should be avoided. Otherwise an isolating valve should be fitted and appropriate backflow prevention devices considered.
- Fittings that are unused should be disconnected at the branch, leaving no dead legs. See figure 12.9.

For further information in a down-to-earth manner, the author recommends the Institute of Plumbing's booklet *Legionnaire's Disease – Good Practice Guide for Plumbers*. Alternatively, guidance can be obtained from the HSE's approved code of practice 'The Control of Legionella Bacteria in Water Systems'.

6.3   Prevention of contamination by cross connection

'Any water fitting conveying rainwater, recycled water or any fluid other than that supplied by the water undertaker, or any fluid that is not wholesome water, shall be clearly identified so as to be easily distinguished from any supply pipe or distributing pipe.'

'No supply pipe, distributing pipe or pump delivery pipe drawing water from a supply pipe shall convey or be connected so that it can convey any rainwater, recycled water or any fluid other than that supplied by the water undertaker, or any fluid that is not wholesome water.'

The two statements above set out the requirements of Paragraph 14 of Schedule 2 of the Water Regulations and quite simply mean that cross connection between wholesome water and unwholesome water is not permitted.

It is quite interesting that the regulations single out rainwater and recycled water. That these are given individual attention is a reflection of the government's view that both rainwater and recycled water could make a useful contribution to water conservation in the future. In its 'Recommendations for requirements to replace water byelaws' the Water Regulations Advisory Committee made the point that as the use of these waters is increased so is the risk of contamination by cross-connection and the risk of backflow is also likely to increase.

Apart from recycled water and rainwater, there are a number of 'unwholesome' waters that could give rise to contamination if cross-connections are made between them and pipes containing wholesome water. These include:

- water supplies and a reserve storage facility for fire fighting;
- water from private wells, springs, etc. (not supplied by the undertaker);
- stored water other than 'protected' drinking water;
- recycled water from industrial equipment or food production processes.

Examples of cross connections are illustrated in figure 6.5 whilst the correct installations for a variety of situations can be seen later in figures 6.27–6.29.

In order to avoid **any potential hazard** and ensure that hazardous connections are not made inadvertently, or otherwise, it is important that all pipes and fittings can be identified to show what is contained within them. Remember, Water Regulations require water fittings containing unwholesome water to be clearly identified. In practice it is best that *all* pipes can be distinguished one from another.

(a) Connections between supply pipe and distributing pipe

No connection shall be made between a supply pipe and a distributing pipe.

(b) Supplies from different sources

Installation not permitted. Correct installation is type AA air gap on supply pipe. See figure 6.7.

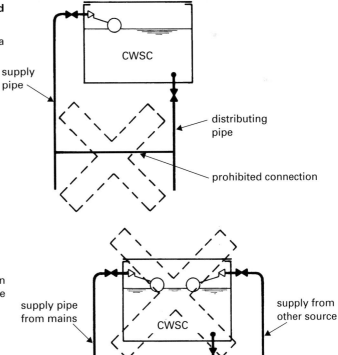

Figure 6.5   Cross connection hazards

(c) Temporary connection to sealed heating system

No closed (sealed) circuit to be connected to a supply pipe unless adequate backflow prevention devices are in place.

Important because primary circuits to heating systems are likely to be contaminated with additives.

This diagram shows an approved method of protection for filling a sealed heating circuit in a house using a 'double check valve arrangement'.

Where the connection is to be a permanent arrangement, the *Water Regulations Guide* recommends the use of a 'type CA backflow preventor with different pressure zones'.

In premises other than a house when backflow risk is more severe, the minimum protection should be a 'type BA backflow preventor with reduced pressure zones' (RPZ valve).

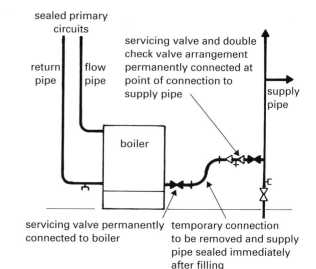

(d) Shower mixer installation

No connection to be made between a supply pipe and a hot or cold distributing pipe.

Water could flow from hot store vessel to the drinking tap if the supply pipe should fail.

May be permitted if feed cistern is protected from contamination and a check valve is fitted to both the hot and cold supplies to the shower.

For correct installation see figure 3.52.

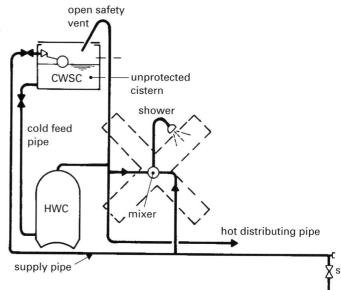

Figure 6.5   continued

- All pipes should be colour coded to show their contents.
- Draw-off taps should be labelled to show those that are suitable for drinking purposes and those that are not.
- Valves and their function should be identified, particularly in industrial and commercial buildings where systems are more complex and unwholesome water more commonly used.
- Accurate drawings should be made, and passed over to the customer, showing where pipes are situated both within buildings and below ground so as to help distinguish them from one another.

See later illustrations, figures 11.71 and 11.72.

It is also important to remember that supply pipes and distributing pipes should not be connected together, even though they may both be supplying water for domestic purposes, unless in certain cases (e.g. shower mixers) appropriate backflow devices are fitted.

Additives to primary hot water or heating circuits

If a liquid (other than water) is used in any type of heating primary circuit, or if an additive, e.g. a corrosion inhibitor, is used in water in such a circuit, the liquid or additive should be non-toxic and non-corrosive.

6.4 Backflow protection

Backflow, which may include both 'backsiphonage' and 'back pressure', may be described as 'the reversal of the flow of water in a water supply system that can lead to contamination of the water supply'.

Backsiphonage may be described as 'backflow caused by siphonage of water from a cistern or appliance back into the pipe which feeds it'. For example, a hosepipe in use when the mains supply is turned off or a severe break in the main occurs causing water to run back into the supply pipe and/or the main.

Back pressure may be described as 'the reversal of flow in a pipe caused by an increase in pressure in the system'. An example of back pressure can be seen in the expansion of water from an unvented hot water heater which is permitted to pass back into the supply pipe.

Fluid risk categories

We became familiar with three categories of backflow risk required by water byelaws which implemented the earlier recommendations of the Backsiphonage Report of 1974.

The Water Supply (Water Fittings) Regulations 1999 now recognize and list, in Schedule 1, five fluid categories based on those developed by the Union of Water Supply Associations of Europe (EUREAU 12) and which are also used in North America and Australia. The five fluid categories represent a range of water qualities depending on how 'drinkable' they are, or how much danger to health they might present.

A comparison between the current fluid categories and the former backflow risk categories is shown in figure 6.6 and a description of the five fluid categories is given in table 6.1 and is taken from Schedule 1 of the Regulations.

Backflow prevention

Appropriate preventative measures should:

- prevent any water returning from any appliance, fitting or process back into the supply or distributing pipe that feeds it; and
- prevent any water returning from a supply pipe back into the water undertaker's main.

It should be remembered that all water taken from a main, whether actually drawn off or still within the water system, is considered suspect and should not be permitted to return to the main.

However, water is permitted to flow back into a supply pipe or distributing pipe where the water has expanded from a hot water system or an instantaneous water heater, but the

Water Byelaws – risk categories

Water Regulations – fluid risk categories

Byelaws		Regulations	
1	Serious health hazard	**5**	Serious health hazard
		4	Significant health hazard
2	Significant or slight health hazard	**3**	Slight health hazard
3	Aesthetic quality is impaired	**2**	Aesthetic quality of water impaired
		1	No health hazard or impairment in its quality

Figure 6.6   Comparison of backflow risk categories and fluid risk categories

Table 6.1   Fluid categories

Fluid category 1
Wholesome water supplied by a water undertaker and complying with the requirements of regulations made under Section 67 of the Water Industry Act 1991(a).

Fluid category 2
Water in fluid category 1 whose aesthetic quality is impaired owing to:
    (a)   a change in its temperature, or
    (b)   the presence of substances or organisms causing a change in taste, odour or appearance, including water in a hot water distribution system.

Fluid category 3
Fluid which represents a slight health hazard because of the concentration of substances of low toxicity, including any fluid which contains:
    (a)   ethylene glycol, copper sulphate solution or similar chemical additives, or
    (b)   sodium hypochlorite (chloros and common disinfectants).

Fluid category 4
Fluid which represents a significant health hazard because of the concentration of toxic substances, including any fluid which contains:
    (a)   chemical, carcinogenic substances or pesticides (including insecticides and herbicides), or
    (b)   environmental organisms of potential health significance.

Fluid category 5
Fluid representing a serious health hazard because of the concentration of pathogenic organisms, radioactive or very toxic substances, including any fluid which contains:
    (a)   faecal material or other human waste,
    (b)   butchery or other animal waste, or
    (c)   pathogens from any other source.

expanded water should be contained and not permitted to flow back to a position where it could be drawn off. Expansion in hot water vessels is discussed in chapter 4.

Backflow can be avoided by good system design and the proper positioning of appropriate backflow prevention devices (see section 6.5). Backflow prevention devices should be placed as near to the point of use as is practically possible, and all devices should be accessible for examination, maintenance, renewal or repair. Non-mechanical means of protection are preferred wherever practicable and will include air gaps, tap gaps, and vented distributing pipes for cistern fed systems.

Guidance to Paragraph 2 of Schedule 2 of the Water Regulations gives the following advice on the installation of mechanical backflow devices:

- they should be readily accessible for inspection, maintenance and renewal;
- devices for protection against categories 2 and 3 should not be located outside buildings. Exceptions are noted for types HA and HUK1 devices that are made specifically for use with existing garden taps in domestic premises;
- devices, particularly vented or verifiable types or those with relief outlets, should not be installed below ground or in chambers below ground;
- where line strainers are installed, e.g. upstream of fluid category 4 prevention devices, a servicing valve should be fitted immediately before the line strainer;
- relief outlets for reduced pressure zone valves should terminate with a type AA air gap at least 300 mm above ground or floor level.

Backflow prevention measures and devices used to prevent backflow should be related to the level of risk, and will vary according to the nature and use of the water supply.

6.5 Backflow prevention devices

A backflow prevention device is defined as 'a device which is intended to prevent contamination of drinking water by backflow'. It may be a mechanical or non-mechanical fitting or arrangement strategically positioned to prevent backflow from occurring.

There are a number of backflow prevention devices in various shapes and forms to suit a range of fluid risk categories. The *Regulator's Specification on Prevention of Backflow* describes these in two tables, the first listing 10 non-mechanical arrangements, and the second showing 14 mechanical devices. The tables, which also indicate the fluid category for which each device is suited, are reproduced here in tables 6.2 and 6.3.

A selection of these arrangements and devices is described and illustrated on the following pages.

Non-mechanical backflow arrangements

Type AA air gap with unrestricted discharge

The type AA air gap satisfies the requirements of BS 6281: Part 1. It has become familiar under water byelaws but now has a new designation under Water Regulations. The type AA air gap gives high-risk protection (fluid category 5) from both backsiphonage and back pressure.

Table 6.2   Schedule of non-mechanical backflow prevention arrangements and the maximum permissible fluid category for which they are acceptable

Type	Description of backflow prevention arrangements and devices	Fluid category for which suited	
		Back pressure	Backsiphonage
AA	Air gap with unrestricted discharge above spill-over level	5	5
AB	Air gap with weir overflow	5	5
AC	Air gap with submerged inlet	3	3
AD	Air gap with injector	5	5
AF	Air gap with circular overflow	4	4
AG	Air gap with minimum size circular overflow determined by measure or vacuum test	3	3
AUK1	Air gap with interposed cistern (for example, a WC suite)	3	5
AUK2	Air gaps for taps and combination fittings (tap gaps) discharging over appliances, such as a wash basin, bidet, bath or shower tray shall not be less than the following: _size of tap or combination fitting_ _vertical distance of bottom of tap outlet above spill-over level of receiving appliance_ not exceeding $G\frac{1}{2}$ 20 mm exceeding $G\frac{1}{2}$ but not exceeding $G\frac{3}{4}$ 25 mm exceeding $G\frac{3}{4}$ 70 mm	X	3
AUK3	Air gaps for taps or combination fittings (tap gaps) discharging over any higher risk domestic sanitary appliance where a fluid category 4 or 5 is present, such as: (a) any domestic or non-domestic sink or other appliance; or (b) any appliance in premises where a higher level of protection is required, such as some appliances in hospitals or other health care premises, shall be not less than 20 mm or twice the diameter of the inlet pipe to the fitting, whichever is the greater	X	5
DC	Pipe interrupter with permanent atmospheric vent	X	5

Notes:
(1)   X indicates that the backflow prevention arrangement or device is not applicable or not acceptable for protection against back pressure for any fluid category within water installations in the UK.
(2)   Arrangements incorporating DC type devices shall have no control valves on the outlet side of the device; they shall be fitted not less than 300 mm above the spill-over level of a WC pan, or 150 mm above the sparge pipe outlet of a urinal, and discharge vertically downwards.
(3)   Overflows and warning pipes shall discharge through, or terminate with, an air gap, the dimension of which should satisfy a type AA air gap.

Table 6.3   Schedule of mechanical backflow prevention arrangements and the maximum permissible fluid category for which they are acceptable

Type	Description of backflow prevention arrangements and devices	Fluid category for which suited	
		Back pressure	Backsiphonage
BA	Verifiable backflow preventer with reduced pressure zone (RPZ valve assembly)	4	4
CA	Non-verifiable disconnector with different pressure zones not greater than 10%	3	3
DA	Anti-vacuum valve (or vacuum breaker)	X	3
DB	Pipe interrupter with atmospheric vent and moving element	X	4
DUK1	Anti-vacuum valve combined with a single check valve	2	3
EA	Verifiable single check valve	2	2
EB	Non-verifiable single check valve	2	2
EC	Verifiable double check valve	3	3
ED	Non-verifiable double check valve	3	3
HA	Hose union backflow preventer. Only permitted for use on existing hose union taps in house installations	2	3
HC	Diverter with automatic return (normally integral with some domestic appliance applications only)	X	2
HUK1	Hose union tap which incorporates a double check valve. Only permitted for replacement of existing hose union taps in house installations	3	3
LA	Pressurized air inlet valve	X	2
LB	Pressurized air inlet valve combined with a check valve downstream	2	2

Notes:
(1) X indicates that the backflow prevention device is not acceptable for protection against back pressure for any fluid category within water installations in the UK.
(2) Arrangements incorporating a type DB device shall have no control valves on the outlet side of the device. The device shall be fitted not less than 300 mm above the spill-over level of an appliance and discharge vertically downwards.
(3) Types DA and DUK1 shall have no control valves on the outlet side of the device and be fitted on a 300 mm minimum type A upstand.
(4) Relief outlet ports from types BA and CA backflow prevention devices shall terminate with an air gap, the dimension of which should satisfy a type AA air gap.

See figures 6.7 and 6.8. This is a pipe arrangement where the supply to a cistern or vessel is arranged as follows:

- the cistern or vessel has an unrestricted overflow to atmosphere;
- the supply discharge and its outlet are not obstructed;
- the water is discharged vertically or not more than 15° from the vertical;
- the vertical distance between the discharge pipe outlet and the spill-over level of the cistern or vessel is *not less* than that shown in table 6.4.

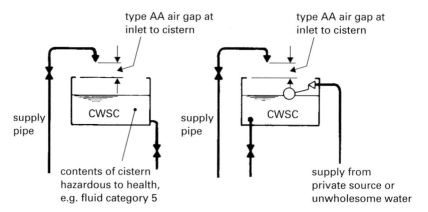

(a) Cistern with hazardous contents   **(b) Cistern with mixed supplies**

Type AA air gap is suitable for use with all risk categories.

Air gap related to size of inlet (see table 6.4).

Flow from inlet to be into air at atmospheric pressure and not more than 15° from the vertical.

Figure 6.7   Examples of type AA air gap at cisterns

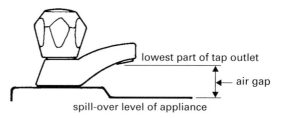

Type AA air gap to be visible, measurable and unobstructed.

Air gap is related to the size of the tap and the category of risk of the appliance (see table 6.4).

Figure 6.8   Example of type AA air gap at draw-off tap

Table 6.4   Air gaps at taps, valves and fittings (including cisterns)

Situation	Nominal size of inlet tap, valve or fitting	Vertical distance between tap or valve outlet and spill-over level of receiving appliance (mm)
Domestic situations fluid category 2/3 (device AUK2)	Up to and including $G\frac{1}{2}$	20
	Over $G\frac{1}{2}$ and up to $G\frac{3}{4}$	25
	Over $G\frac{3}{4}$	70
Non-domestic situations fluid category 4/5 (device AUK3)	Any size inlet pipe	Minimum diameter 20 mm, or twice the diameter of the inlet pipe, whichever is the greater

Its principal merit is that it has no moving parts and cannot readily be destroyed by vandals. It is accepted throughout the world.

Type AB air gap with weir overflow

The type AB air gap (see figure 6.9) is suitable for high-risk (category 5) situations and is particularly useful where there is a need to protect the contents of the storage vessel from ingress of contaminants such as dust and insects, e.g. feed and expansion cisterns in industrial or commercial premises.

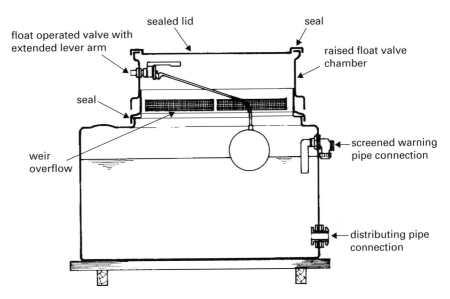

Weir overflow and warning pipe screened to prevent ingress by insects, dust, etc.

All joints on cistern sealed to prevent contamination.

Screened cistern vent, and open vent connection not shown.
These should be sealed to prevent contamination to the cistern water.

Figure 6.9   Example of type AB air gap with weir overflow

Type AG air gap with minimum size circular overflow

The type AG air gap arrangement satisfies the requirements of BS 6281: Part 2. See figure 6.10. In a cistern or other similar vessel, which is open at all times to the atmosphere, the vertical distance between the lowest point of discharge and the critical water level should be one of the following:

(1)   sufficient to prevent backsiphonage of water into the supply pipe;
(2)   not less than the distances shown in table 6.4.

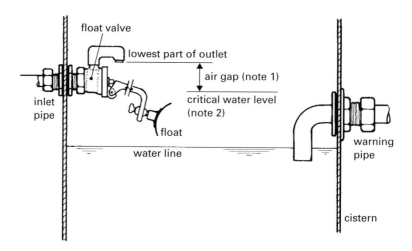

Note 1   The air gap is related to the size of the inlet and is the minimum permitted vertical distance between the 'critical' water level and the lowest part of the float valve outlet (see table 6.4).

Note 2   The critical water level is the highest level the water will reach at the maximum rate of inflow, i.e. float removed.

Note 3   Type AG air gaps should comply with the requirements of BS 6281: Part 2.

Figure 6.10   Type AG air gap (type B air gap to BS 6281: Part 2)

In storage cisterns the type AG type air gap, as shown in figure 6.10, is impractical. Its critical water level, which determines the difference in fixing heights between the float-operated valve and the warning pipe, cannot be readily calculated. The critical water level will vary from one installation to another depending on the inlet pressure and the length and gradient of the warning pipe.

In most cases it is expected that water undertakers will accept the simpler arrangement seen in figure 6.11.

The AUK1 air gap with interposed cistern

The AUK1 air gap with interposed cistern is used to 'disconnect' supply pipes and distributing pipes from appliances and vessels that might cause a backflow risk and is effective against:

(a)   backsiphonage risks of any fluid category; and
(b)   back pressure risk up to fluid category 3.

Common examples can be seen in WCs and urinals where the flushing cistern acts as the interposed cistern (see figure 6.12). Further examples in commercial and industrial situations are shown later in this chapter.

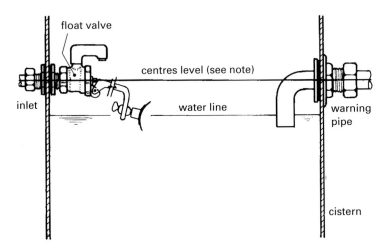

This arrangement is acceptable if:
- the cistern complies with Water Regulations;
- the float-operated valve is of the reducing flow type.

A reducing flow type float valve is one which gradually closes as the water level in the cistern rises, e.g. diaphragm float valve to BS 1212: Parts 2 or 3.

In this cistern the critical water level is assumed to be level with the centre line of the float valve body.

Figure 6.11 Acceptable alternative to the type AG air gap

Type AUK1 air gap with interposed cistern illustrated here in WC.

The AUK1 air gap should conform with type AG and comply with BS 6281: Part 2.

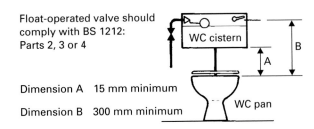

Figure 6.12 Type AUK1 air gap with interposed cistern

Type DC pipe interrupter with permanent atmospheric vent

A pipe interrupter (see figure 6.13) will admit air into a system, without the use of moving or flexible parts, to prevent backflow of water when a vacuum occurs. Whilst this device contains no moving parts, it can be subject to vandalism (by blockage of the airways). This device is suitable for protection from fluid category 5 backsiphonage risks, but cannot be used to protect against back pressure.

Typical use is with pressure flushing valves to WCs and urinals, where air is pulled in to break the siphon if a backflow condition occurs.

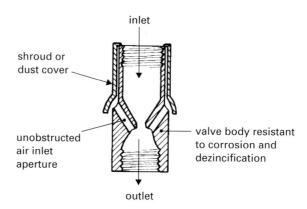

Valve should produce vacuum on outlet side and must discharge vertically downwards.

Must be fitted at least 300 mm above overflowing level of appliance served or 150 mm in the case of a urinal.

No tap or valve to be installed downstream (outlet size).

Pipe downstream not to be reduced in size.

Length of pipe downstream to be as short as possible.

Pipe interrupter to be readily accessible for repair.

Pipe interrupter to comply with BS 6281: Part 3.

Figure 6.13   Example of pipe interrupter – type DC

Mechanical backflow arrangements

Type BA verifiable backflow preventer with reduced pressure zone valve

The **reduced pressure zone valve** (or RPZ as it is better known), is relatively new to the backflow scene and is designed to meet all but the very highest backflow risks, being rated to fluid category 4 for both back pressure and backsiphonage.

The type BA, RPZ valve is shown in figure 6.14 whilst installation details are shown in figure 6.15. Figure 6.16 shows how the valve works.

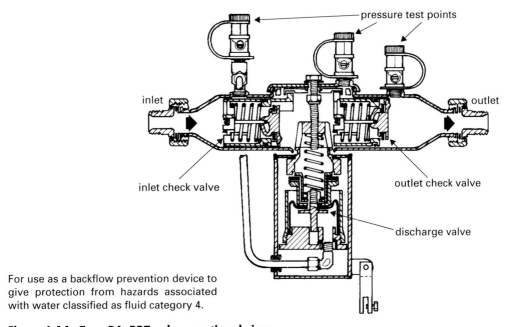

For use as a backflow prevention device to give protection from hazards associated with water classified as fluid category 4.

Figure 6.14   Type BA, RPZ valve – sectional view

Type CA non-verifiable disconnector with difference between pressure zones not greater than 10%

The non-verifiable disconnector works by venting the intermediate pressure zone of the valve to the atmosphere when the incoming pressure drops to 10% of the pressure at the outlet of the device. This valve will give protection against back pressure and back-siphonage risks up to fluid category 3.

Type DA anti-vacuum valve or vacuum breaker

An anti-vacuum valve or vacuum breaker (see figure 6.17) is a mechanical device with an air inlet which remains closed when water flows past it, but which opens to admit air if there is a vacuum in the pipe. The vacuum breaker must close once pressure and flow return to normal.

Upstands are required for anti-vacuum valves to work to maximum effect, and provide additional protection should the protection at point of use fail (see figure 6.18).

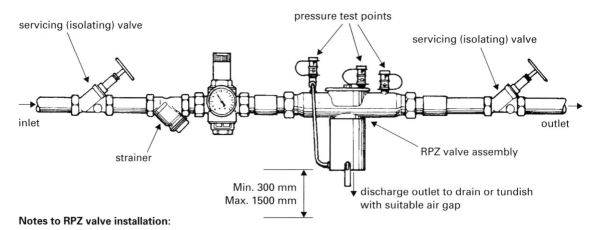

Notes to RPZ valve installation:

Valve to be installed in suitable position, free from flooding, free from effects of frost and in a secure, lockable cabinet. It should not be installed above any electrical equipment.

The assembly should be readily accessible except for closure of secure cabinet doors.

Assembly to be installed above ground in horizontal position (unless otherwise approved) with an in-line strainer downstream of the BA device.

Clearance to be allowed for test equipment to be fitted and to permit maintenance; minimum 200 mm in front of the device and at rear, 50 mm space for sizes up to DN 50 or 100 mm for devices above DN 50.

Discharge outlet to be at least 300 mm above floor level and maximum distance from floor of 1500 mm.

Adequate drainage to be provided, and air gap arranged between relief valve outlet and tundish/drain. No drinking water supply to be drawn off after (downstream of) an RPZ assembly.

RPZ valve assemblies will cause a drop in pressure and may not be suitable for use on low pressure supplies.

Assemblies to be flushed out and sterilized after installation and before use.

Assembly to be checked following installation to ensure relief valves function correctly.

Assembly to be 'site tested' before use and at regular intervals after installation by an accredited tester and test certificate to be issued after each test.

Records should be kept of all installation and maintenance procedures. Copies to be retained by both installer and the customer.

Testing should be carried out at not more than 12 month intervals.

Accredited installers and testers need to be approved by the water supplier.

Water supplier to be notified and approval given before RPZ valve is installed, and supplier should be notified of any maintenance tests carried out.

For more comprehensive information refer to *Information and Guidance Note No. 9-03-02* issued by the WRC Evaluation and Testing Centre, or from the water supplier.

Figure 6.15   Example of RPZ assembly

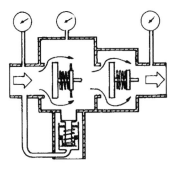 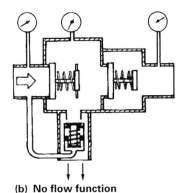 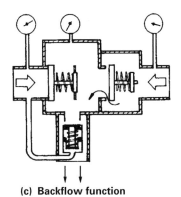

(a) Normal flow-through function

(b) No flow function

(c) Backflow function

For normal function with flow-through valve, both check valves are open, and discharge valve is closed.

Both check valves are closed and discharge valve is open.

Upstream check valve is closed and downstream check valve is open. Discharge valve is open.

Figure 6.16   How the RPZ valve works

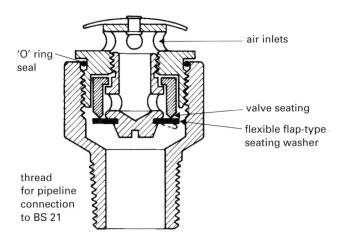

air inlets

'O' ring seal

valve seating

flexible flap-type seating washer

thread for pipeline connection to BS 21

Air inlet remains closed when water inside the device is at or above atmospheric pressure, and opens to admit air under vacuum conditions at the valve inlet.

Diagram shows terminal type; in-line type also available.

Needs regular opening and shutting to keep it operable.

Must be same size as pipe to which it is connected.

No control valve to be fitted downstream.

Failure of these valves to close would mean waste of water and possible damage to building.

Satisfies Water Regulations if it complies with BS 6282: Part 2 and there is no control valve downstream.

Must be fitted at least 300 mm above overflow level of appliance served.

Figure 6.17   Type DA anti-vacuum valve, atmospheric type

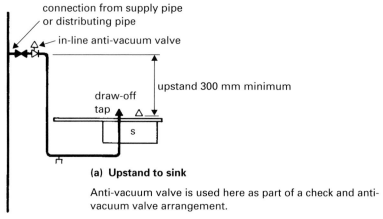

(a) **Upstand to sink**

Anti-vacuum valve is used here as part of a check and anti-vacuum valve arrangement.

Upstand is necessary for anti-vacuum valve to be effective.

Upstand 300 mm for cisterns and fixed appliances.

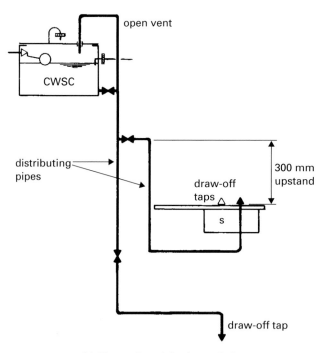

(b) **Upstand to sink, cistern fed**

Open vent shown here is an alternative to the check and anti-vacuum valve.

Upstand is needed for the vent to be fully effective.

Where vent is used, upstand should be at least 300 mm.

Branch not to rise above its connection.

Figure 6.18   Upstands to anti-vacuum valves

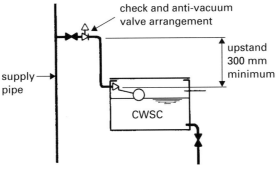

(c) Upstand to cistern

Upstand 300 mm above cistern and fixed appliances.

Figure 6.18   continued

Type EB non-verifiable single check valve

A check valve (see figure 6.19) is a mechanical device which, by means of a resilient 'elastic' seal or seals, permits flow of water in one direction only and is closed when there is no flow. The check valve is suitable for low-risk fluid category 2 only. Typical use is on an inlet to a domestic water softener.

The check valve should:
- be resistant to corrosion and dezincification;
- operate satisfactorily at temperatures up to 65°C;
- when shut, prevent any flow from inlet to outlet where water pressure does not exceed 10 mb;
- comply with BS 6282: Part 1.

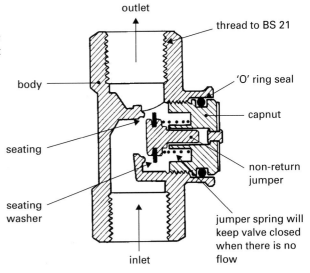

Figure 6.19   Check valve to BS 6282: Part 1

Type ED non-verifiable double check valve assembly

This comprises two check valves with a test cock fitted between them (see figure 6.20).

Double check valve assembly should comply with BS 6282: Part 1.

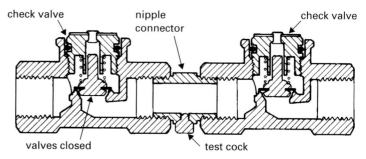

Figure 6.20   Double check valve assembly

Type DUK1 combined check and anti-vacuum valve

This is shown in figures 6.21 and 6.22 and consists of a single check valve upstream combined with an anti-vacuum valve. It is suitable for category 3 backsiphonage risk and category 2 back pressure risk.

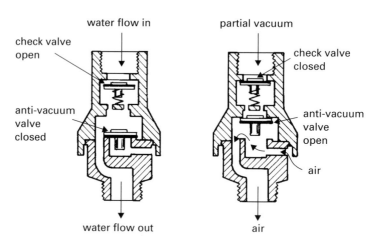

Must be used with an upstand for anti-vacuum valve to be effective.

Should comply with BS 6282: Part 4.

Figure 6.21   Combined check and anti-vacuum valve assembly

Categories of backflow risk

Table 6.5 lists appliances which fall into each of five fluid risk categories. The table lists minimum backflow protection measures, but a higher grade device may be used if more convenient. For situations not listed, the water undertaker should be consulted.

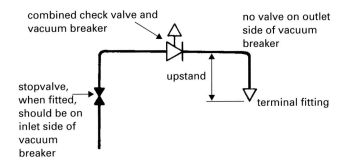

Check valve to be fitted upstream of vacuum breaker.

Figure 6.22   Use of combined check valve and anti-vacuum valve assembly

Table 6.5   Determination of fluid categories with examples

(a) Determination of fluid category 1

Fluid category 1:	Wholesome water supplied by a water undertaker and complying with the requirements of regulations made under Section 67 of the Water Industry Act 1991.
Example:	Water supplied directly from a water undertaker's main

(b) Determination of fluid category 2

Fluid category 2:	Water in fluid category 1 whose aesthetic quality is impaired owing to: (a) a change in its temperature; or (b) the presence of substances or organisms causing a change in its taste, odour or appearance, including water in a hot water distribution system.
Examples:	Mixing of hot and cold water supplies Domestic softening plant (common salt regeneration) Drink vending machine in which no ingredients or carbon dioxide are injected into the supply or distributing inlet pipe Fire sprinkler systems (without anti-freeze) Ice-making machines Water-cooled air conditioning units (without additives)

(c) Determination of fluid category 3

Fluid category 3:	Fluid which represents a slight health hazard because of the concentration of substances of low toxicity, including any fluid which contains: (a) ethylene glycol, copper sulphate solution or similar chemical additives; or (b) sodium hypochlorite (chloros and common disinfectants).
Examples:	Water in primary circuits and heating systems (with or without additives) in a house Domestic washbasins, baths and showers Domestic clothes and dishwashing machines Home dialysing machines Drink vending machines in which ingredients or carbon dioxide are injected Commercial softening plant (common salt regeneration only) Domestic hand-held hoses with flow controlled spray or shut-off control Hand-held fertilizer sprays for use in domestic gardens Domestic or commercial irrigation systems, without insecticide or fertilizer additives, and with sprinkler heads not less than 150 mm above ground level

Table 6.5 continued

(d) Determination of fluid category 4

Fluid category 4:	Fluid which represents a significant health hazard due to the concentration of toxic substances, including any fluid which contains: (a) chemical, carcinogenic substances or pesticides (including insecticides and herbicides); or (b) environmental organisms of potential health significance.
Examples:	*General* Primary circuits and central heating circuits in other than a house Fire sprinkler systems using anti-freeze solutions *House gardens* Mini-irrigation systems without fertilizer or insecticide application, such as pop-up sprinklers or permeable hoses *Food processing* Food preparation Dairies Bottle-washing apparatus *Catering* Commercial dishwashing machines Bottle-washing apparatus Refrigerating equipment *Industrial and commercial installations* Dyeing equipment Industrial disinfection equipment Printing and photographic equipment Car washing and degreasing plants Commercial clothes washing plants Brewery and distillation plant Water treatment or softeners using other than salt Pressurized fire-fighting systems

Table 6.5   continued

(e) Determination of fluid category 5

Fluid category 5:	Fluid representing a serious health hazard because of the concentration of pathogenic organisms, radioactive or very toxic substances, including any fluid which contains: (a) faecal matter or other human waste; or (b) butchery or other animal waste; or (c) pathogens from any other source.
Examples:	*General* Industrial cisterns Non-domestic hose union taps Sinks, urinals, WC pans and bidets Permeable pipes in other than domestic gardens, laid at or below ground level, with or without additives Grey water recycling systems *Medical* Any medical or dental equipment with submerged inlets Laboratories Bedpan washers Mortuary and embalming equipment Hospital dialysis machines Commercial clothes washing plant in health care premises Non-domestic sinks, baths, washbasins and other appliances *Food processing* Butchery and meat trades Slaughterhouse equipment Vegetable washing *Catering* Dishwashing machines in health care premises Vegetable washing *Industrial and commercial installations* Industrial and commercial plant, etc. Mobile plant, tankers and gulley emptiers Laboratories Sewage treatment and sewage cleansing Drain cleaning plant Water storage for agricultural purposes Water storage for fire-fighting purposes *Commercial agricultural* Commercial irrigation outlets below or at ground level and/or permeable pipes, with or without chemical additives Insecticide or fertilizer applications Commercial hydroponic systems

Note   The list of examples shown above for each fluid category is not exhaustive.

6.6   Secondary or zone backflow protection

This is used on a supply pipe or distributing pipe where there is an increased risk of backflow because of the following:

- exceptionally heavy use;
- presence of a hazardous substance;
- possibility of internal backflow in buildings of multiple occupation, e.g. flats or tall buildings.

Secondary or zone backflow protection should be provided in addition to those prevention devices installed at points of use and is required when:

- supply or distributing pipes convey water to two or more separately occupied premises;
- premises are required to provide a storage cistern capacity for 24 hours or more of normal use or premises receiving intermittent supply.

Acceptable arrangements for wholesite or zone backflow protection are shown in figure 6.23.

(a) Secondary or zone backflow protection on supply pipes to separate premises

Single or double check valve depending on level of risks.

No need for secondary protection at lowest level.

No part of the branch pipe to be higher than its connection to the common supply pipe.

Connection of branch pipes to be at least 300 mm above the overflowing level of highest appliance served.

Secondary protection will protect premises at one level from those at another level, e.g. flats or shops with flat over.

Zone protection will protect one part of a building from another part, e.g. industrial or commercial premises.

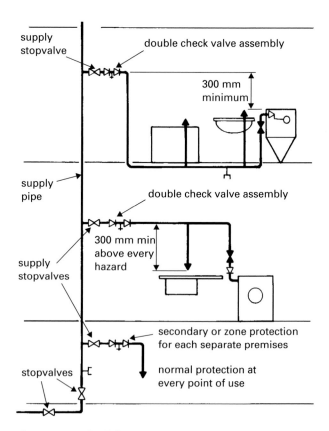

Figure 6.23   Acceptable arrangements for secondary or zone backflow protection

(b) Secondary or zone backflow protection on distributing pipes

No part of the branch pipe to be higher than its connection to the common distributing pipe.

Single or double check valve depending on level of risk.

No need for secondary protection at lowest level.

Connection of branch pipe to be at least 300 mm above the overflowing level of highest appliance served.

Secondary protection will protect premises at one level from those at another level.

Zone protection will protect one part of a building from another part.

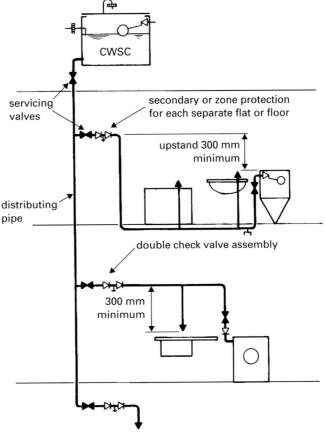

(c) Secondary or zone backflow protection on distributing pipes using a vent pipe

Vent will admit air to prevent backflow.

No need for secondary protection at lowest level.

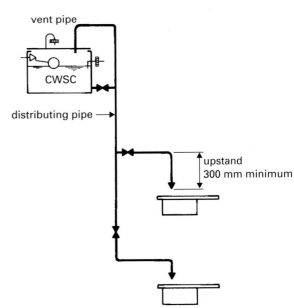

Figure 6.23 continued

6.7   Application of backflow prevention devices

In general, backflow can be prevented by good system design combined with the use of back-flow devices or arrangements chosen to suit the category of risk for which they are designed.

Water Regulations advise that where possible backflow protection should be achieved without the use of mechanical devices. For example, point-of-use protection may include the 'tap gap' or 'air gap' as seen in table 6.4 and figures 6.7 and 6.8. In the case of cistern-fed appliances, zone protection may be provided in the form of permanently vented distributing pipes. In many high-risk situations an air gap is the only device permitted.

Backflow prevention arrangements and devices should be positioned and installed so they are readily accessible for inspection, repair and renewal. It is important that all backflow devices are regularly checked and maintained in good working condition. This applies particularly to mechanical devices.

Bidets and WCs adapted as bidets

Bidets are an obvious contamination risk that come within fluid category 5 of backflow risk. There are two types of bidet: the over-rim type, and the ascending spray or submerged inlet type. Each is protected differently. The former may be used with its taps under mains pressure, the latter is required to be supplied from a storage cistern.

Over-rim types should be arranged with type AUK2 air gaps in accordance with table 6.4 between the tap outlets and the spill-over level of the appliance (see figure 6.24). It is an offence to attach a hand-held flexible spray or similar fittings to taps or bidets where connection is made directly from the supply pipe.

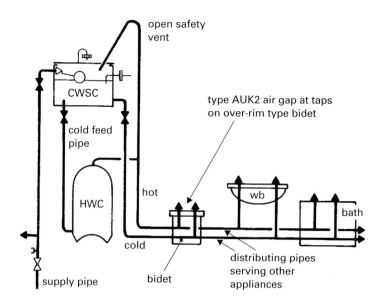

Over-rim type bidets may be supplied from distributing pipe or supply pipe provided type AUK2 air gap is maintained.

Figure 6.24   Over-rim type bidet

Ascending spray type bidets, and those having hand-held sprays attached, present a more serious risk than over-rim types and are not permitted to be connected directly from a supply pipe. When making connections to these the following points should be borne in mind:

- They should be supplied with hot and cold water via a break cistern.
- Both hot and cold connections should be supplied from independent, dedicated distributing pipes that do not supply any other appliance, except in the following cases:
  - (a) a common cold distributing pipe that serves WC or urinal flushing cistern only in addition to the bidet;
  - (b) where the bidet is the lowest appliance in the premises and there is no likelihood of other fittings being fitted at a lower position at a later date and there is no spray attachment fitted, e.g. in a bungalow with all rooms at one level or in a house with a bidet at the lowest floor level.
- The branch pipe connection to the bidet should be at least 300 mm above the spill-over level of the bidet or 300 mm above the highest point that a spray attachment might reach.

Figure 6.25 shows typical acceptable arrangements.

WCs and urinals

WC pans and urinals are within fluid risk category 5 and present a serious risk of backflow whatever their situation, whether in a house or in industrial or commercial premises. There are two suitable types of backflow device for these:

(1) an **interposed cistern (type AUK1)**, which means a siphonic or non-siphonic flushing cistern, may be used in any type of premises; or

(2) a **pipe interrupter with permanent atmospheric vent (type DC)**, fitted to the outlet of a manually operated pressure flushing valve, may be connected to a supply pipe or distributing pipe (but not in a house). There should be no other obstruction between the outlet of the pipe interrupter and the flush pipe connection to the appliance.

   The pipe interrupter should be positioned so that its vent outlet is at least:
   - 300 mm above the overflowing level of the WC pan, or
   - 150 mm above the urinal bowl being served.

Taps and shower outlets to sanitary appliances

All single outlet taps, combination taps and fixed shower heads should discharge above the appliance, terminating with an air gap as shown in figure 6.8. In domestic premises a tap gap (AUK2) should be used, whilst all other premises should maintain the more stringent AUK3 air gap.

Sinks in both domestic and non-domestic situations are considered to be a fluid category 5 risk and as such the minimum protection is the type AUK3 air gap. However, this is not generally a problem as sinks need additional space for access to work and for the filling of buckets and other appliances (see figure 2.1). Where appliances such as baths and wash

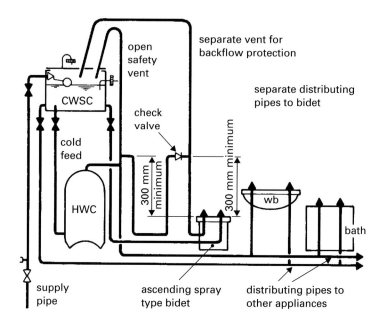

Before fitting a bidet with an ascending spray or flexible hose, the water undertaker MUST be notified (Regulation 5)

(a) Bidet supplied from separate dedicated distributing pipes

Precautions required to prevent backflow from bidet through cylinder to appliances:
- vent to atmosphere;
- check valve fitted downstream of vent;
- 300 mm minimum upstand.

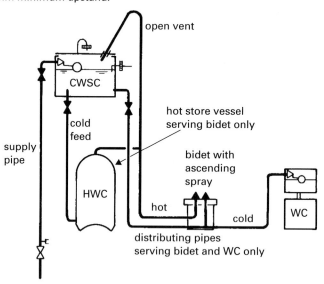

(b) Hot store vessel feeding bidet only

Connection between WC and bidet permitted because this creates no additional hazard.

No special precautions required where hot store vessel feeds the bidet only.

Figure 6.25   Ascending spray type bidet

basins in domestic premises have submerged inlets they are considered to be a category 3 risk and both hot and cold inlets should be supplied through type EC or type ED double check valves. Appliances for non-domestic use will have a higher category of risk and will need backflow protection to suit the risk. For instance, appliances in hospitals are a category 5 risk.

Washing machines and dishwashers

Household machines have backflow protection to fluid category 3 built in during manufacture. Before installing an appliance of this kind reference should be made to the *Water Materials and Fittings Directory* in which they will be listed if they are approved under the Water Regulations Advisory Scheme. Commercial machines such as those used in laundromats or similar premises are a category 4 risk, whilst clothes washing plant or equipment in health care establishments are fluid risk category 5.

Drinking water fountains

These should be designed so that there is a minimum 25 mm air gap between the water delivery nozzle and the spill-over level of the bowl. Additionally, the nozzle should be screened or shrouded to prevent mouth contact.

Outside taps and garden supplies

Backflow protection for hose taps depends on the level of risk for the individual application. Any backflow prevention device used should be fitted inside a building where it will not be subjected to frost damage.

- Hosepipes held in the hand for garden and other uses should be fitted with a *self-closing mechanism* at the hose outlet.
- In a house situation, any garden tap to which a hose connection can be made should be fitted with a *double check valve* positioned inside a building. This will give adequate protection if the hose is held in the hand or where the hose outlet is fixed and provides a permanent air gap (see figure 6.26(a)).
- Mini-irrigation systems and porous hoses used in a house garden require a *double check valve* as minimum protection and additionally a *pipe interrupter with moving element (type DB)* fitted at the connection of the hose and at least 300 mm above the highest water outlet in the system (see figures 6.26(b) and 6.26(c)).
- The *double check valve* is also considered sufficient protection for hand-held hoses used for spraying fertilizers or domestic detergents in household garden situations.
- For more severe risks, such as the application of insecticides, that are considered to be a fluid risk category 5 even in domestic premises, higher levels of backflow control are needed.

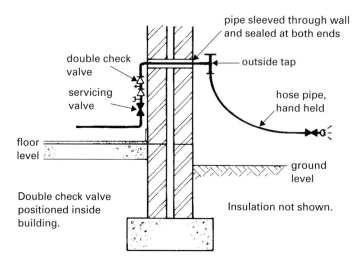

(a) Hose union tap, and tap with hose pipe held in the hand

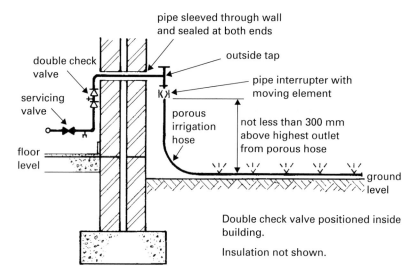

(b) Porous hose or mini-irrigation system – ground level or sloping away from the building

Figure 6.26   Backflow protection to external taps in houses

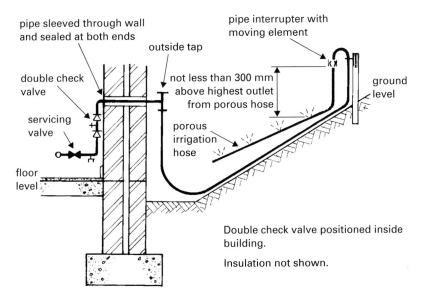

pipe sleeved through wall and sealed at both ends

pipe interrupter with moving element

outside tap

double check valve

not less than 300 mm above highest outlet from porous hose

ground level

servicing valve

porous irrigation hose

floor level

Double check valve positioned inside building.

Insulation not shown.

(c) Porous hose or mini-irrigation system – ground rising away from the building

Figure 6.26   continued

Existing garden taps

In a house situation a hose union tap fitted before the Water Supply (Water Fittings) Regulations 1999 came into force (1 July 1999) may be fitted in one of three ways. Either:

(1)   the existing hose union tap should have a *double check valve* installed inside a building; or
(2)   the tap should be replaced by one that incorporates a *double check valve arrangement (type HUK1)*; or
(3)   a *hose union backflow preventer (type HA)*, or a *double check valve*, should be fitted to the outlet of the tap.

Outside taps and systems in commercial premises

Taps used for non-domestic applications generally present a higher risk than those in domestic premises. Backflow protection should be provided to suit the level of risk and the application, e.g. commercial, horticultural or industrial applications. For example, soil watering and irrigation systems such as permeable hoses are considered to be a fluid risk category 5 and should be supplied only through one of the following backflow prevention devices:

• Type AA air gap with unrestricted discharge;
• Type AB air gap with weir overflow;
• Type AUK1 air gap with interposed cistern.

There may also be a need to provide additional zone protection.

Backflow protection in industrial/commercial premises

There are many industrial and trade premises where risk of contamination may be present, and where backflow protection may be required in addition to any zone backflow protection that is provided. Examples of some of these are shown in the following pages, and include improvements to backflow protection in existing fire sprinkler systems.

Figure 6.27 shows an example of where a property is supplied with water for both domestic and non-domestic purposes.

Spill-over level of appliances served must be at least 300 mm below the invert level of the warning/overflow pipe, and at least 15 mm below the base of the cistern.

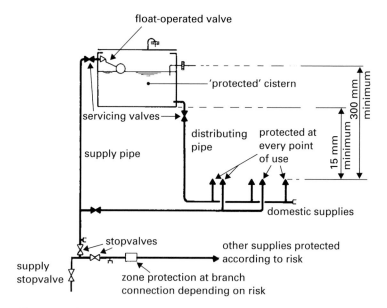

Pipes used for domestic purposes should not be connected to pipes for any other purposes.

Figure 6.27   Supplies for domestic and other purposes

Figure 6.28 shows examples of the prevention of contamination by backflow or cross connection within industrial, commercial, trade, research, educational, medical and similar establishments, in addition to any zone backflow protection that might be required.

In premises where drinking water and non-drinking water supplies are made available, they should be clearly identified as 'drinking water', 'non-drinking water' or 'fire-fighting water', as appropriate. Hoses intended for drinking water should be used only for that purpose and should be marked 'not for cleaning purposes'.

Water regulations require that 'any fluid that is not wholesome water shall be clearly identified so as to be easily distinguished from any supply pipe or distributing pipe'.

Figure 6.29 shows the prevention of contamination by backflow or cross connection within agricultural and similar establishments (in addition to any wholesite or zone backflow protection).

Diagram shows type AA air gap.

Could also use type AB or type AD air gap.

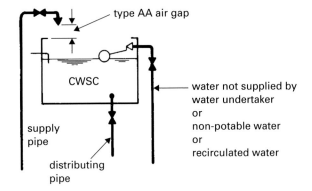

(a) **Water supplied by undertaker and non-potable water from other sources**

Diagram shows type AA air gap.

Could also use type AB or AD air gap.

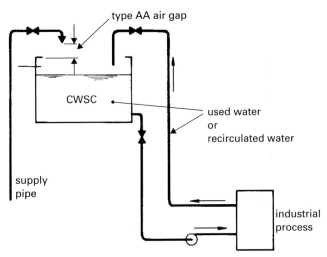

(b) **Re-used or recirculated water**

Diagram shows type AB air gap.

Could also use type AA or type AD air gap.

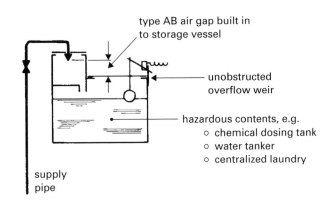

(c) **Supplies to fixed or mobile appliances in industrial processes**

Figure 6.28   Backflow protection in industrial and commercial installations

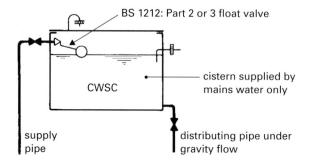

(a) Supplies from storage

Spill-over level of appliances served must be at least 300 mm below the invert level of the warning/overflow pipe, and at least 15 mm below the cistern base.

Where pump is fitted to distributing pipe, cistern must be fitted with type AA, AB or AD air gap arrangement.

Supplies to static or mobile appliances, e.g. crop sprayer, to be connected via type AA or AB air gap or properly connected break cistern serving no other appliance.

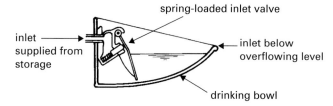

(b) Supply to animal drinking bowl from storage

Drinking bowl without adequate air gap or shrouded inlet should be supplied from storage using an interposed cistern with a dedicated distributing pipe that serves only similar fluid category 5 (high-risk) appliances.

If inlet has type AA air gap and is shrouded from mouth contact, the drinking bowl could be supplied direct from the supply pipe or distributing pipe.

Further backflow protection may be needed if other appliances are supplied from same source.

Figure 6.29   Backflow protection in agricultural and horticultural installations

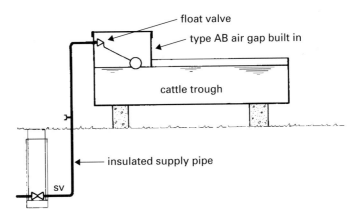

float valve

type AB air gap built in

cattle trough

insulated supply pipe

SV

(c) Cattle trough connected to supply pipe

Figure 6.29   continued

Fire protection systems

The scope of this book does not permit detailed discussion of fire protection systems in commercial and industrial premises, which are subject to the rules of the Loss Prevention Council. However, fire protection to domestic and residential buildings is discussed in chapters 13 and 14. Before commencement of any fire installation the local water undertaker should be consulted, particularly in respect of requirements for the prevention of contamination.

Sprinkler systems are commonly installed, particularly in high-risk situations. They are fitted for emergency fire protection only and should be used for no other purpose. Water in them may become stagnant and create contamination risks, particularly where substantial volumes are stored in ground level or elevated cisterns.

Wet sprinkler systems without additives, first-aid fire hose reels and hydrant landing valves are fluid risk category 2 and require the minimum protection of a *single check valve* only.

Wet sprinkler systems with additives, and systems containing hydro-pneumatic pressure vessels, are considered to be in fluid risk category 4, and will require either a *verifiable backflow preventer (RPZ valve)* or an air gap, type AA, AB or AD.

Where a common supply pipe serves a fire protection system and a supply pipe for drinking water and domestic purposes, the fire supply should be connected immediately on entry to the building and appropriate backflow protection should be fitted close to the point of connection.

Figure 6.30 shows backflow protection in fire sprinkler systems.

Wet sprinkler systems without additions are fluid risk category 2.

Systems with additives, e.g. antifreeze, will need protection for category 4 risk.

Sprinkler controls not shown.

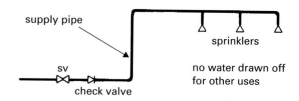

(a) System direct from supply pipe

Connection to sprinkler system to be taken from supply pipe at or near to point of entry.

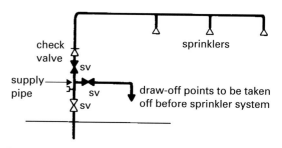

(b) Wet system with draw-offs

Where cistern supplies only the sprinkler system, it will require a type AA or type AB air gap for category 5 risk.

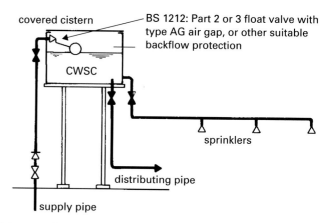

(c) Wet system from storage

Cistern supplied from mains only.

Water for other purposes to be drawn separately from storage cistern.

Separate cistern for fire purposes preferred.

An uncovered cistern must supply the sprinklers only.

Figure 6.30   Backflow protection in fire sprinkler systems

Type AA air gap shown. Could use type AB air gap.

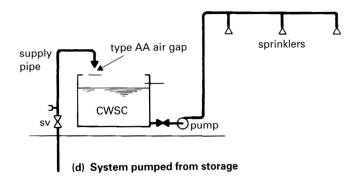

(d) System pumped from storage

Type AA air gap shown. Could use type AB air gap.

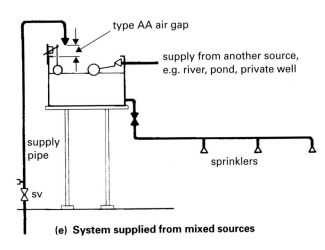

(e) System supplied from mixed sources

Figure 6.30   continued

Chapter 7
Frost precautions

Water temperatures in systems will vary from time to time depending on a number of factors, for example ambient temperature of the surrounding air, temperature of the incoming water supply, provision of insulation and heat to the building and its pipework systems, much of which, in turn, will be affected by weather conditions and the time of the year.

Temperature changes in systems are inevitable and must be accepted provided they do not become too extreme. In extreme cases protective measures may need to be taken.

Water installations should be protected against the effects of:

- frost and the formation of ice in pipes and fittings;
- loss of heat from hot water pipes, fittings and storage vessels;
- heat gains in cold water pipes, fittings and storage vessels;
- condensation on the surface of pipes;
- thermal stress.

Heat losses from hot water cylinders, pipes and fittings can cause serious waste of heat energy. Pipes and fittings should be insulated to prevent this. In the case of cylinders and tanks a lagging jacket should be fitted, or preferably hot water storage vessels that are factory insulated should be used. This subject is covered more fully in chapter 8.

Heat gains in cold water pipes can lead to an increase in bacterial activity and subsequently to a deterioration in water quality. This can be offset by using insulation to the standards shown later in table 7.2. Where hot water pipes are run in close proximity to each other, the cold pipe should be placed below the hot one to avoid the effect of rising heat.

Condensation can occur where cold pipes pass through areas of relatively high humidity. The moisture formed may lead to some corrosion of metal pipes or in extreme cases, damage to the building structure.

7.1 Protection from frost

When water loses heat and its temperature drops below freezing point it turns to ice. Upon freezing its volume increases by approximately 10%. This results in:

- damage to pipework and fittings due to the greater volume of ice compared to water;
- the risk of explosion should hot water apparatus be put into operation when part of the system is blocked with ice.

The temperature of water supplied is quite low. In winter, just a small reduction in temperature will cause freezing.

The best precaution against freezing of water services in buildings is the obvious one of keeping the inside of the building continuously warm by the provision and maintenance of adequate heating. This will be easier to achieve and more economical if the building has been designed and constructed so as to minimize the loss of heat through its structure. When the whole building is not heated, or where heating is only intermittent, the localized heating of water pipes and fittings or the heating of their immediate surroundings may suffice. Localized heating, such as trace heating, in conjunction with a frost thermostat, should be used only in addition to other forms of frost protection, or where those are unsuitable.

Location of pipes

Pipes should be located so as to minimize the risk of frost damage, avoiding areas that are difficult to keep warm, especially the following:

- unheated parts of the building, e.g. roof spaces, cellars and underfloor spaces, garages or outhouses;
- areas where draughts could occur, e.g. near windows, air bricks, ventilators, external doors or under suspended floors;
- cold surfaces, e.g. chases and ducts in outside walls and places where pipes come into direct contact with outside walls;
- any exterior position above ground.

If it is not possible to avoid these locations then protection should be provided.

Although some plastic pipes and cisterns are flexible and not easily fractured by ice formation, such formation stops the supply of water. Therefore precautions are still necessary.

7.2   Protection of pipes and fittings

Pipes underground

These should be laid at least 750 mm below the ground surface bearing in mind any expected changes in ground levels (see figure 7.1). This minimum requirement will be sufficient in most cases, but where more severe weather can be expected, then greater depth of cover can be provided up to a maximum of 1350 mm.

The insulating value of the soil will be affected by its nature and water-retaining capacity as well as the degree of exposure of the site. Therefore extra depth of cover may be required to suit local conditions; see figure 7.2.

Stopvalves below ground

Underground stopvalves should not be brought up to a higher level merely for ease of access, but should remain at the same depth as the pipe.

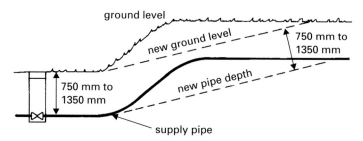

Where ground is to be levelled, ensure pipe is maintained at full depth of cover. This is important, especially on new installations, where ground is often levelled after services are laid.

Minimum depth of cover must be maintained over whole length of pipe.

Figure 7.1   Pipes in uneven ground

Pipes entering buildings

Pipes should enter buildings at the same depth as laid (see figures 7.3 and 7.4). Where it is not possible to maintain a minimum depth of 750 mm, pipes should be insulated. See also figure 10.3.

Pipes and fittings outside buildings

Where possible these should be laid underground. Where this is not possible sufficient protection against the effects of frost should be provided. See figures 7.5 and 7.6.

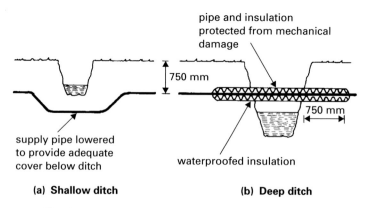

Pipe insulated where impractical to take it below deep ditch.

Figure 7.2   Pipes passing under or through ditches

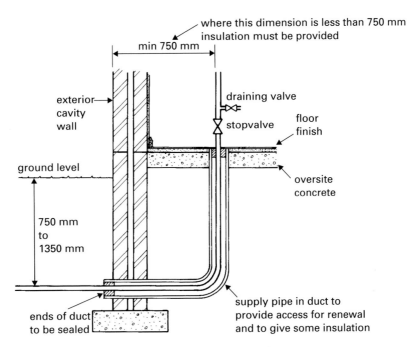

where this dimension is less than 750 mm
insulation must be provided

min 750 mm

exterior
cavity
wall

draining valve

stopvalve

floor
finish

ground level

oversite
concrete

750 mm
to
1350 mm

ends of duct
to be sealed

supply pipe in duct to
provide access for renewal
and to give some insulation

Figure 7.3   Pipes entering buildings – solid floor construction

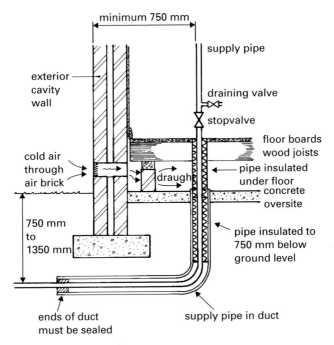

minimum 750 mm

supply pipe

exterior
cavity
wall

draining valve

stopvalve

cold air
through
air brick

draught

floor boards
wood joists

pipe insulated
under floor
concrete
oversite

750 mm
to
1350 mm

pipe insulated to
750 mm below
ground level

ends of duct
must be sealed

supply pipe in duct

Shallow foundations as shown here would not be normal in new buildings,
but where this situation does occur the pipe beneath the foundation should
be protected from settlement by the building.

Figure 7.4   Pipes entering buildings – suspended timber floor

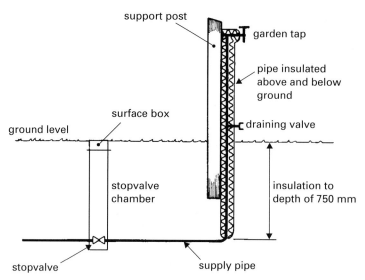

Insulation to be waterproofed throughout and protected from damage.

Note   Backflow protection device needed if hose pipe is to be connected.

Figure 7.5   Pipe rising from below ground to garden stand-pipe

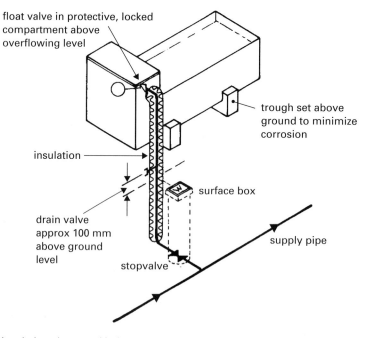

Insulation above and below ground, water proofed and protected from damage by animals.

Figure 7.6   Pipe rising from below ground to cattle trough

Pipes and cisterns in or above roof spaces

When considering the insulation of roof spaces and pipes and components within roof spaces (see figure 7.7), full advantage should be taken of any heat rising from rooms. Roof spaces are notoriously cold and draughty places and the cause of many frozen pipes in the UK. This has not been helped by Building Regulations which require ventilation of roof spaces to prevent condensation. Consideration should also be given to the following:

- ceiling insulation should be omitted below cisterns;
- insulation should be provided all over any cistern and include any pipes rising to connect with it;
- pipes within roof spaces, including warning and overflow pipes, should wherever possible be fixed below any ceiling insulation;
- any pipes other than those situated as above should be adequately insulated.

If a cistern is sited above the roof of a building, it must be protected by installing it, together with its inlet and outlet pipes, in an insulated enclosure either provided with its own means of heating or opening into some heated part of the building itself.

7.3   Draining facilities

It should be possible to drain down all parts of a hot or cold water system including its pipes, fittings, components and appliances, so that pipes not in use can be emptied in cold weather. An empty pipe cannot suffer frost damage. For pipes to be adequately drained, allowance must be made for the entry of air. Usually this will be done through taps and vent pipes but in some special cases air inlet valves will be needed.

To assist draining:

- pipes should be laid to slight and continuous falls to draining valves at low points;
- draining taps should have means for connection of a hose pipe;
- cisterns, cylinders and tanks should be fitted with draining taps unless they can be drained through pipes leading to a draining tap elsewhere.

Draining taps must be of the screw-down pattern having a removable key and must comply with BS 2879 (see figure 7.8).

BS 6700 recommends that all pipes and fittings be drainable. However, it is the author's view that there are some exceptions to this rule. Draining taps should not be fitted within sealed ducts or below floors or below ground where they may be inaccessible (see figure 7.9). More importantly, they should not be fitted where they might become submerged and create a contamination risk (see figure 7.10).

Where a building is divided into parts, the pipework should be arranged so that each part can be isolated and drained without affecting the other parts. A stopvalve and drain tap should be located in a convenient position close to where the pipe enters the building or part of the building. Unless the stopvalve is installed within a normally heated building, it should be protected against frost damage.

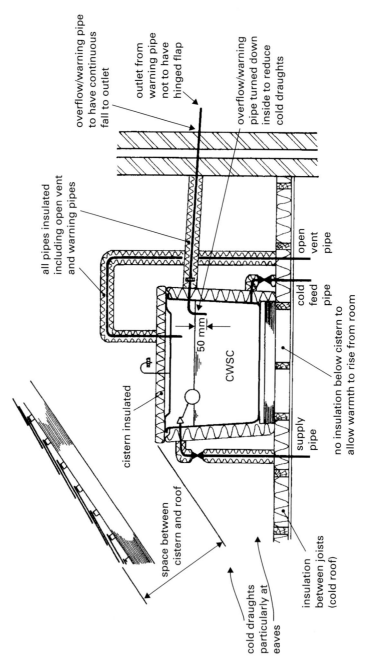

Figure 7.7 Insulation of pipes and cisterns in roof space

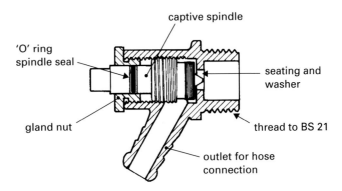

Figure 7.8 Drain cock to BS 2879

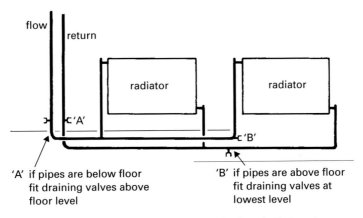

'A' if pipes are below floor
fit draining valves above
floor level

'B' if pipes are above floor
fit draining valves at
lowest level

For pipe drops below floor level where the floor is likely to be covered by carpets and furniture it may be better to fit the draining valves above, but near the floor (alternative A).

Alternative B is more suitable where pipes do not drop below floor level.

Figure 7.9 Draining arrangements – pipe drops below floors

Every external standpipe, livestock watering appliance, garden tap, garage tap, or similar water fitting should, where possible, be supplied through a stopvalve located in a convenient position within a normally heated building. Where this is not possible, the stopvalve should be protected against frost damage and a draining tap fitted in an appropriate position at low level.

Pipes and fittings in any part of a building that is unheated or unused in winter, including any water closet, garage or conservatory, or any other outbuilding, should be arranged to enable them to be isolated and drained separately.

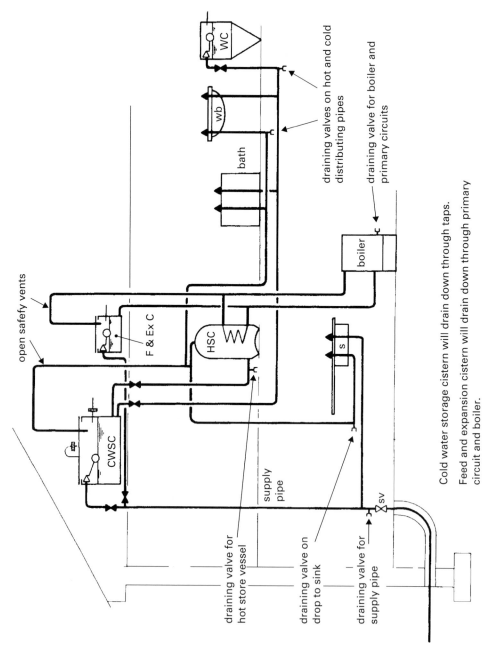

Figure 7.10 Hot and cold water system showing draining valve positions

7.4 Insulation against frost damage

The thickness of insulation used to protect an installation is dependent on the following factors:

- the type of insulation used and its thermal conductivity value (see table 7.1);
- the reason for the insulation, whether to protect against frost damage, heat gains, heat losses, or condensation;
- whether the pipes, fittings and components are in heated or unheated premises, or whether they are indoors or outdoors;
- the requirements of Water and Building Regulations (see also chapter 8).

Table 7.1 Examples of insulating materials

Thermal conductivity W/(m.K)	Material
Less than 0.020 0.021 to 0.035 0.04 to 0.055 0.055 to 0.07	Rigid phenolic foam Polyurethane foam Corkboard Exfoliated vermiculite (loose fill)

The Water Regulations Guide discusses the background to the criteria used in BS 5422 for calculating insulation, and considers two conditions, 'normal' and 'extreme', where differing thicknesses of insulation might apply.

Normal conditions would include water fittings installed within buildings in areas that are not subjected to draughts from outside the building. Examples given are cloakrooms, storerooms, utility rooms, roof spaces where pipes are below the ceiling insulation, and unheated parts of otherwise heated commercial buildings.

Extreme conditions include water fittings installed inside unheated or marginally heated buildings, or below suspended floors, cold roof spaces or other areas where draughty conditions are likely.

Insulation tables to suit a wide range of contingencies for various types of building usage can be found in BS 5422. The Standard gives calculated figures for the protection of copper and steel pipes in both domestic and commercial premises. Two of these tables relating to protection against frost in domestic buildings are reproduced in table 7.2.

Table 7.2 gives figures based on the thickness of insulation needed in normal or extreme conditions to delay frost for up to 8 or 12 hours respectively. Commercial insulation materials should be chosen to give at least the thicknesses shown in the table. However, for small diameter pipes it may not be possible to obtain insulation as thick as that stated in the table. To achieve good frost protection where insulation thicknesses given in table 7.2 cannot be obtained, it may be advisable to choose one of the following options:

- select an insulation material with a lower thermal conductivity;
- increase the pipe size; or
- provide supplementary heat to maintain the water temperature, e.g. trace heating.

The description 'extreme conditions' does not include water fittings installed outside buildings. Water pipes located outside buildings in open air, or less than 750 mm below ground,

Table 7.2   Minimum insulation thickness to guard against freezing for domestic cold water systems for up to 8 hours (*12 h*)

Outside diameter (mm)	Inside diameter (mm)	Insulation thickness (mm)									
		Specified condition 1					Specified condition 2				
		λ = 0.020	λ = 0.025	λ = 0.030	λ = 0.035	λ = 0.040	λ = 0.020	λ = 0.025	λ = 0.030	λ = 0.035	λ = 0.040
Copper pipes to BS EN 1057											
15.0	13.6	11 (20)	15 (30)	20 (43)	26 (62)	34 (88)	12 (23)	17 (35)	23 (53)	31 (78)	41 (113)
22.0	20.2	6 (9)	7 (12)	9 (16)	11 (20)	13 (24)	6 (10)	8 (14)	10 (18)	12 (23)	15 (28)
28.0	26.2	4 (6)	5 (8)	6 (10)	7 (12)	9 (14)	4 (7)	6 (9)	7 (11)	8 (13)	10 (16)
35.0	32.6	3 (5)	4 (6)	5 (7)	6 (9)	7 (10)	4 (5)	4 (7)	5 (8)	6 (10)	7 (11)
42.0	39.6	3 (4)	3 (5)	4 (6)	5 (7)	5 (8)	3 (4)	4 (5)	4 (6)	5 (7)	6 (9)
54.0	51.6	2 (3)	3 (4)	3 (4)	3 (5)	4 (6)	2 (3)	3 (4)	3 (5)	4 (5)	4 (6)
76.1	73.1	2 (2)	2 (3)	2 (3)	3 (4)	3 (5)	2 (2)	2 (3)	2 (3)	3 (4)	3 (4)
Steel pipes to BS 1387											
21.3	16.1	9 (15)	12 (21)	15 (29)	19 (38)	24 (50)	10 (18)	14 (26)	18 (35)	23 (48)	29 (64)
26.0	21.7	6 (9)	7 (12)	9 (15)	11 (18)	13 (22)	6 (10)	8 (13)	10 (17)	12 (21)	15 (26)
33.7	27.3	4 (7)	5 (8)	7 (10)	8 (12)	9 (15)	5 (7)	6 (9)	7 (12)	9 (14)	10 (17)
42.4	36.0	3 (5)	4 (6)	5 (7)	5 (8)	6 (10)	3 (5)	4 (6)	5 (8)	6 (9)	7 (11)
48.3	41.9	3 (4)	3 (5)	4 (6)	5 (7)	5 (8)	3 (4)	4 (5)	4 (6)	5 (7)	6 (9)
60.3	53.0	2 (3)	3 (4)	3 (5)	4 (5)	4 (6)	2 (3)	3 (4)	3 (5)	4 (6)	4 (7)
76.1	68.8	2 (2)	2 (3)	2 (3)	3 (4)	3 (4)	2 (3)	2 (3)	3 (4)	3 (4)	3 (5)

Key:

Specified conditions 1: water temperature 7°C; ambient temperature –6°C; evaluation period 8 h (*12 h*); permitted ice formation 50%; normal installation, i.e. inside the building and inside the envelope of the structural insulation

Specified conditions 2: water temperature 2°C; ambient temperature –6°C; evaluation period 8 h (*12 h*); permitted ice formation 50%; extreme installation, i.e. inside the building but outside the envelope of the structural insulation

λ = thermal conductivity at mean temperatures of insulation (W/m^{2}°C)

Bracketed figures shown in *italics*, e.g. (*12*), provide for 12 hours' protection from freezing.

Note 1   Thicknesses given are calculated specifically against the criteria noted in the table and may not satisfy other design requirements.

Note 2   Some of the insulation thicknesses given are too large to be applied in practice. These are included to show that in some cases, particularly in smaller sizes, a material of given thermal conductivity may not be able to provide the required degree of frost protection under the design conditions. Therefore, to increase the degree of frost protection it may be necessary to increase the pipe size, select insulation with a lower thermal conductivity or use some means of putting heat back into the system.

or in other areas where they may be subjected to harsh weather conditions, may need additional insulation over and above that given in table 7.2. For these applications where an increased insulation value should be applied, the advice of insulation manufacturers should be sought.

Where necessary, insulating material must be resistant to, or protected from, mechanical damage, rain, moist atmosphere, subsoil water and vermin. If the insulation is not water resistant or is in vulnerable areas, such as in open air or below ground, the insulation should be protected with a suitable waterproof material to keep the insulation dry.

Where pipes are fixed near to inside surfaces of outside walls there is no need to insulate them against the effects of frost. There is, however, a need to insulate against heat gains in cold pipes, heat losses in hot pipes, and as a means of preventing condensation. It is important that these pipes are not fixed in direct contact with the outside wall; rather, they should be secured with spacer clips or, if saddle clips are used, on to a pipe board (see figure 7.11).

Space should be allowed around pipes for the insulation, which should be continuous over pipes and fittings with space left only for the operation of valves.

Manufacturers' literature will give details of the application and use of insulating materials in various situations.

Insulation materials should be resistant to moisture vapour because if water is absorbed its insulation value will be lost. Where the insulation is not water resistant, a vapour barrier should be applied to its outer surfaces.

Readers should note that no amount of insulation is able to prevent a pipe from freezing. It can, however, slow down the heat loss and delay the effects of frost.

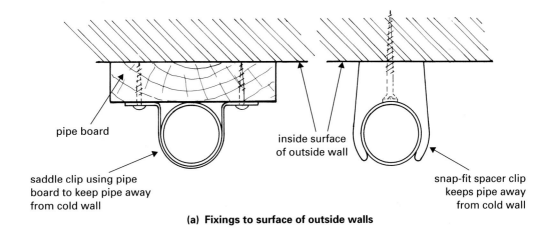

(a) **Fixings to surface of outside walls**

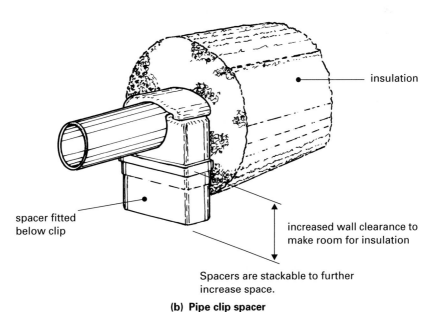

(b) **Pipe clip spacer**

Figure 7.11   Fixings to the inside of outside walls

Local heating

Local heating methods may be used in areas where other forms of protection are impractical or unsuitable, for example in unheated roof spaces or garages. More commonly, local heating is seen to be used in conjuction with a frost thermostat in tank rooms and pump rooms (see figure 2.41). Local heating, in conjunction with a frost thermostat, should only be used when other forms of frost protection are unsuitable.

Trace heating

Trace heating can be used as a form of local heating to prevent frost damage in pipes and fittings during periods of extreme cold. Trace heating should conform to BS 6351-1 and should be used in addition to any insulation provided and *not* as a substitute. Trace heating is a simple low-cost way of putting heat where it is needed to prevent frost damage (figure 7.12).

Trace heating consists of continuous, thin, flexible cables that are easily fitted to vulnerable pipes and fittings. Trace heating cable carries an electrical current, and because of its electrical resistance, it gets hot and will provide gentle heat to the pipe it is required to protect.

Self-limiting (self-regulating) tape controls its own heat output according to the temperature around it. The material from which the cable is made has the ability to vary its electrical resistance and thus the heat given off as the ambient temperature rises and falls, e.g. if the temperature of the pipe rises the heat output from the cable will drop.

As part of the system a thermostat should be used, pre-set to suit the particular frost risk. The tape should be fitted so that it has maximum thermal contact with the pipe and should be fixed using temperature rated insulation tape. Trace heating should be installed prior to any insulation being fitted to prevent heat being lost and to make sure any heat energy produced is only used to protect the pipe and its contents.

The electrical connections should be made and tested by a competent person and the electrical installation should comply with BS 7671.

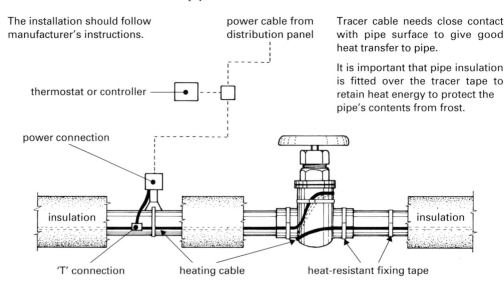

Figure 7.12   Use of trace heating for frost protection

Chapter 8
Water economy and energy conservation

The 1987 edition of BS 6700 suggested a relationship between water usage and energy consumption, based on the cost of supplying water and disposing of effluent. It also suggested that metering will produce considerable saving in water consumption.

There is a need to conserve water, because as we consume more, sources become increasingly costly and difficult to find. It is, therefore, in the interest of the consumer, and of concern to the plumber, that water is not consumed unduly.

With regard to metering and the cost of water supply, the author only partly agrees with the previous standard's philosophy. Whilst metering will most likely help to reduce the consumption of water, the cost to the consumer must increase because of the difficulties of reading and maintaining meters, and the additional cost of purchasing and installing meters and ancillary materials. Under the present system, consumers on higher rateable values and low usage who opt for meters do receive smaller water bills but, it is feared, at the expense of the majority whose properties are on lower rateable values, or those with large families.

As metering becomes more widespread or mandatory, then the costs of installing, maintaining and reading meters must in the long run cause a general increase in payments for water. At the same time, the energy consumed in manufacturing, installing and maintaining meters and their associated components is likely to exceed the energy savings from reduced water consumption.

Designers of water systems should be constantly aware of the need to keep water usage and energy consumption to a minimum. The following notes have this need in mind.

8.1 Water economy

Water byelaws have always been written in order to prevent or reduce wastage or excessive use of water. Under current Water Regulations which came into effect on 1st July 1999 similar requirements apply. Water is costly to produce, and during hot dry spells it is becoming increasingly difficult to maintain a constant supply. BS 6700 was written with water economy in mind and suggests a number of ways in which savings can be made and Water Regulations now take this a step further.

Leakages

Approximately 30% of all the potable water produced in Britain is lost through waste, undue consumption or misuse. Much of this is due to leakages underground from mains and service pipes. Water undertakers carry out waste detection programmes to reduce this loss, but because of the age and condition of many of our underground pipes, it is a continuing problem which can only be solved by regular testing and monitoring of internal and external pipework systems. (See chapter 12 which includes methods for the detection of leakages in metered and unmetered supply pipes.)

Additionally:

(1) Warning pipes from cisterns should discharge where they can readily be seen, e.g. over a doorway where the discharge may cause a nuisance.
(2) Ponds and pools should be built so that water loss is kept to a minimum. They should not lose more than 3 mm depth of water per day, after taking rainfall and evaporation into consideration. Ponds and pools must be fitted with an impervious lining or membrane. Any pond or pool installation of more than 10 000 l capacity and designed to be replenished by automatic means must be notified to the water undertaker before the installation is carried out.

Flushing of WCs and urinals

WC flushing accounts for about 25% of all domestic water used in buildings. Urinals consume water in large quantities if not properly controlled.

The Water Supply (Water Fittings) Regulations 1999 have introduced new requirements for the flushing of WCs and urinals. In brief, these are aimed at bringing us into line with European practices and encouraging new and innovative flushing arrangements with the result that they introduce to this country a number of changes to previous practices.

WC flushing cisterns

To reduce the amount of water used in flushing, Water Regulations now require that sizes of cisterns used for the flushing of WCs be reduced compared to those previously permitted.

WC cisterns are permitted with non-siphonic flushing devices. Typical of these are those used on the continent with valve type actuators using gravity flow rather than the old familiar siphonic arrangement.

Cisterns with siphonic flushing arrangements are still permitted but *all* WC cisterns are required to meet the strict performance criteria set out in the '*Water Regulator's Specification for the performance of WC suites*'. Cisterns, when flushed, must be capable of clearing the contents of the pan effectively using a single flush and flushing volumes for WC cisterns are reduced from 7.5 l to a maximum of 6 l.

The dual-flush cistern is back in fashion. Manufacturers are encouraged to produce dual-flush cisterns that are easier and more positive to use than previous ones. Dual-flush cisterns are required to give a maximum full flush of 6 l and a lesser flush volume of two-thirds that of the full flush, i.e. 4 l.

Also permitted are 'pressure flushing cisterns' which use incoming water pressure to compress air which in turn is used to increase the pressure of water available for flushing the WC pan.

Implementation of WC flushing arrangements under the Water Regulator's Specification is shown in table 8.1. Figures 8.1–8.3 illustrate the new WC flushing arrangements.

WC pans are designed to suit the associated cistern to ensure effective cleansing. Care should be taken that the correct cistern or pan is fitted when either is renewed otherwise effective pan clearance may not be achieved.

Table 8.1   Maximum permitted capacity of WC cisterns

Type of appliance	Use	Maximum permitted volume (litres)	Restrictions
Single flush and dual flush	Domestic	6	Flushing by cistern only; siphonic and non-siphonic devices permitted
Single flush and dual flush	Non-domestic	6	Flushing by cistern or by pressure flushing valve; siphonic and non-siphonic devices permitted

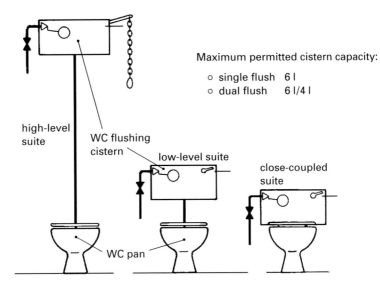

Maximum permitted cistern capacity:

o single flush   6 l
o dual flush   6 l/4 l

high-level suite

WC flushing cistern

low-level suite

close-coupled suite

WC pan

Figure 8.1   Capacity of WC flushing cisterns

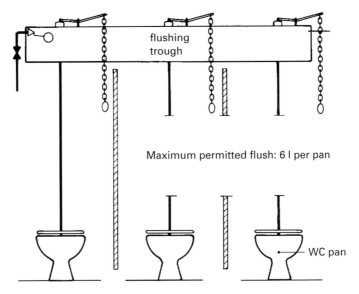

flushing trough

Maximum permitted flush: 6 l per pan

WC pan

Figure 8.2   Capacity of WC trough cisterns

Pressure flushing valves

As an alternative to the flushing cistern, 'pressure flushing valves' are approved for the flushing of both WCs and urinals. Flushing valves have been used on the continent for many years and have been extensively used in ships where the sea provided more than ample water for flushing. However, flushing valves are *not* permitted for the flushing of WCs in private dwellings, or anywhere that a minimum flow rate of 1.2 l/s cannot be achieved at the appliance. See figures 8.3 and 8.4.

Flushing arrangements for urinals

Water Regulations require that automatic flushing cisterns should supply a maximum of 10 l per hour, per bowl, stall, or 700 mm length of slab, delivered not more than three times per hour (see figure 8.5).

Additionally, supply pipes to urinal flushing cisterns are required to be fitted with devices that will control the supply during periods when the urinal is not in use. There are two methods of controlling the supply: time control (see figure 8.6), and hydraulic control (see figures 8.7 and 8.8).

Where a time control is used, the pipe supplying the urinal cistern should also be fitted with a lockable control valve.

Alternatively, individual bowls or stalls may be user operated as required, e.g. by a manual chain pull (see figure 8.9) or push button operation or infra-red sensor or similar. A multiple-style urinal installation is shown in figure 8.10.

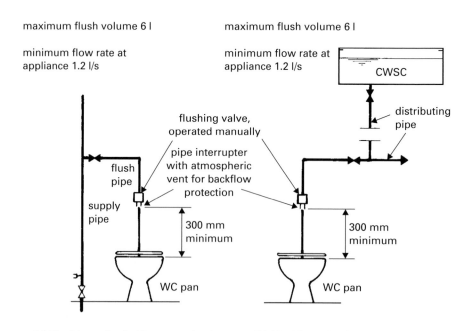

(a) **Flushing valve fed from supply pipe**   (b) **Flushing valve fed from distributing pipe**

Figure 8.3   Pressure flushing valves for WCs

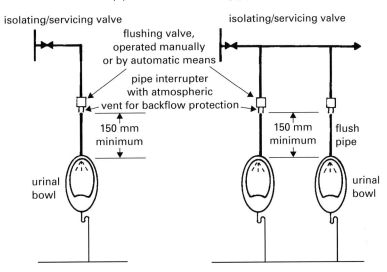

Pressure flushing valves may be supplied from the supply
pipe or from distributing pipes

(a) **Pressure flushing valve serving
single urinal bowls**

(b) **Pressure flushing valve serving
range of urinal bowls**

Figure 8.4   Pressure flushing valves for urinals

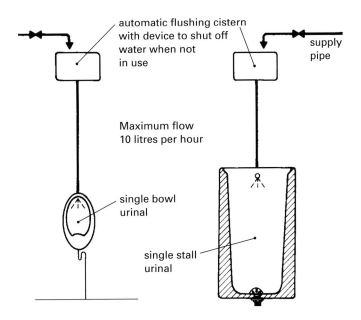

Figure 8.5   Urinal flushing cisterns – single appliances

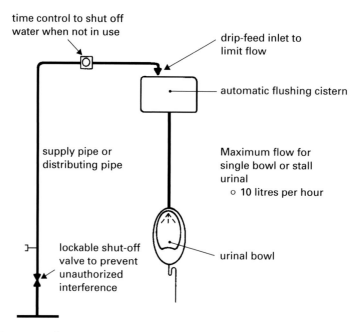

Figure 8.6   Time control to automatic flushing cistern

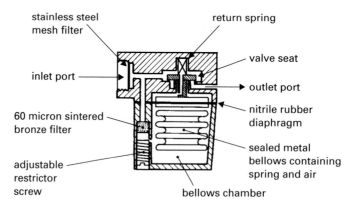

Figure 8.7   Hydraulic flow control device for automatic flushing cistern

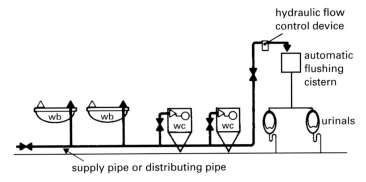

hydraulic flow control device

automatic flushing cistern

urinals

supply pipe or distributing pipe

No use of appliance = no flow through flow control device.

Draw-offs at other appliances will create pressure variations causing the hydraulic flow control to open, thus permitting a small quantity of water to pass into the automatic flushing cistern.

Figure 8.8   Use of hydraulic flow control device

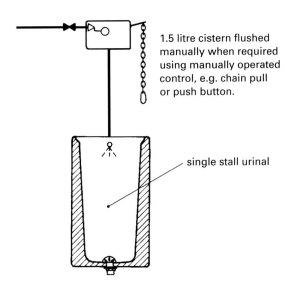

1.5 litre cistern flushed manually when required using manually operated control, e.g. chain pull or push button.

single stall urinal

Figure 8.9   Single urinal with manual flushing control

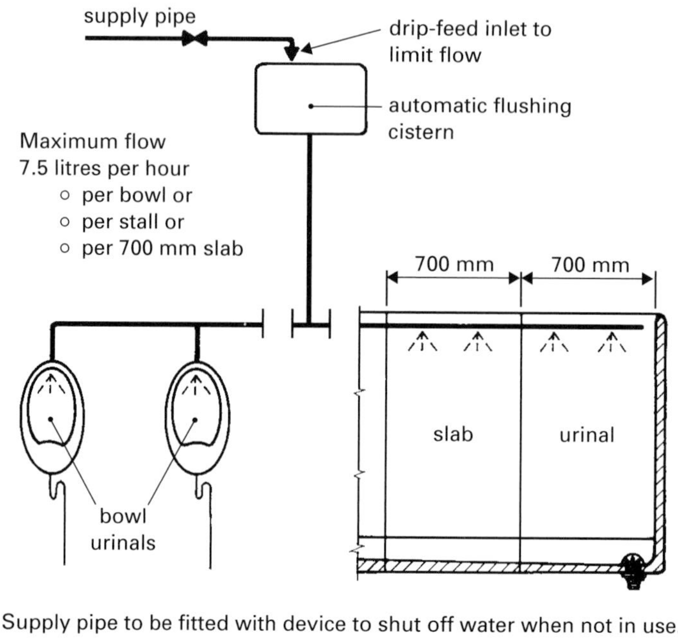

Supply pipe to be fitted with device to shut off water when not in use.

Figure 8.10 Multiple urinal installation

Requirements for the maximum flushing volumes of water to be used in the flushing of urinals are shown in table 8.2.

It may not always be feasible to size the automatic flushing cistern for multiple urinals according to the number of bowls, stalls or slabs, or to give the age-old recommended three flushes per hour.

Table 8.3 provides for a variation of flushing arrangements based on the formula:

$$\text{Time interval (min)} = \frac{\text{cistern capacity (l)} \times \text{time in minutes (60)}}{\left(\dfrac{\text{maximum flush requirement}}{7.5\ \text{l per bowl}}\right) \times \left(\begin{array}{c}\text{number of bowls, stalls}\\ \text{or 700 mm width of slab}\end{array}\right)}$$

For example, using a 13.5 l cistern to flush five urinal bowls:

$$\text{Time interval} = \frac{\text{cistern capacity (13.5 l)} \times \text{time (60)}}{\text{maximum flush (7.5 l per bowl)} \times \text{number of bowls (5)}}$$

$$= \frac{13.5 \times 60}{7.5 \times 5}$$

$$= 21.6\ \text{mm}$$

The reduction in quantity of water used should lead to the reduction of limescale build-up in appliances and drains, and consequent reduction in odours.

Table 8.2   Maximum permitted volumes of water for flushing urinals

Appliance	Maximum volume
For a single bowl or stall supplied from an automatic flushing cistern	10 l/hour
For more than one appliance supplied from an automatic flushing cistern	7.5 l/hour per bowl, stall, or per 700 mm width of slab
For individual bowls or stalls, user operated, and supplied via manual chain pull or button operated from cistern or pressure flushing valve	1.5 l per flush as required

Table 8.3   Volumes and flushing intervals for urinals

Number of bowls, stalls, or per 700 mm of slab	Volume of automatic flushing cistern l				Maximum fill rate l/h
	4.5 l	9 l	13.5 l	18 l	
	Shortest period (minutes) between flushes				
1	27	54	81	108	10
2	18	37	54	72	15
3	12	24	36	48	22.5
4	9	18	27	36	30
5	7.2	14.4	21.6	28.2	37.5
6	6	12	18	24	45

Water supplies for WC and urinal flushing

For a great many years the water industry resisted the direct connection of WC pans to supply and distributing pipes and strict measures were used, i.e. an interposed cistern (flushing cistern), to guard against the possibility of backflow from these appliances to the main or to other water fittings.

Under Water Regulations, pressure flushing cisterns for WCs and pressure flushing valves for both WCs and urinals may be supplied from a supply pipe or a distributing pipe, providing appropriate backflow prevention devices are in place, the device being a 'pipe interrupter with permanent atmospheric vent'.

Waste plugs

Waste plugs should be fitted to all baths, basins, sinks or similar appliances, except where delivery is less than 3.6 l/min and the appliance is designed not to have a plug, e.g. basins with spray taps, and shower trays. Appliances for medical or veterinary purposes are also excepted.

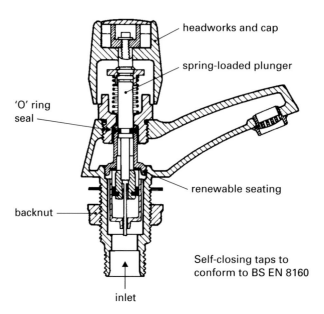

Figure 8.11   Non-concussive self-closing tap

Self-closing taps

Self-closing taps should be of the non-concussive type (see figure 8.11) and be capable of closing against 2.6 times the working pressure. These taps are very effective when new, but tend to fail in the open position after a period in use so they should only be used in buildings where regular maintenance and inspection can be ensured.

Spray taps and aerators

Spray taps can provide savings of up to 50% in both fuel and water. However, they have several disadvantages:

• they should not be used where basins are subject to heavy fouling by grease or dirt;
• they require regular maintenance;
• they are only suitable for hand rinsing;
• the heads may block in hard water areas;
• since self-cleansing velocities of waste water may not be achieved, residues may build up in waste pipes.

Aerators may reduce consumption by reducing the flow rate, and when compared with spray taps will give improved flow pattern. They have a minimum operating pressure that should be checked against available pressure before installation.

Showers

Showers are generally said to use less water than baths. However, the reduced consumption is often offset by more frequent use. Also, with the arrival of pressurized hot water systems, water consumption for showering will increase.

Washing troughs and fountains

Fittings serving these appliances should be capable of discharging to individual units without discharging to other units. A unit means a 600 mm length of straight trough or of the perimeter of a round appliance.

Domestic appliances

Maximum amounts of water used per complete cycle of operations:

Clothes washer without water-using tumbler dryer:	27 l per kg of washload for standard 60°C cotton cycle
Clothes washer with water-using tumbler dryer:	48 l per kg of washload for standard 60°C cotton cycle
Dishwasher:	4.5 l per place setting

Other economy measures

The following measures may also improve water economy:

- protection from mechanical damage and corrosion, especially underground (trench preparation and backfill, pipe depth);
- use of approved fittings, components and pipes;
- adequate frost precautions;
- use of spray taps.

8.2   Grey water and recycled rainwater

Building Regulations encourage the use of grey water (recycled waste water) and rainwater. In Approved Document H2, guidance is given on the construction and use of tanks for the storage of grey water or rainwater for reuse within buildings. This guidance does not apply to water butts used for the garden.

Grey water and rainwater storage tanks should:

- not allow leakage of the contents or ingress of subsoil water to occur, and should be ventilated;
- be fitted with an anti-backflow device on any overflow connected to a drain or sewer to prevent contamination of the stored grey water or rainwater in the event of surcharge in the drain or sewer;

- be provided with access for emptying and cleaning. Access covers should be of durable quality having regard to the corrosive nature of the tank contents. The access should be lockable or otherwise engineered to prevent personnel entry.

Additionally, where grey water or rainwater is used to augment mains water, precautions should be taken to protect the mains water supply from contamination (see chapter 6).

8.3 Energy conservation

It is a requirement of Building Regulations that reasonable provision is made for the conservation of fuel and power in buildings. This requirement includes:

- provision for the use of high energy condensing boilers in both new and replacement installations;
- the use of effective control systems on hot water and space heating systems; and
- the use of insulation to prevent undue loss of heat from hot water storage vessels, pipes, fittings and other components.

Building Regulations also promote competent persons self-certification schemes in a bid to encourage installers to become more competent in their work and conform to the energy conservation requirements of Part L1A (new dwellings) and L1B (existing dwellings).

Note Energy conservation in buildings other than dwellings is dealt with in Parts L2A and L2B of the Building Regulations and is outside the scope of this book.

Every hot store vessel should be fitted with a thermostat to keep the hot water at the required temperature, and a time switch that will shut off the heat source when there is no demand. Energy can also be saved by reducing the quantity of water heated, bearing in mind the methods used to heat or control the energy input to the storage vessel.

Hot store vessels should be of adequate capacity without being oversized and heating methods should, where possible, enable a reduced quantity of water to be heated when desired. Examples of this include the use of double element or twin element immersion heaters in electrically heated systems and economy valves on gas circulators. Hot water storage vessels must be fitted with adequate thermal insulation to limit heat loss from the surface area of the vessel.

Pipes to and from hot water apparatus and central heating components should be insulated unless they are designed to contribute to space heating. It is particularly important that any secondary circulation circuits are well insulated. Table 3.4 gives maximum lengths permitted for hot water distributing pipes without insulation. It is recommended that hot water distributing pipes should be as short as possible to minimize heat energy loss. Maximum rates of heat loss from hot water pipes should not exceed those shown in table 8.4 and thickness of insulation for practical purposes should equal or be better than that shown in table 8.5.

Trace heating used for the heating of hot water pipes should be of the self-regulating type, and the system should comply with BS 6351-1.

Use of electrical booster pumps for hot and cold water systems should be minimized. The energy needed to boost water pressure in a building is about 0.02 kW/m^3 per metre of lift. Therefore, where mains pressure is insufficient to supply the upper floors of a building, it is advisable and more energy efficient to use mains pressure to the limit of its supply.

Table 8.4 Maximum permitted rate of energy loss from pipes

Pipe diameter (OD) mm	Maximum permissible heat loss W/m^2
8	7.06
10	7.23
12	7.35
15	7.89
22	9.12
28	10.07
35	11.08
42	12.19
54	14.12

Note: Figures above are based on pipe temperature of 60°C and still air temperature of 15°C.

Table 8.5 Insulation thickness to control heat loss for domestic hot water systems and central heating installations in potentially unheated areas

Outside diameter of copper pipe upon which thickness is based (mm)	Insulation thickness (mm)					Heat loss Ambient temperature −1°C	
	$\lambda =$ 0.025	$\lambda =$ 0.030	$\lambda =$ 0.035	$\lambda =$ 0.040	$\lambda =$ 0.045	Hot water at 60°C	Central heating at 75°C
10	10	16	22	31	44	6.8 W/m	8.6 W/m
12	12	18	26	36	49	7.3 W/m	9.2 W/m
15	15	22	31	42	58	7.8 W/m	9.7 W/m
22	19	26	35	47	62	8.2 W/m	10.2 W/m
28	21	28	38	49	64	9.0 W/m	11.3 W/m
35	22	30	39	51	64	10.0 W/m	12.6 W/m
42	23	31	41	52	65	11.0 W/m	13.8 W/m
54	25	33	42	53	65	12.8 W/m	16.0 W/m
Cylinders	35	42	50	58	67	38.2 W/m	47.9 W/m

$\lambda =$ thermal conductivity at mean temperature of insulation (W/m^2 C).
Note: Heat loss relates to the specified thicknesses and temperatures.

8.4 Building Regulations and energy conservation

Part L of the Building Regulations is used by government to improve the energy efficiency in buildings. More efficient hot water and heating systems will cut fuel consumption and reduce the amount of CO_2 being discharged to atmosphere. Requirements to limit heat losses from buildings using improved insulation will in turn provide energy savings in both new and existing buildings.

In hot water services the provisions of Parts L1A and L1B aim to make hot water (and heating) systems in dwellings become more efficient by:

- ensuring that only the most efficient boilers are installed;
- limiting heat losses from hot water pipes and storage vessels;
- providing for suitable time and temperature controls;
- introducing work methods that ensure appliances and system controls are properly commissioned and maintained;
- promoting competent persons self-certification schemes to encourage installers to become more competent in their work and conform to the energy conservation requirements of the Regulations;
- requiring installers to provide information to occupiers of buildings that will enable them to operate heating appliances and systems more efficiently;
- notifying local building control bodies and occupants that installations conform fully with Building Regulations.

Scale control

Building Regulations identify the need to sustain energy efficiency in domestic hot water heating systems by the control of scale in systems and provision of water treatment. Part L refers to the *Domestic Heating Compliance Guide* which requires that in areas of the country served with hard water (in excess of 200 mg/l as $CaCO_3$) provision should be made to treat the feed water to water heaters and the hot water circuit of combination boilers to reduce accumulation of limescale and the consequent reduction in energy efficiency.

Formation of scale in boilers and water heaters is due to 'hardness' in water and is caused by the presence of dissolved minerals, mainly calcium and magnesium, and associated anions, bicarbonate, sulphate and chloride. When hard water is heated, bicarbonates decompose and calcium carbonate is deposited in the heater and associated pipework. This 'furring' of pipes and heaters can cause blockage and equipment failure, but also coats heating surfaces, effectively insulating them so that the efficiency of the heater is impaired.

Treatments to control limescale would include the fitting of a base exchange water softener, or a scale inhibitor, or continuous dosing of the system with a recognized chemical inhibitor. These are discussed briefly in chapter 2.

Boiler efficiency

Boilers installed in dwellings are required to be energy efficient and have a minimum SEDBUK rating. SEDBUK (seasonable efficiency of domestic boilers in the UK) is a measure of the energy efficiency of the boiler based on the percentage of fuel actually burned.

The SEDBUK rating applies to boilers fitted in new buildings and to those installed to replace existing ones. The minimum efficiency requirements for boilers are shown in table 8.6 and will vary depending on the type of fuel used by the boiler. It should be noted that back boilers may be installed with SEDBUK ratings 2% points lower than those specified for normal boilers. Boilers running on solid fuel should have efficiency ratings as recommended in the HETAS certification scheme.

Whilst it may be possible in unusual circumstances for less efficient boilers to be installed in existing dwellings, these installations will be rare and installers will need to show proof that the premises are unsuitable for the preferred high energy boilers.

For boilers installed in buildings other than dwellings, energy efficiency needs to be calculated to give a maximum permissible carbon emission factor.

Table 8.6   Minimum SEDBUK ratings for boilers installed in dwellings

Central heating system fuel	Minimum SEDBUK efficiency rating
Natural gas (mains supplied)	78%
Liquefied petroleum gas (LPG)	80%
Oil	85%
Oil (combination boilers)	82%

Note: For boilers not SEDBUK rated an appropriate seasonal efficiency can be taken from the government's standard assessment procedure for energy rating of dwellings (SAP).

Minimum provision for the control of hot water and heating systems in both new and existing dwellings

Systems in new dwellings or complete replacement systems in existing dwellings should meet the requirements of Part L in full. It is also recommended that controls for existing installations should be updated to meet the standards for new systems but only if it is practical and economic to do so.

Boiler interlock

Boiler interlock is required where systems supply domestic hot water as well as space heating. Its purpose is to ensure that the boiler and pump are switched off when there is no demand for either hot water or space heating. Boiler interlock cannot be controlled by thermostatic radiator valves (TRVs) alone.

In a solid fuel boiler, boiler interlock is not required unless recommended by the manufacturer of the boiler.

If an electrically heated central heating boiler also heats the hot water then boiler interlock will be required.

Boiler control

For solid fuel boilers thermostatic control is needed to control the burning rate of the boiler according to the boiler water temperature. It may not be possible to switch off the heat output completely but a boiler thermostat will effectively reduce heat output to a minimum. In most cases a room thermostat will shut off the circulating pump, which will indirectly cause the boiler to operate at minimum output.

In electrically heated systems the boiler should have flow temperature control and be capable of modulating the power input to the primary water.

Space heating zones

Dwellings having total usable floor area up to 150 m² should be divided into at least two space heating zones each with its own independent temperature control. One of these should control the living area. Where the floor area is greater than 150 m², each heating zone should have its own independent time control and its own independent temperature control.

In single-storey open-plan dwellings where the living area accounts for more than 70% of the total floor area, sub-zoning of temperature control is not necessary.

In existing dwellings where only the boiler is replaced, the space heating system may be controlled as one zone.

Water heating zone

A hot water service zone should be provided in addition to the space heating zones mentioned above. A separate hot water zone is not required where water is heated instantaneously, e.g. in a combination boiler.

Time control for space and water heating

The following methods of time control would be suitable:

1. a full programmer with separate timing to each circuit;
2. a separate timer for each circuit including the hot water circuit;
3. a programmable room thermostat to each heating circuit, with a separate timer for the hot water circuit.

In gas and oil fired systems where the total usable floor area is more than 150 m^2 the following timing methods may be used:

1. multiple heating zone programmers;
2. a single multi-channel programmer;
3. programmable room thermostats;
4. a separate timer for each circuit;
5. a combination of (3) and (4) above.

The above methods should also provide a control for the hot water circuit.

Where the hot water is produced instantaneously, such as in a combination boiler, time control is only required for the space heating zone.

In an existing system, where only the hot water cylinder is being replaced and there is no separate time control for the hot water circuit, it would be reasonable to have one timing device to control both hot water and space heating.

Solid fuel systems that have automatic ignition may be able to take advantage of the more sophisticated control systems.

Temperature control of space heating

Separate temperature control should be provided for each heating zone within a dwelling. Suitable methods include:

1. a room thermostat or programmable room thermostat for each zone (circuit);
2. a room thermostat or programmable room thermostat in the main zone, plus individual radiator controls, e.g. thermostatic radiator valves (TRVs) on all radiators in the other zones;
3. a combination of (1) and (2) above.

In an existing gas or oil system where the hot water circuit is on gravity circulation and only the hot water cylinder is being replaced, a thermo-mechanical cylinder thermostat should be installed as a minimum provision.

Temperature control of the hot water system

The temperature of stored hot water should be controlled using a cylinder thermostat and either a zone valve or three-port valve. In dwellings of more than 150 m² total floor area where there is more than one hot water circuit, each circuit should have separate timing and temperature controls such as:

- a multiple heating zone programmer for each circuit;
- a single multi-channel programmer to serve two or more circuits;
- a separate timer for each circuit.

Non-electric hot water controllers are not suitable and should not be used.

A zone valve is not suitable for the control of circuits to thermal stores but a second pump could be used in place of the zone valve.

In an existing system, a thermo-mechanical cylinder thermostat could be installed as a minimum provision.

8.5   Insulation to meet Building Regulations

Building Regulations require that hot water (and heating) pipes, fittings and storage vessels be insulated. These requirements, which aim to minimize energy use in buildings, apply to both new and replacement systems.

New systems

Guidance in documents L1A and L1B advises that pipes for hot water and heating in dwellings should be insulated in accordance with the *Domestic Heating and Hot Water Compliance Guide*. In turn it is suggested that further guidance on insulation can be obtained from BS 5422 or from the BRE Report No. 262 *Thermal insulation: avoiding risks*. The *Domestic Heating Design Guide* gives the following insulation requirements:

- Primary circulation pipes for hot water and heating should be insulated throughout their length, except where practical constraints when penetrating joists and other structural elements make this provision impractical.
- Primary and secondary pipes should be insulated wherever they pass outside the heated living space and further frost protection may be needed where pipes pass through unheated parts of the premises.
- Pipes connected to hot water storage vessels (including the vent pipe) should be insulated for at least 1 m from their connection to the cylinder (or they should be insulated up to the point where they become concealed).
- Secondary circulation pipes should be insulated throughout their length.

Additionally the insulation should be labelled to show it meets the level of heat loss compliance given in the guide. Maximum permitted heat loss from insulated hot water and heating pipes is given in table 8.4.

Replacement systems

In existing systems, where a boiler or hot water storage vessel is replaced, any pipes that are accessible, or exposed as part of the work, should be insulated. Insulation should, where practicable, meet the standards set out for new systems and on completion should be labelled to show that it meets the level of heat loss compliance given in table 8.4, which is reproduced from the *Domestic Heating Compliance Guide*.

Tables for the environmental insulation thickness for hot water pipes in domestic premises are given in BS 5422 and are reproduced in table 8.5. Examples of suitable insulation materials are given in table 7.1.

Insulation standards for hot water storage vessels

Cylinders used for the storage of hot water for domestic use are required under the *Domestic Heating Compliance Guide* to be insulated to the standards set out in BS 1566-1 and should not exceed the value given by the formula:

$$1.6(0.2 + 0.051V^{2/3}) \text{ kWh per 24 hours}$$

where

V is the nominal cylinder capacity in litres.

The *Domestic Heating Compliance Guide* gives two exceptions to this rule: primary stores and unvented systems heated primarily by electricity. Unvented systems heated primarily by electricity should not exceed the value given by the formula:

$$1.28(0.2 + 0.051V^{2/3}) \text{ kWh per 24 hours}$$

where

V is the nominal cylinder capacity in litres.

An example of a calculation using the above formula is set out below and the graph in figure 8.12 shows maximum permitted heat loss against cylinder capacities. The graph, based on the above formulae, can be used to check that cylinders supplied meet the required insulation standard. Within the graph, an example of its use for a cylinder of 117 l capacity will give a comparison with the calculation below.

Example of calculation for an unvented hot water vessel of 117 l capacity heated by electricity.

$$\begin{aligned}
\text{Maximum heat loss} &= 1.28(0.2 + (0.051(117^{2/3}))) \\
\text{(kWh per 24 hours)} &= 1.28(0.2 + (0.051(117^{0.6667}))) \\
&= 1.28(0.2 + (0.051 \times 23.921)) \\
&= 1.28(0.2 + 1.21996) \\
&= 1.28 \times 1.41996 \\
&= 1.82 \text{ kWh per 24 hours}
\end{aligned}$$

Primary (thermal) store vessels should be insulated to the requirements of Section 4.3.1 or 4.3.2 of the Water Heater Manufacturers Association *Performance specification for thermal stores*. Two formulae are given, with or without pipework.

Insulation for thermal stores with pipework should not exceed the value given by the formula:

$$3.6 + 0.002(S_V - 120) \text{ kWh per 24 hours}$$

where

S_V is the nominal cylinder capacity in litres.

Insulation for thermal stores without pipework should not exceed the value given by the formula:

$$2.8 + 0.002(S_V - 120) \text{ kWh per 24 hours}$$

where

S_V is the nominal cylinder capacity in litres.

For thermal stores heated primarily by electricity, the standing heat loss should be 15% less than that given for thermal stores heated from other energy sources.

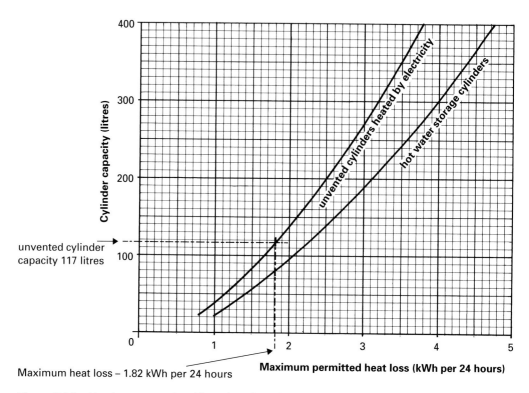

Figure 8.12   Maximum permitted heat loss from hot water cylinders

Chapter 9
Noise and vibration

Water Regulations require water fittings to be constructed of materials that are resistant to damage from vibration.

Noise is caused by vibration and is generally seen (or heard) to be a nuisance to building occupiers, but vibration associated with pipework noise can also at times cause damage to pipes and fittings leading to leakage within the system.

It is unfortunate that this important aspect of water installations has been omitted from BS 6700, despite its earlier inclusion in the 1987 edition. Because of its importance, the author has decided to continue its inclusion in this book.

In water systems many materials are susceptible to vibration and will transmit or even accentuate noises produced. Most system noises, which many operatives explain away as unavoidable, can be avoided by better design and workmanship.

9.1 Flow noises

Pipework noise becomes significant at water velocities over 3 m/s. It is important therefore that systems are designed to keep water velocities below 3 m/s by increasing the pipe size, as necessary. The causes of some common flow noises are shown in figures 9.1, 9.2 and 9.3.

friction, vibration and noise where water rubs
against pipe walls

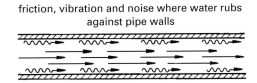

Where velocities are below 3 m/s noise is not significant.

Figure 9.1   Flow noise in pipes

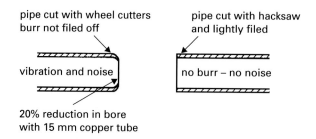

pipe cut with wheel cutters
burr not filed off

pipe cut with hacksaw
and lightly filed

vibration and noise

no burr – no noise

20% reduction in bore
with 15 mm copper tube

Burrs left on pipe ends after cutting will add to turbulence in pipes
and increase noise levels.

Figure 9.2   Effect on water flow of burrs on pipe ends

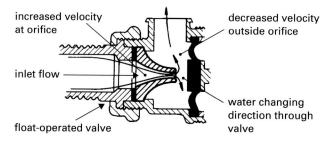

increased velocity
at orifice

decreased velocity
outside orifice

inlet flow

water changing
direction through
valve

float-operated valve

Changes in flow direction, and drop in pressure through valve
may lead to cavitation which will further increase flow noise. Can
be reduced by lowering flow velocity.

Figure 9.3   Noise caused by water flow through narrow apertures

Cavitation

Cavitation can be simply described as wear or erosion of the internal surfaces of pipes and fittings caused by turbulent water flow.

When cavitation occurs, water flow noise increases. Cavitation is not common in pipework as it usually only occurs at water velocities of 7 m/s to 8 m/s in elbow fittings. However, it can occur through reduced pressures at the upper parts of systems which incorporate long pipe drops.

Outlet fittings generally incorporate abrupt changes of direction in water flow, and there is a sudden drop in pressure at the outlet side of the seating of taps and float valves. These conditions are ideal for cavitation, and are the major cause of noise in these fittings. Although figure 9.4 shows a bib tap to BS 1010, this is a problem common to most types of taps and valves.

Cavitation noises can be controlled by reducing the pressure drop across the valve seating. For example, if the inlet pressure is reduced and the tap is opened more fully it will operate more quietly. Similarly, a float valve with a lower inlet pressure and larger orifice will operate with less noise.

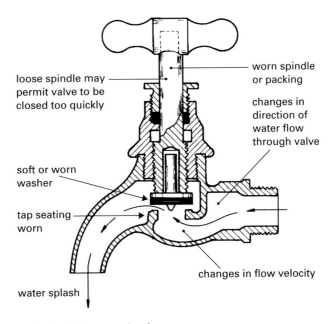

loose spindle may permit valve to be closed too quickly

worn spindle or packing

changes in direction of water flow through valve

soft or worn washer

tap seating worn

changes in flow velocity

water splash

Figure 9.4 Causes of noise in taps and valves

Fast flow through narrow apertures

Noises in float valves and stopvalves can often mean that the waterway is becoming blocked with particles carried along in the water (see figure 9.5). To avoid this, stopvalves should always be left in the fully open position so that particles can pass through and out of the system.

In the past, a good way to reduce the problem of blockage, and also reduce general flow noise in float valves, was to fit an equilibrium float valve. The use of the Portsmouth type valve is restricted under the Water Regulations, but there are diaphragm type valves shown in the *Water Fittings and Materials Directory* in sizes from 1″ (G1) upwards. There is also a ceramic disc type listed that provides full flow during fill and is designed to reduce the possibility of water hammer noise.

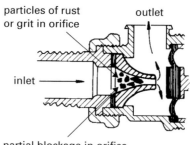

particles of rust
or grit in orifice

outlet

inlet

partial blockage in orifice

Any obstruction in a valve which significantly reduces the bore
will create noise. Commonly found in:
   o stopvalves that are partially closed;
   o stopvalves that are partially blocked;
   o float valves partially blocked (as illustrated).

It is the mistaken belief of many that to partially close a stopvalve
will reduce pressure. In fact, it will create noise and unnecessary
wear to the valve.

Figure 9.5   Noise caused by obstruction in valves

9.2   Water hammer noise

Valve closure

Sudden valve closure will cause shock waves to be transmitted along pipes with a loud
hammering noise. Rapid closure can be prevented by regular maintenance, making sure
that packing glands are correctly adjusted and spindles are not loose. Other precautions
include the restriction of velocities and avoidance of long straight pipe runs. Limiting water
velocities to 3 m/s will not, in itself, reduce water hammer, but will help reduce the magni-
tude of pressure peaks produced.

Solenoid valves and self-closing taps

These often cause water hammer noise. Use non-concussive types, properly and regularly
maintained.

Vibration of the cistern wall

This is very common in copper cisterns (see figure 9.6). It causes the float valve to open and
shut in rapid succession causing water hammer, or sometimes a noise similar to that of an
electric motor. The cure is to stabilize the cistern wall with a strap around it, and to fix the
supply pipe securely to prevent it and the connected float valve from vibrating.

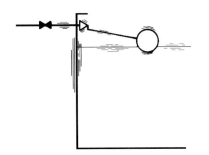

Vibrating cistern wall will move float body and create noise.

Figure 9.6   Vibration of cistern wall

Float valve oscillation

This causes a regular series of loud banging noises which occur when the valve is almost closed. This type of noise can be most disturbing to occupiers of buildings, and could well cause damage to pipes and fittings. There are a number of reasons for it, and BS 6700 (1987 edition) suggested that the commonest cause is the formation of waves on the surface of the cistern water. The author does not agree entirely, and believes the waves are a result of the oscillation and not the cause. However, the problem is there whichever is the case!

For the float valve to operate correctly (see figure 9.7), its closing force (float buoyancy) must overcome the incoming force of the water pressure. If the incoming pressure and the buoyancy force nearly balance, the conditions for oscillation are ideal. A shock wave against the washer will cause it to open slightly. The float is depressed into the water and subsequently bounces back to reclose the valve abruptly, creating a further shock wave which in turn rebounds to open the valve again; thus a cycle is established.

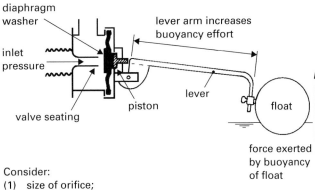

Consider:
(1)   size of orifice;
(2)   length of lever arm;
(3)   diameter of float.

As the cistern water level rises, the closing force exerted on the piston by the float and lever arm must close the diaphragm washer against the incoming force of the inlet pressure.

Figure 9.7   Prevention of float valve oscillation

Many plumbers suggest fitting a damping plate to the float or the lever arm, which will 'dampen down' the effects of the shock waves, or alternatively fitting baffle vanes in the cistern to prevent surface waves affecting the float. In practice these often seem a little 'Heath Robinson'.

Another way of improving flotation is simply to increase the lever arm length, thus achieving better leverage, or to fit a larger float. One could also reduce the size of the orifice seating. Although this might lead to more flow noise, it may be preferable to water hammer.

Tap washer oscillation

This is usually associated with worn, split or soft washers, which vibrate violently as water passes them (see figure 9.8). The noise can be extremely loud but the cure is simple – change the washer!

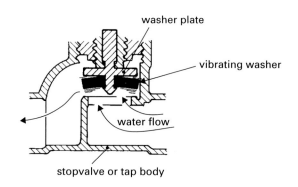

Worn, split or soft washers will vibrate violently.

Figure 9.8   Tap and valve washer oscillation

9.3   Other noises

Pumps

These will not cause excessive noise if well designed, unless flow exceeds the pump rating, or the static pressure is insufficient. Noise transmission from pumps can be reduced by:

- using rubber hose isolators between pump and pipework;
- isolating pipework from building structures with resilient inserts fitted in brackets;
- the use of a hydraulic–acoustic filter tuned to the unwanted frequency.

Boilers

Noise in boilers may occur due to nucleate boiling at localized hot spots. Noise associated with scale formation and rust deposits may be controlled by chemical or physical means or by water softening.

Splashing noises

These occur when water strikes the water surface in cisterns. Silencer tubes screwed into float valve outlets are now prohibited because of the risk of backsiphonage. Some float valves are fitted with collapsible silencer tubes, which are permitted. Others have outlets designed to reduce splashing noise.

Many taps are produced with flow correctors or aerators, which help to reduce water impact noise and splashing into sinks and other appliances. Metal sinks of pressed stainless steel increase splashing sounds, and some treatment in addition to that provided by the manufacturers may be worth considering.

Thermal movement

In hot pipes this causes creaks, squeaks and more impulsive sounds. The use of resilient pipe clips, brackets or pads between pipes and fittings will provide sufficient flexibility to cope with expansion and contraction. Long straight runs are a particular problem for which expansion joints or loops should be considered.

Air or vapour bubbles

These commonly cause flow noise. As a result of poor design and operation of hot water systems, bubbles are formed in hot water cylinders or heaters. Systems should be designed to avoid general or localized boiling and to allow removal of air when filling.

Gases

These are formed by corrosive action in primary circuits and can also cause noise when pumped around systems. This can normally be prevented in primary circuits by the use of corrosion inhibitors.

9.4   Noise transmission and reduction

Noise is transmitted to the listener from its source in a number of ways, including direct airborne transmission, through building structures (see figure 9.9), along pipes, and through the water contained in the pipe.

In metal pipes noise is transmitted with little loss, but plastics can reduce noise transmission depending on the pipe material and its thickness. For lengths of between 5 m and 20 m, the reduction is approximately 1 dB/m to 2.5 dB/m.

The insertion of metal-bellows vibration isolators can reduce transmissions by 5 dB to 15 dB, whilst reinforced rubber hose isolators can be even better.

Resilient mountings can help isolate a storage cistern from its supporting structure, preventing the transmission of noise through ceilings into habitable rooms.

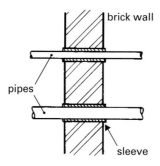

Small flexible pipe
  o little noise
Large rigid pipe
  o vibration transmitted to wall

Brick will not transmit appreciable sound from small flexible pipes, but larger, more rigid pipes such as steel will induce some vibration.

(a) Solid walls

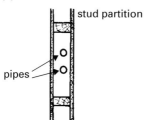

Hollow partition will act as a sound box to transmit and increase pipe noises.

Lightweight structures vibrate readily, transmitting and accentuating sound from pipes. To prevent this, use lightweight flexible pipes (copper or plastics) and flexible vibration isolating clips.

(b) Hollow walls

Figure 9.9 Structure-borne noises

Chapter 10
Accessibility of pipes and water fittings

BS 6700 recommends that all pipes and fittings must be readily accessible for inspection, repair and renewal.

Additionally, Water Regulations state that: 'No water fitting shall be embedded in any solid wall or floor'.

However, the degree of accessibility will depend largely upon:

- how strictly the regulations are enforced by the water undertaker;
- the personal opinion of the designer, installer and building engineer;
- cost considerations of installation and maintenance of accessible installations using ducts, chases, access panels, etc.;
- the likelihood of routine inspection and maintenance being carried out;
- the consequences of leakages from inaccessible parts of the pipework;
- the reliability of joints, resistance of pipes and joints to internal and external corrosion and flexibility of pipes when being passed through chased ducts or sleeves;
- the importance of aesthetic considerations, on the one hand where pipes are surface mounted and, on the other, the consequences of breaking into, repairing and otherwise spoiling expensive decorations and floor finishings.

10.1 Pipes passing through walls, floors and ceilings

No pipe may be installed in the cavity of an external cavity wall, other than where it has to pass through from one side to the other (see figures 10.1 and 10.2).

Pipes entering buildings

Pipes entering buildings should be located in a sleeve or duct that will permit the pipe to be readily replaced. Each end of the duct should be sealed using a non-hardening,

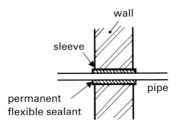

Pipes passing through walls or floors should be in a sleeve to permit movement relative to wall or floor.

Sleeves intended to carry a water pipe should not contain any other pipe or cable.

Figure 10.1   Pipe sleeved through wall

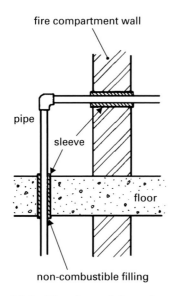

Sleeves should:
- o permit ready removal and replacement of pipes;
- o be strong enough to resist any external loading exerted by the wall or floor;
- o be sealed with fire-resistant material that will accommodate thermal movement.

Figure 10.2   Pipe sleeved through wall and floor

non-cracking, water-resistant material for a length of 150 mm to prevent access by water, gas or vermin.

An example of pipes entering buildings is given in figure 10.3.

Pipes in walls, floors and ceilings

No pipe should be buried in any wall or floor, or under any floor, unless arranged as shown in figure 10.4. Where pipes are enclosed as shown it is preferred that there are no joints enclosed. Pipes should be wrapped to prevent corrosion and permit thermal movement.

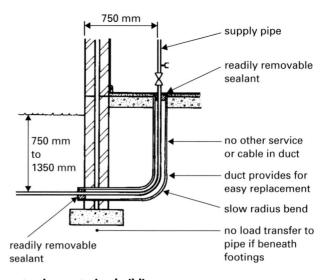

Figure 10.3   Access to pipes entering buildings

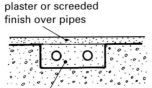

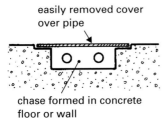

Closed-circuit pipes may be screeded or plastered over provided they are in a proper chase, and can be readily exposed for repair or replacement and are wrapped in impervious tape.

(a) Pipes plastered or screeded over

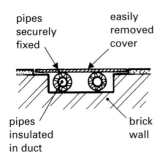

Pipes other than closed circuits must be in a duct that is reasonably easy to expose for examination, repair or replacement, without causing any structural damage.

(b) Pipes under removable cover

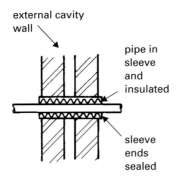

Pipe to be insulated if ducted in an outside wall or in any wall in an unheated building.

(c) Pipes in external walls

external cavity
wall

pipe in
sleeve
and
insulated

Sleeve passing through cavity walls must pass right through the cavity and the pipe must be insulated. Other than this arrangement no pipes should be laid in a cavity.

sleeve
ends
sealed

(d) Pipes passing through external walls

Figure 10.4 Pipes in chases and ducts

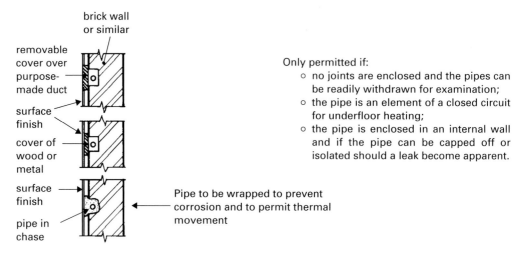

Only permitted if:
- no joints are enclosed and the pipes can be readily withdrawn for examination;
- the pipe is an element of a closed circuit for underfloor heating;
- the pipe is enclosed in an internal wall and if the pipe can be capped off or isolated should a leak become apparent.

(e) Pipes enclosed within internal walls

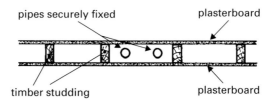

Pipes may be enclosed in internal walls of the timber studded type.

(f) Pipes in stud partition

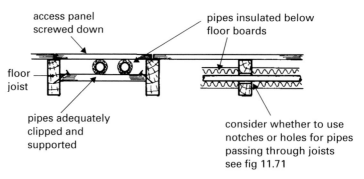

Pipes under suspended floors should be avoided. Where unavoidable they must be insulated and access panels formed in the floor for examination and repair.

Access panels should not affect the stability of the floor.

Boards removable at intervals of not more than 2 m and at every joint for inspection of whole length of pipe.

(g) Pipes under timber floors

Figure 10.4   continued

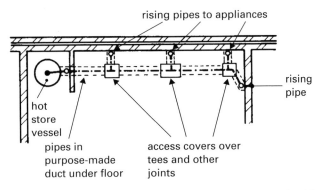

Where pipes are enclosed in a solid floor, access covers should
be provided at tees and joints.

(h) Typical ducting arrangements

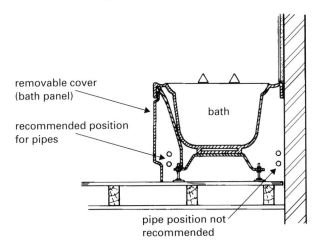

Although BS 6700 recommends that pipes are positioned in front
of a bath, the back position gives better fixing for pipes and easier
access for bath replacement. However, joints behind bath should
be avoided.

(i) Pipes concealed under baths

Figure 10.4   continued

10.2   Stopvalves

Stopvalves above ground should be positioned so as to be readily accessible for examination, maintenance and operation. They should be accessible for use without the need for hand tools to remove any access cover. It is preferable that the valve is not behind any cover.

Stopvalves on underground pipes should be enclosed within a pipe guard or chamber with surface box to provide access for shutting off with a metal stopvalve key (see figure 10.5).

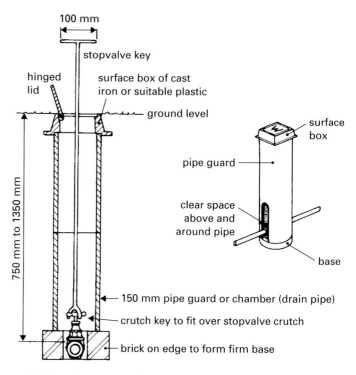

(a) Chamber construction

Surface box to be suitable for relevant traffic loading, e.g. heavy or light grade.

Stopvalves and pipes more than 1350 mm deep will be deemed to be inaccessible.

Surface box to conform to BS 5834-2 and be of a grade to suit its location and traffic loading.

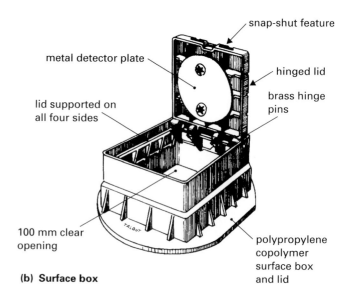

(b) Surface box

Figure 10.5   Access to below ground stopvalves

10.3   Water storage cisterns

These should be positioned so that they are easy to inspect, clean and maintain (see figure 10.6).

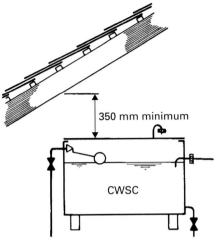

Clear space required all round for inspection of cistern and pipes.

Internal access must be provided for cleaning and maintenance.

Combination units and cisterns in cupboards require 225 mm minimum space.

350 mm minimum

CWSC

(a) Small cisterns

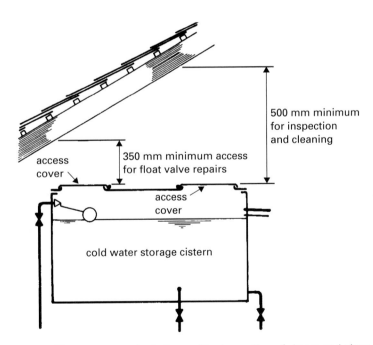

500 mm minimum for inspection and cleaning

350 mm minimum access for float valve repairs

access cover

access cover

cold water storage cistern

Clear space required all round for inspection of cistern and pipes.

(b) Large cisterns

Figure 10.6   Access to cisterns

Chapter 11
Installation of pipework

Pipes and fittings need to be chosen and installed to suit their purpose and the conditions in which they are to be situated.

Pipes should be laid or fixed:

- to avoid frost, mechanical damage or corrosion;
- so that they do not leak, cause undue noise or permit any contamination of water contained in them;
- to comply with the requirements of relevant British or European Standards or be approved under the UK Water Fittings Testing Scheme or another equivalent testing scheme.

When jointing pipes take care to ensure that:

- joints are mechanically sound and clean inside;
- pipes are cut squarely and all burrs removed and distorted ends rounded;
- cutting tools are in good condition to limit distortion;
- joints comply with British or European Standards and are listed in the *Water Fittings and Materials Directory*;
- when applying heat the risk of fire is eliminated, and the operator does not breathe any harmful fumes given off from soldering or welding processes;
- only approved jointing materials or compounds are used;
- joints, clips and fittings are compatible with the pipe material and will not cause corrosion.

Pipes and components should be handled with care. Damage caused during installation can seriously affect the life and performance of the system.

Bending of pipes should be carefully carried out using purpose-made equipment to avoid deformation or damage. Avoid crimping, kinking and restriction of the pipe bore which can cause damage to the pipe bore, or loss of water flow, or create additional flow noise in the system. Any damaged pipes should be discarded. Pipes of copper, stainless steel and black low carbon steel are ideally suited to bending. Pipes of galvanized steel must *not* be bent as this will damage the protective zinc coating.

11.1 Steel pipes

There are three grades of low carbon steel pipes to BS 1387: heavy, medium and light, each of which is obtainable with or without galvanization inside and out. For hot and cold services within buildings, only galvanized pipes are permitted, as follows:

Heavy (identified by a red band painted around the pipe near its ends) – for use underground where, in addition to galvanization, other forms of protection should be used to guard against exterior corrosion, e.g. a bituminous coating. Exposed threads should be painted.

Medium (identified by a blue band) – for general use above ground only.
Light (identified by a brown band) – permitted on some fire fighting installations.

Jointing of galvanized steel tubing for water services is achieved by screwing the pipe and fitting together, pipe joints being made on site using hand or electrically powered threading machines. After jointing, any exposed threads should be painted or, if underground, treated with a suitable bitumen or corrosion-preventing coating complying with BS 5493. Where pipelines change direction, manufactured bends should be used. Welded or brazed joints should not be used and pipes should not be bent, as these processes will damage the pipe's galvanized coating. It is extremely dangerous to weld or braze galvanized tube because the zinc fumes are highly toxic.

Low carbon steel pipes and fittings should be protected from corrosion, particularly in damp or otherwise corrosive conditions such as below ground. Proprietary pipe wrap materials are available for this purpose, and should be applied so that none of the pipework remains exposed.

It is advisable to consult water suppliers before installing any low carbon steel to establish whether the water supply will cause excessive corrosion.

A selection of joints for use with low carbon steel tube can be seen in figures 11.1–11.3, whilst further information on jointing and fixing is given in tables 11.1 and 11.2.

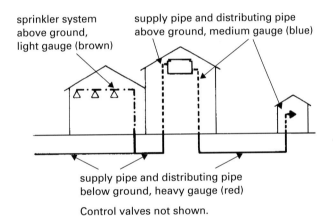

Figure 11.1   Grades of steel pipe and their uses

Table 11.1   Thread engagement lengths for steel pipe to BS 1387

Nominal size of pipe inside diameter mm	Thread length mm
15	13
20	15
25	17
32	19
40	19
50	24
80	30
100	36
150	40

Table 11.2   Maximum spacing of fixings for internal steel piping

Nominal size of pipe inside diameter mm	Spacing on horizontal run m	Spacing on vertical run m
15	1.8	2.4
20	2.4	3.0
25	2.4	3.0
32	2.7	3.0
40	3.0	3.6
50	3.0	3.6
80	3.6	4.5
100	3.9	4.5
150	4.5	5.4

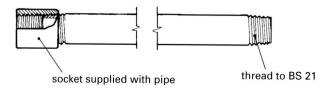

socket supplied with pipe   thread to BS 21

(a) Pipe–standard length 6.4 m

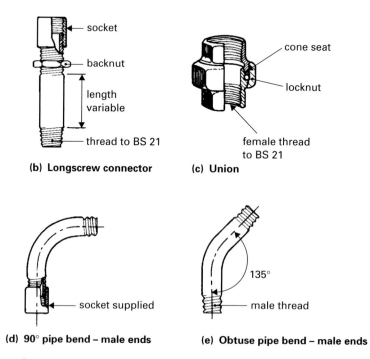

socket

backnut

length variable

thread to BS 21

cone seat

locknut

female thread to BS 21

(b) Longscrew connector   **(c) Union**

socket supplied

135°

male thread

(d) 90° pipe bend – male ends   **(e) Obtuse pipe bend – male ends**

Figure 11.2   Selection of fittings for steel tubes to BS 1387

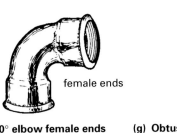

(f) 90° elbow female ends

female ends

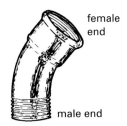

female end

male end

(g) Obtuse bend – male and female

(h) Tee 90° – equal

1 ——— 2

3

Numbers indicate method of specifying outlets:
- state 'in line' outlets first;
- state larger end first.

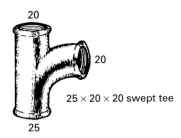

20

20

25 × 20 × 20 swept tee

25

(i) Swept or pitched tee

Figure 11.2 continued

(a) Saddle clip of galvanized steel

(b) Screw-on clip of galvanized cast iron

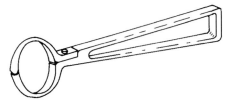

(c) Clip of galvanized cast iron for building in

Figure 11.3 Pipe fixings for steel tubes

11.2   Copper pipes

Copper tube is ideally suited to use in hot and cold water supplies. It is resistant to external corrosion in most soil conditions and to internal corrosion from the majority of water supplies. Copper is high on the electrochemical scale and should not be connected directly to other metals, particularly galvanized steel, unless the other metal is resistant to, or protected from, the effects of galvanic (electrolytic) action. In areas where unacceptable green staining occurs or electrolytic corrosion is promoted, alternative materials should be chosen, or water treatment should be considered.

Only copper tube to BS EN 1057 should be used. This standard includes those grades of tube that were formerly covered by BS 2871.

Copper tube for water services to BS EN 1057 is graded according to its hardness in the range of sizes shown in table 11.3.

Tables 11.4 and 11.5 give a range of pipes to BS EN 1057 and show how the former BS 2871 tube has been absorbed into this new European Standard. Typical fixings for copper tube are illustrated in figures 11.4–11.6 and spacing recommendations are given in table 11.6.

Copper tube is also available with a polyethylene coating to add protection from external corrosion. Joints would need wrapping with a suitable adhesive tape after testing. In some areas this may be necessary to prevent corrosion to underground pipes. The polyethylene coating should be coloured blue when the tube is to be used underground.

Jointing methods for copper tubes

Copper fittings may be manufactured from copper or copper alloys such as brass or gunmetal. Copper is used for capillary solder fittings only. Fittings made from copper alloys should comply with the requirements of BS EN 1254.

For underground use, and in situations where the water is capable of causing dezincification, fittings are required to be dezincification resistant. Dezincification is a form of electrolytic corrosion in which the zinc content of the brass is corroded away, leaving the copper behind and the remaining brass in a porous and weakened condition.

Table 11.3   Copper tube to BS EN 1057

| Material temper | | Range of sizes (OD) | | Standard lengths available |
EN hardness number	Common term	mm from	mm to	m
R220	Annealed	6	54	10 to 20
R250	Half hard	6	159	3 and 6
R290	Hard	6	267	3 and 6

OD = outside diameter.

Annealed (soft temper) tube is used in coils for microbore heating circuits, smaller connections to taps and in a heavier wall thickness for underground use.

Half hard tube is obtainable in straight lengths in thicknesses suitable for above and below ground use. It is ideally suited to bending by machine or by hand.

Hard tube is obtainable in straight lengths. Because of its hardness it will not readily stretch and should not be bent.

Table 11.4   R250 half hard copper tube to BS EN 1057 in straight lengths

OD mm	Wall thickness								
	0.6	0.7	0.8	0.9	1	1.2	1.5	2	2.5
6	X		Y						
8	X		Y						
10	X		Y						
12	X		Y						
15		X			Y				
22				X		Y			
28				X		Y			
35						X	Y		
42						X	Y		
54						X		Y	
66.7						X		Y	
76.1							X	Y	
108							X		Y
133							X		
159								X	

Notes
(1)   X indicates tube to former BS 2871: Part 1, Table X.
(2)   Y indicates tube to former BS 2871: Part 1, Table Y.

OD = outside diameter.

Table 11.5   R220 annealed copper tube to BS EN 1057 in coiled lengths

OD mm	Wall thickness					
	0.6	0.7	0.8	0.9	1	1.2
6	W		Y			
8	W		Y			
10		W	Y			
12			Y			
15					Y	
22						Y
28						Y

Notes
(1)   W indicates tube to former BS 2871: Part 1, Table W.
(2)   Y indicates tube to former BS 2871: Part 1, Table Y.

OD = outside diameter.

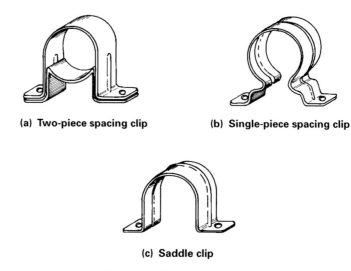

(a) Two-piece spacing clip   (b) Single-piece spacing clip

(c) Saddle clip

Figure 11.4   Copper fixing clips for small diameter copper tubes

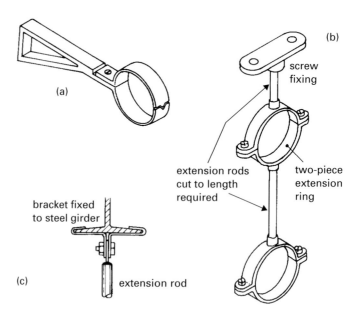

(a) Built-in clip

(b) Two-piece pipe ring with extension rod and backplate

(c) Alternative fixing for pipe ring

Figure 11.5   Brass fixing clips for copper tubes

Table 11.6   Maximum spacing of fixings for internal copper and stainless steel piping

Type of piping	Nominal size of pipe, outside diameter mm	Spacing on horizontal run m	Spacing on vertical run m
Copper tube to	15	1.2 (1.8)*	1.8 (2.0)*
BS EN 1057	22	1.8 (2.4)	2.4 3.0)
R250 half hard,	28	1.8 (2.4)	2.4 (3.0)
R290 hard	35	2.4 (2.7)	3.0 (3.0)
and	42	2.4 (3.0)	3.0 (3.6)
stainless steel to	54	2.7 (3.0)	3.0 (3.6)
BS EN 10312	76	3.0 (3.0)	3.6 (3.6)
	108	3.0 (3.6)	3.6 (4.5)
	133	3.0 (3.9)	3.6 (4.5)
	159	3.6 —	4.2 —

Note   * Figures for stainless steel tube are shown in brackets.

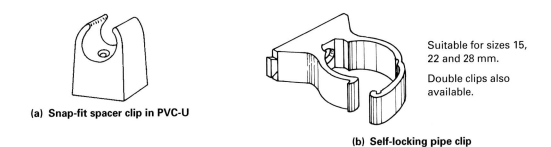

(a) Snap-fit spacer clip in PVC-U

Suitable for sizes 15, 22 and 28 mm.

Double clips also available.

(b) Self-locking pipe clip

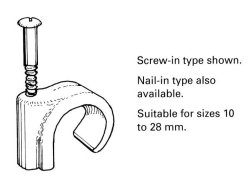

Screw-in type shown.

Nail-in type also available.

Suitable for sizes 10 to 28 mm.

(c) Side fixing saddle clip

Figure 11.6   Clips of plastic for copper tube

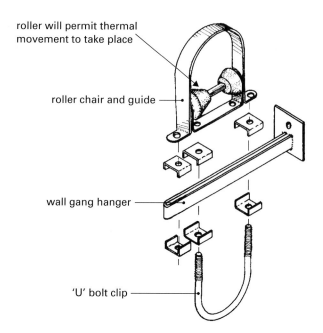

roller will permit thermal
movement to take place

roller chair and guide

wall gang hanger

'U' bolt clip

Figure 11.7   Pipe hanger fixings for large diameter pipes

Gunmetal, made of copper and tin, contains no zinc and is, therefore, immune to dezinci-fication. It is, however, expensive to manufacture because each individual fitting must be individually cast. Brass on the other hand can be hot stamped into shape before machining.

Duplex brass has traditionally been used for the manufacture of fittings, but it was susceptible to dezincification and its use was permitted only above ground in those areas not likely to be affected by dezincification. Duplex brass has now been replaced by dezincification-resistant (DZR) brass for the production of the vast majority of copper alloy fittings. DZR brass fittings may be marked with one of the following symbols:

CR or DRA

Suitable methods for the jointing of copper tube include: compression fittings, capillary solder joints, push-fit and press-fit fittings, brazing and braze welding. These are described below. See also table 11.8.

Compression fittings – type A (non-manipulative)

These are the most usual type of compression fitting for use above ground. They must not be used below ground. See figures 11.8 and 11.9.

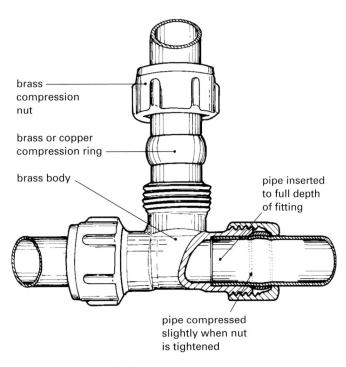

brass compression nut

brass or copper compression ring

brass body

pipe inserted to full depth of fitting

pipe compressed slightly when nut is tightened

For use above ground only.

Also suitable for stainless steel tube.

Cut pipe ends squarely and file off burrs.

No manipulation of pipe ends needed.

Do not overtighten brass backnut.

Jointing compounds not required.

(a) Assembly of compression fittings – type A (non-manipulative)

Figure 11.8   Compression fittings – type A

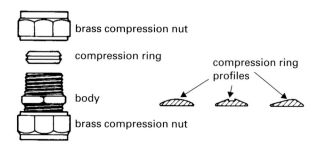

Rings from one fitting are not compatible with fittings from another manufacturer.

(b) Straight coupling

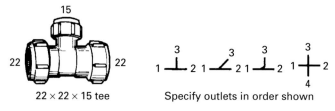

22 × 22 × 15 tee

Specify outlets in order shown

(c) Tee with reduced branch

(d) Elbow – copper to male iron

(e) Straight swivel coupling
Supplied with one fibre washer

Figure 11.8   continued

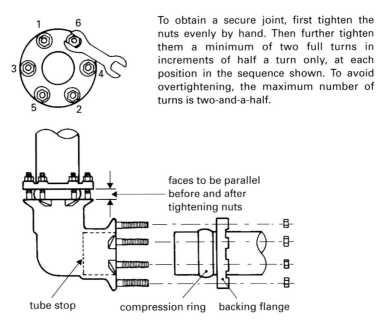

To obtain a secure joint, first tighten the nuts evenly by hand. Then further tighten them a minimum of two full turns in increments of half a turn only, at each position in the sequence shown. To avoid overtightening, the maximum number of turns is two-and-a-half.

Figure 11.9   Large diameter compression fitting – type A

Compression fittings – type B (manipulative)

These require the flaring of pipe ends and give extra security against joints pulling apart in extreme conditions. This is especially important underground, where any resulting leakage could well go undetected. See also figures 11.10 and 11.11.

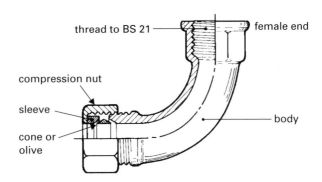

Figure 11.10   Type B compression fitting for copper tube – female iron to copper bend

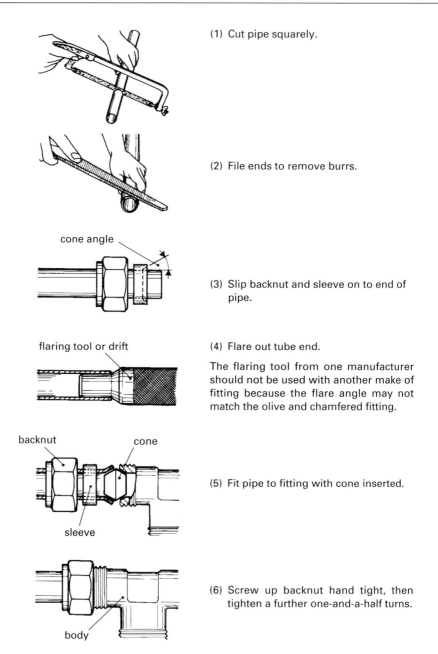

(1) Cut pipe squarely.

(2) File ends to remove burrs.

(3) Slip backnut and sleeve on to end of pipe.

(4) Flare out tube end.

The flaring tool from one manufacturer should not be used with another make of fitting because the flare angle may not match the olive and chamfered fitting.

(5) Fit pipe to fitting with cone inserted.

(6) Screw up backnut hand tight, then tighten a further one-and-a-half turns.

Figure 11.11   Method of assembly of type B compression fittings

Capillary fittings

These are made in two types, as illustrated in figure 11.12. They are suitable for use both above and below ground and are made in a range of copper or brass materials.

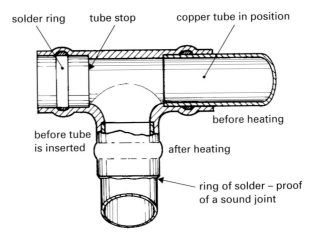

Extra solder should not be added as there is already sufficient within the fitting.

(a) Integral solder ring type

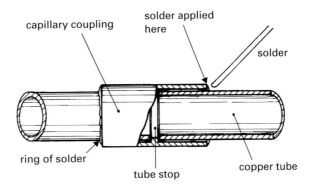

Because the end feed fitting contains no solder, this must be applied to the joint as it is heated.

(b) End feed type

Figure 11.12   Capillary soldered joints

The soldering method is as follows:

(1) Cut the end square and remove burrs.
(2) Thoroughly clean inside of fitting and outside of pipe with proprietary sandpaper (not emery board) or cleaning cloth.
(3) Apply suitable flux to prevent oxidation and assist cleaning (just a thin smear to both surfaces).
(4) Assemble joint.
(5) Apply heat with blow torch or electrical tongs.
(6) Add solder to joint area (end feed type only) and continue to apply heat until solder flows around and into joint. For solder ring fittings apply heat until solder appears and forms a ring around the end of the fitting (no extra solder needed).
(7) Leave joint undisturbed to cool.
(8) Remove flux residues or they may corrode the pipe.

Solders have traditionally been lead/tin alloys, the use of which is prohibited by Water Regulations in favour of lead-free solders, e.g. tin/silver alloy No. 28 or 29 or tin/copper alloy No. 23 or 24 to BS EN 29453.

Push-fit and press-fit joints

The risk associated with the use of heat in the making of pipe joints is increasingly being highlighted. On some sites the use of heat producing tools such as the blowlamp may not be permitted because of the fire risk. For the above reasons, the use of soldering, brazing and braze welding is likely to diminish.

Whilst we have used heat-free joints for many years, over the past few years new methods have been developed. Push-fit and press-fit joints were introduced for use with copper tube in the mid to late 1990s and are now firmly placed as suitable alternatives to traditional jointing methods.

Push-fit joints are available in sizes from 10 mm to 54 mm and can be used for hot, cold and central heating applications. Joints rely on an 'O' ring for a good seal along with a stainless steel grab ring that will prevent the joint from pulling apart under pressure. The joint is quickly and easily made and can be dismantled for re-use using a simple disconnection tool applied to the joint collar.

It is important that, prior to jointing, the pipe end is cut squarely, de-burred and chamfered to avoid damage to the 'O' ring seal when inserting the tube into the collar. The pipe should be pushed fully home with a twisting action until a distinct 'click' can be heard denoting that it has reached the tube stop. Examples of these joints can be seen in figure 11.13.

This type of joint can also be applied to pipes of stainless steel, polybutylene (PB) and cross-linked polyethylene (PEX).

The **press-fit joint** also relies on an 'O' ring seal but the seal is completed using a mechanical compression tool to squeeze the joint and its 'O' ring to make a positively interlocked joint. Pipes should be cut squarely, de-burred and inserted fully into the joint socket. A depth gauge is available to help mark the tube to show when the pipe is fully inserted. No paste, flux or lubricant is needed (see figure 11.14).

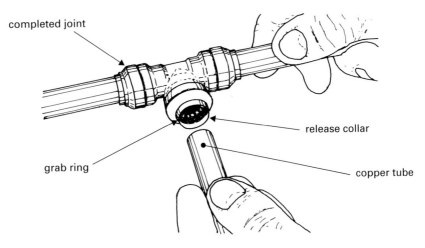

Jointing procedure

(1) Cut pipe squarely and clean off burrs.
(2) Insert tube through release collar until it rests against the grab ring.
(3) Push tube firmly with twisting motion until a 'click' is heard indicating it has reached the tube stop.
(4) Pull on the tube to make sure the joint is secure.

(a) Making a push-fit joint t

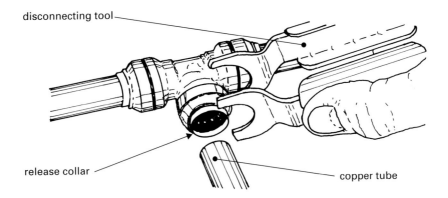

Disconnection procedure

(1) Position the disconnecting tool forks around the fitting. One fork around the neck of the fitting and the smaller one around the pipe.
(2) With one hand squeeze the disconnecting tool to compress the release collar and with the other hand twist out the tube. Use thumb as a lever to assist disconnection.
(3) Check the fitting before attempting to reconnect.

(b) Disconnecting a push-fit joint t

Figure 11.13   Push-fit joints for copper tube

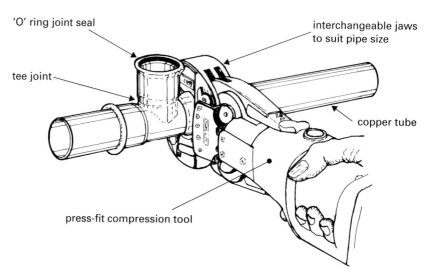

Figure 11.14   Making a press-fit joint to copper tube

Press-fit joints can be used on pipes ranging from 12 mm to 108 mm in diameter and are suitable for:

- cold water applications up to 16 bar pressure at 20°C; and
- hot water installations up to 6 bar pressure at 110°C.

With correctly sized jaws fitted, the press-fit tool must fit squarely over the joint ring, before commencing the compression cycle. The joint will be completed when the jaws fully enclose the fitting, at which time the power will cease automatically.

When designing the pipework system, space must be allowed to engage and operate the clamping jaw around the fitting (see figures 11.15 and 11.16).

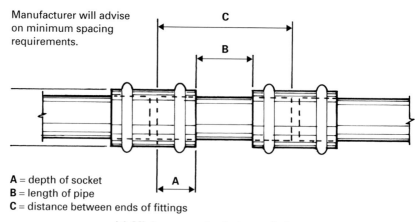

(a) Minimum spacing between fittings

Figure 11.15   Space requirements around pipes for operation of press-fit tool

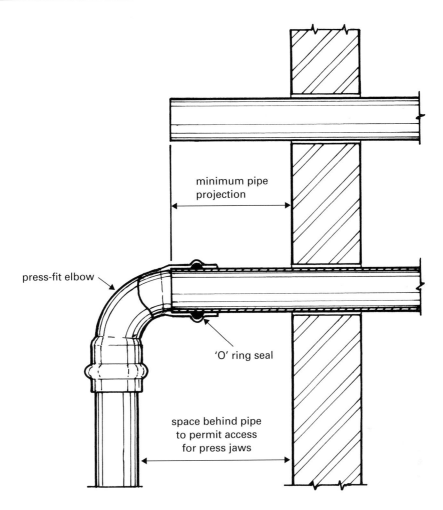

A pipe projecting through a wall must be long enough to permit the press-fit jaws to locate the fitting without hindrance.

Pipe running alongside wall needs space for press-fit jaws to locate over fitting.

The minimum projections and spaces required will be specified by the press-fit tool manufacturer.

(b) Press-fit joint to copper tube protruding through wall

Figure 11.15   continued

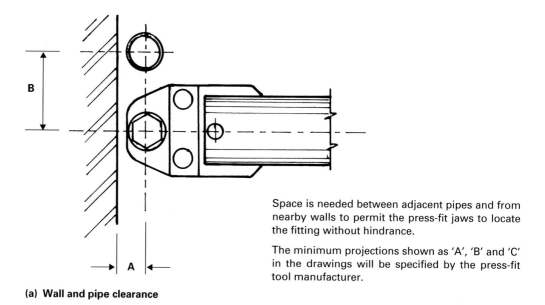

Space is needed between adjacent pipes and from nearby walls to permit the press-fit jaws to locate the fitting without hindrance.

The minimum projections shown as 'A', 'B' and 'C' in the drawings will be specified by the press-fit tool manufacturer.

(a) Wall and pipe clearance

(b) Corner and pipe clearance

Figure 11.16   Space requirements for operation of press-fit tool near walls

Press-fit is a fast and efficient method that should reduce jointing and installation times. Any cost savings are to some extent offset by the initial outlay of the press-fit tool that can be electrically powered either by mains supply or by battery. Alternatively there are manual versions available for use where there is no power supply on site or no means to recharge batteries. These more economic manually operated tools are perhaps more suited to smaller installations and repair work.

Brazed joints

Brazing is a capillary joint which uses a filler metal that has a melting point above 450°C but which is lower than the melting point of the metals being joined. The method of jointing is similar to that of soft soldering but it is carried out at higher temperatures. In brazing, the parent metals are not melted, rather they are 'wetted' or 'tinned' by elements within the filler metal.

Brazing is ideally suited to joints that need to resist high pressures or temperatures, and in the past has been used extensively in hospitals where specifications have demanded the use of capillary joints without a lead content. Joints may be hand-made using specialized equipment or by the use of copper or copper alloy fittings, similar in appearance to capillary soldered joints. Further information on brazed joints is given in figure 11.17 and tables 11.7 and 11.8. Figure 11.17 shows a hand-formed joint to copper tube.

Brazing filler metals for use with copper tube and fittings should conform to BS EN 1044 and are of two main types:

(1)   Copper-silver-zinc alloys (Cu-Ag-Zn) that require the use of a flux; and
(2)   Copper-phosphorus-silver alloys (Cu-P-Ag). Copper to copper joints using this filler metal can be made without the need for a flux. For connections between copper and other copper-rich metals a flux will be needed.

Table 11.7   Minimum depths of overlap for brazed joints

Pipe diameter (mm)	15	22	28	35	42	54	67	76	108
Joint depth (mm)	8	10	12	15	18	20	22	25	30

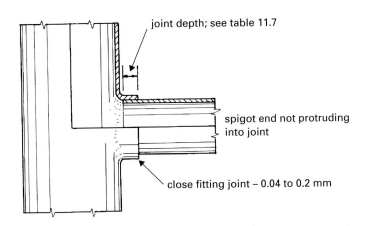

Figure 11.17   Hand-made brazed joint to copper tube

Table 11.8   Brazing alloys

(a) Type AG: Silver brazing filler metals

EN ISO 3677 type	Nominal percentage composition				Melting range °C	
	Silver (Ag)	Copper (Cu)	Zinc (Zn)	Tin (Sn)	Solidus °C	Liquidus °C
AG 101	59–61	22–24	12–16	2–4	620	685
AG 102	55–57	21–23	15–19	4–6	620	655
AG 103	54–56	20–22	20–24	1.5–2.5	630	660
AG 104	44–46	26–28	23.5–27.5	2–3	640	680
AG 105	39–41	29–31	26–30	1.5–2.5	650	710
AG 106	33–35	35–37	25.5–29.5	2–3	630	730
AG 107	29–31	35–37	30–34	1.5–2.5	665	755
AG 108	24–26	39–41	31–35	1.5–2.5	680	760
AG 201	62–64	23–25	11–15	0	690	730
AG 202	59–61	25–27	12–16	0	695	730
AG 203	43–45	29–31	24–28	0	675	735
AG 204	29–31	37–39	30–34	0	680	765
AG 205	24–26	39–41	33–37	0	700	790
AG 206	19–21	43–45	34–38	0.05–0.25	690	810
AG 207	11–13	47–49	38–42	0.05–0.25	800	830
AG 208	4–6	54–56	38–42	0.05–0.25	820	870

Notes
(1)   This table contains filler metals from BS EN 1044 that do not contain cadmium. Fillers containing cadmium can produce highly toxic fumes and are not recommended.
(2)   Zinc is also toxic and alloys containing zinc should be used with care and the work area well ventilated.

(b) Type CP: Copper–phosphorus brazing filler metals

EN ISO 3677 type	Nominal percentage composition				Melting range		
	Silver (Ag)	Copper (Cu)	Zinc (Zn)	Tin (Sn)	Solidus °C	Liquidus °C	Minimum brazing temperature °C
CP 101	Remainder	6.6–7.5	17–19	0	645	645	650
CP 102	Remainder	4.7–5.3	14.5–15.5	0	645	800	700
CP 103	Remainder	7.0–7.6	5.5–6.5	0	645	725	690
CP 104	Remainder	5.7–6.3	4.5–5.5	0	645	815	710
CP 105	Remainder	5.9–6.7	1.5–2.5	0	645	825	740
CP 201	Remainder	7.5–8.1	0	0	710	770	720
CP 202	Remainder	6.6–7.4	0	0	710	620	730
CP 203	Remainder	5.9–6.5	0	1.8–2.2	710	890	760
CP 301	Remainder	5.6–6.4	0	6.5–7.5	690	825	740
CP 302	Remainder	6.4–7.2	0		650	700	700

Notes
(1)   Copper–phosphorus fillers may be used on copper to copper joints without the need for a flux.
(2)   These metals should not be used with ferrous metals or copper alloys containing nickel.

Filler metals can be obtained with or without a flux coating and in the following sizes:

- preferred lengths 500 mm and 1000 mm;
- preferred diameters (in mm) 1.0, 1.5, 2.0, 2.5, 3.0 and 5.0.

A selection of filler metals from BS EN 1044 can be seen in table 11.8. The table can be used by quoting a series of numbers from the table depending on the required characteristics of the filler metal. For example:

- for a silver brazing filler rod quote EN 1044-AG 103;
- for a copper–phosphorus filler rod quote EN 1044-CP 101.

Note   These are examples, not recommended filler alloys. Selection will depend on the properties and application, e.g. quick flowing, low melting point, etc.

In Table 11.8(a) the author has selected only filler alloys that have no cadmium content. Filler metals containing cadmium can produce highly toxic fumes and are not recommended. Zinc is also toxic and alloys containing zinc should be used with care and the work area well ventilated. It should be noted that any fumes from brazing alloys and their fluxes can be hazardous and should not be breathed in.

Fluxes for brazing are of two types and should conform to BS EN 1045:

(1)   For high temperature brazing (over 750°C), a borax based flux is advised.
(2)   For low temperature brazing (below 750°C), an alkali fluoride flux may be used.

Braze-welded joints

Braze welding is described in BS 699 as 'the jointing of metals using a technique similar to fusion welding and a filler metal with a lower melting point than the parent metal, but neither using capillary action as in brazing nor intentionally melting the parent metal'.

Braze (bronze) welded joints (see figure 11.18) are very strong, but are costly to produce and require the use of a high degree of skill.

Fluxes used should be borax based and should be cleaned off after use. Filler metals will be of brass with the addition of silicon and tin. Two suitable metals are listed in BS 1453. See table 11.9.

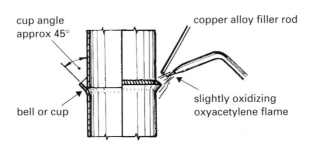

Figure 11.18   Braze-welded butt joint

Table 11.9   Recommended filler metals for braze-welding copper joints

Type	Nominal percentage composition (by weight)					
BS 1453	Copper	Zinc	Silicon	Tin	Magnesium	Iron
C2	57 to 63	Balance	0.02 to 0.5	Optional to 0.5	0	0
C4	57 to 63	Balance	0.15 to 0.3	Optional to 0.5	0.05 to 0.25	0.1 to 0.5

11.3   Stainless steel pipes

Stainless steel pipe to BS EN 10312 is similar to copper in use. Its external dimensions are the same as copper. Jointing methods are generally similar to those for copper tube but stainless steel pipes should *not* be joined by soft soldering. Although stainless steel and copper pipes may generally be mixed, joining small copper areas to large stainless steel areas should be avoided.

Stainless steel pipe is much more rigid than copper and therefore needs a little more accuracy in bending. Bending is done using the copper pipe bending machine and should not be attempted using a bending spring.

Pipe diameters range from 15 mm to 159 mm. Spacing for fixings is given in table 11.6.

Jointing methods include:

(1) compression fittings of stainless steel, or copper alloy to BS EN 1254-2;
(2) joints for silver soldering and brazing to BS EN 1254-2 or BS EN 1254-4;
(3) push-fit or press-fit joints;
(4) anaerobic adhesive bonding (up to 85°C) (see figure 11.19). Anaerobic adhesive jointing is not permitted where the pipe may be:
 (a) embedded in a wall or floor;
 (b) enclosed in a chase or duct; or
 (c) in a position where access is restricted.

Note   For methods (1) to (3) see jointing methods for copper tube. See also figure 11.10.

A selection of stainless steel joints for stainless steel pipes is shown in figure 11.20.

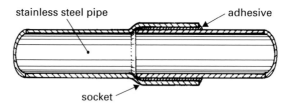

Method for anaerobic adhesive bonding
(1) Check fit between tube and fitting.
(2) Ensure bond areas are grease-free using solvent degreaser.
(3) Abrade bonding surfaces with medium emery cloth (80 grit).
(4) Apply ring of adhesive to leading edge of pipe and slip pipe into fitting.
(5) Allow:
  ○ one minute for curing,
  ○ one hour to withstand static pressure,
  ○ one day for full strength.

Figure 11.19   Anaerobic adhesive bonded joint for stainless steel tubes

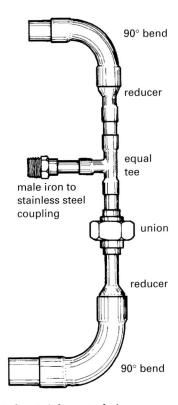

Figure 11.20   Selection of joints for stainless steel pipes

11.4   Plastics pipes

There is a wide range of plastics pipes available for use in hot and cold water supply installations. These are considered separately in the following pages. It is important that the
various plastics are recognized individually as separate materials, e.g. polyethylene (PE),
unplasticized polyvinyl chloride (PVC-U), chlorinated polyvinyl chloride (PVC-C), cross-
linked polyethylene (PE-X), etc. Plastics pipework systems for pressure applications are not
automatically inter-compatible. Each material has its own range of properties and jointing
methods, and is often suited to different applications and should be selected accordingly.

There is one area, however, where all the plastics materials considered here can, to some
degree, be considered together, and that is their reaction to the effect of heat. Coefficients
of expansion in plastics are generally about ten times greater than for metals, so much more
attention should be given to allowances made for thermal movement.

Plastics pipes have relatively low softening and melting points, much lower than metals.
They should therefore not be used where heat may cause them to become softened and
weakened. Plastics pipes used for hot water must be capable of withstanding temperatures
of 100°C or more.

Plastics pipes are flexible and may tend to droop between fixings, particularly where clips are too widely spaced or where subjected to heat.

It is generally recommended that plastics pipes are not threaded because threading reduces the pressure resistance of the pipe. If threaded joints are desired, the manufacturer should first be consulted as to the advisability of their use.

Where jointing materials or packing are required, a PTFE (polytetra-fluoroethylene) tape is recommended. Oil-based compounds should not be used as they may deteriorate the plastics material of the pipe or its fittings. Steel grips or stilsens that may damage the pipe should not be used on plastics pipes and fittings.

Polyethylene pipes (PE)

BS 6700 recommends that the requirements of CP 312-3 be followed when installing and jointing polyethylene pipes for the supply of drinking water. Polyethylene pipes should conform to BS EN 12201-2 and fittings to BS EN 12201-3 or BS 5114. Compression fittings of copper alloy should conform to BS EN 1254.

Polyethylene has proved to be an excellent material for cold water installations, particularly when used below ground. It is corrosion resistant, easy to lay and simple to joint. It can be obtained in long lengths permitting supply pipes to be laid with the minimum of joints. Its flexibility allows it to be bent around obstacles and threaded through ducts into buildings. It must not be used for hot water installations.

When laying polyethylene pipelines below ground, it is advisable to bed and cover the pipe with a selected soil or granular material such as pea shingle, to prevent damage from stones and flints, and to avoid deformation of the pipe when backfilling the trench.

In the past polyethylene has been known to be permeable to gas, and is liable to be damaged by contact with oils or oil-based products.

For use in construction work, polyethylene tube is manufactured in two colours:

- Black medium-density polyethylene tube to BS 6730 is made for above ground use. The tube material has a carbon black pigment added to prevent the passage of light through the pipe wall.
- Blue medium-density polyethylene tube to BS 6572 is made for use below ground, or in positions where the tube is fully protected from sunlight, e.g. within ducts. Its blue colour makes it easy to identify below ground. It is important for the safety of installers and end users that pipes (and their contents) can be readily identified.

It is expected that these standards will be replaced by BS EN 12201.

In addition to the tubes currently manufactured to BS 6730, BS 6570 or BS EN 12201, there are other polyethylene pipes still in use that were installed to meet various previous standards. Table 11.10 gives information on some of these. When connecting to existing pipelines, care should be taken that any fittings and inserts used are correct for the dimensions of the particular pipe.

Fixing supports for polyethylene pipes

Because polyethylene is a flexible material, pipes above ground should be supported continuously, or at least be fixed to the maximum spacing requirements shown in table 11.11.

Table 11.10   Types and grades of polyethylene pipes

Type of piping	Grade	Maximum working pressure	Type of piping	Nominal size, outside diameter mm	Approximate bore mm	Maximum working pressure bar
Low-density polyethylene (Type 32) to BS 1972 and high-density polyethylene (Type 50) to BS 3284	Class B (light)	6	Blue medium-density polyethylene to BS 6572 for below ground use only and black medium-density polyethylene to BS 6730 for above ground use only	20	15	
	Class C (medium)	9		25	20	
				32	25	12
	Class D (heavy)	12		50	40	
				63	50	

Preferred lengths: straight pipes 6 m, 9 m, 12 m.   Coils 50 m, 100 m, 150 m.

Note   BS 1972 and BS 3284 are withdrawn. These are shown because of their extensive use in the past.

Table 11.11   Maximum spacing of fixings for internal polyethylene pipes

Type of piping	Nominal diameter of pipe (DN)	Spacing on the horizontal run m	Spacing on the vertical run m
Low-density polyethylene to BS 1972 (withdrawn) (nominal size in inches)	$\frac{3}{8}$	0.30	0.60
	$\frac{1}{2}$	0.40	0.80
	$\frac{3}{4}$	0.40	0.80
	1	0.40	0.80
	$1\frac{1}{4}$	0.45	0.90
	$1\frac{1}{2}$	0.45	0.90
	2	0.55	1.10
	$2\frac{1}{2}$	0.55	1.10
	3	0.60	1.20
	4	0.70	1.40
Black polyethylene (PE) pipe to BS EN 12201-2 (nominal size in mm)	20	0.50	0.90
	25	0.60	1.20
	32	0.60	1.20
	50	0.80	1.50
	63	0.80	1.60
SDR 11 polyethylene (PE) pipe to BS EN 12201-2 (nominal size in mm)	16	0.60	1.20
	20	0.70	1.40
	25	0.80	1.60
	32	0.90	1.80
	40	1.00	2.00
	50	1.00	2.00
	63	1.10	2.20
	90	1.30	2.60
	110	1.50	3.00
	125	1.60	3.20
	160	1.80	3.60

Jointing methods for polyethylene pipes

A variety of jointing methods is available for polyethylene pipes and fittings are produced from both plastics and metals. Copper alloy compression fittings for use with PE pipe should comply with BS EN 1254-3 and BS 5114 for other joints. For smaller pipes, mechanical joints are predominantly used, and include both push-fit and compression fittings. These are illustrated in figures 11.21–11.27. For larger sizes, mechanical joints may be used and, particularly for mains pipes, thermal fusion jointing is popular.

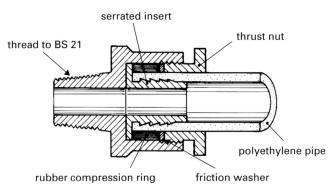

Suitable for smaller sizes only (up to 63 mm).

Figure 11.21   Compression fitting for polyethylene pipes

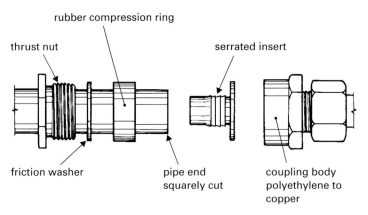

This illustration and the instructions for making the joint are for a compression joint manufactured by Talbot. Similar fittings are available from several other manufacturers.

(1)   Assemble parts in sequence shown.
(2)   Using a hide mallet or similar, knock insert into pipe end until its flange touches pipe face.
(3)   Push rubber compression ring and friction washer against flange and locate assembly in fitting body.
(4)   Screw in thrust nut until hand tight, then tighten fully with spanner (one-and-a-half to two turns).

Figure 11.22   Making the compression joint

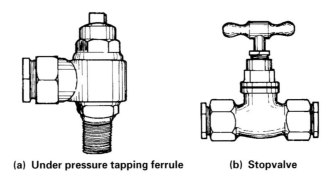

(a) Under pressure tapping ferrule   (b) Stopvalve

Figure 11.23   Examples of fittings for polyethylene pipes

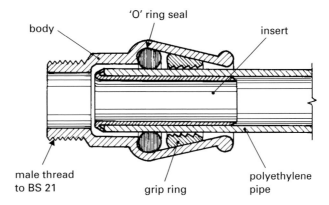

Figure 11.24   Push-fit joint for polyethylene pipe

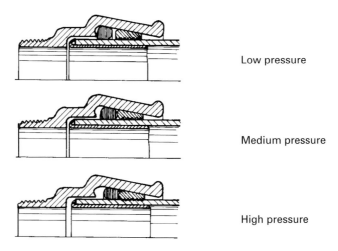

Low pressure

Medium pressure

High pressure

As water pressure increases the components tighten to create a pressure seal.

Figure 11.25   How the push-fit joint works

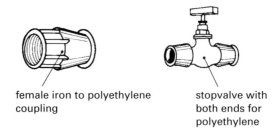

female iron to polyethylene
coupling

stopvalve with
both ends for
polyethylene

These fittings are available in a variety of sizes from
20 mm to 180 mm

Figure 11.26   Examples of small diameter fittings

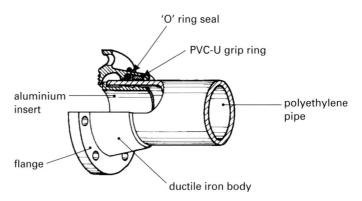

'O' ring seal

PVC-U grip ring

aluminium
insert

polyethylene
pipe

flange

ductile iron body

Figure 11.27   Example of large diameter fitting – flange adaptor

Push-fit joints

These are shown in figures 11.24–11.27.

Thermal fusion joints

There are three main techniques currently in use for the making of thermal fusion joints: socket fusion, butt fusion and electrofusion. These are shown in figures 11.28–11.30. It is important that heat input is strictly controlled to give correct fusion temperatures. Operatives should be fully trained in thermal fusion techniques.

Thermal fusion joints are suitable for large diameter polyethylene (PE) pipes, but are not popular for small diameters because other methods of jointing are so much easier. They are also suitable for polypropylene (PP) pipes.

(a) Socket fusion

Pipes may be joined using a range of polyethylene fittings. Pipe ends are heated to soften mating surfaces using a heating element, after which the pipe end is pushed fully into the socket and held firmly until fused and cooled down.

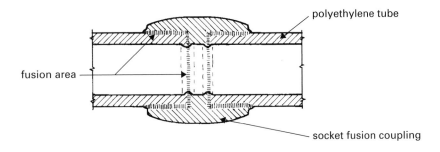

(b) Butt fusion

Used to join pipe ends only. Jointing is achieved by softening ends using a flat plate heating element, following which the pipe ends are pressed firmly together to form one solid mass. Popular method for the jointing of water mains.

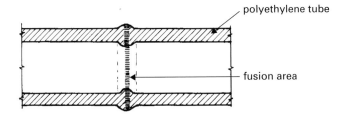

(c) Electrofusion

Perhaps the most reliable method but more expensive. A range of fittings are available, e.g. tees, bends, couplings, tapping saddles, etc. Fittings have an electric heating coil built in to each socket. The pipe is inserted into the socket, an electrical connection is made and the heat automatically applied. The joint is allowed to cool after the correct fusion temperature has been reached.

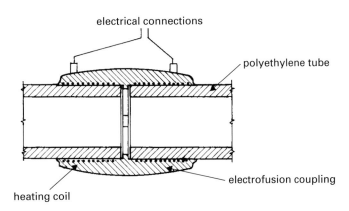

Figure 11.28   Thermal fusion jointing methods

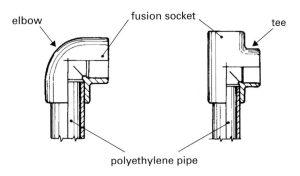

Figure 11.29   Thermal fusion joints

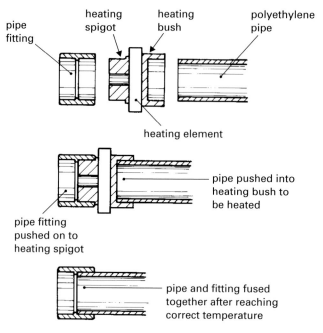

Also suitable for polypropylene pipe.

Pipes should only be joined with compatible fittings, i.e. pipe and fittings in polyethylene or pipe and fittings in polypropylene.

Correct heat is essential.

Follow manufacturers' instructions for jointing method.

Figure 11.30   Making socket fusion joints

Unplasticized polyvinyl chloride (PVC-U) pipes

PVC-U (unplasticized polyvinyl chloride) is an excellent material for cold water pipes for temperatures up to 20°C, above which its mechanical properties are reduced. PVC-U pipes become increasingly brittle as temperatures are lowered. Particular care should be taken when handling them at temperatures below 5°C.

Available in three pressure ranges and a variety of diameters, from 10 mm to 600 mm (see table 11.12), PVC-U is used in large quantities for water mains but is not so popular for smaller services. It is equally suited to installations above and below ground, but only for cold water applications.

Pipes and fittings of PVC-U should conform to BS EN 1452 or to BS 3505 or, if for industrial use, to BS 3506. They should be installed and used in accordance with the recommendations of BS CP 312-2.

Where mechanical joints are made using copper alloy fittings these should be dezincification resistant or immune. If there is adequate access, in positions above ground, solvent cement joints can be used.

Table 11.12   Sizes and pressure ratings of PVC-U pipe to BS 3505

Pressure class	Pressure rating at 20°C bar	Range of nominal sizes, inside diameter inches	mm
C	9	2 to 24	50 to 600
D	12	$1\frac{1}{4}$ to 18	32 to 450
E	15	$\frac{3}{8}$ to 16	10 to 400

Fixings for PVC-U pipes

Examples of fixings and their spacing are shown in figure 11.31 and table 11.13.

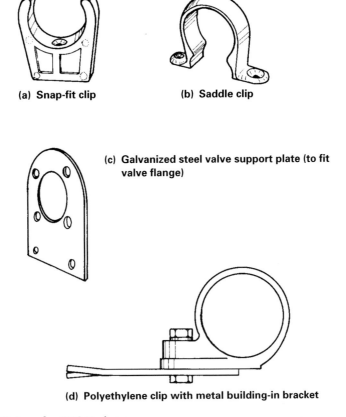

(a) Snap-fit clip   (b) Saddle clip

(c) Galvanized steel valve support plate (to fit valve flange)

(d) Polyethylene clip with metal building-in bracket

Figure 11.31   Fixings for PVC-U pipes

Table 11.13   Maximum spacing of fixings for internal PVC-U and PVC-C pipes

Type of piping	Nominal size of pipe inches	mm	Spacing on horizontal run m	Spacing on vertical run m
PVC-U to BS 3505 (Figures are for normal ambient temperatures below 20°C. For temperatures above 20°C the pipe manufacturer should be consulted)	$\frac{1}{4}$	10	0.6	1.1
	$\frac{1}{2}$	13	0.7	1.3
	$\frac{3}{4}$	20	0.7	1.4
	1	32	0.8	1.6
	$1\frac{1}{4}$	32	0.9	1.7
	$1\frac{1}{2}$	40	1.0	1.9
	2	50	1.1	2.2
	3	75	1.4	2.8
	4	100	1.6	3.1
	6	150	1.9	3.7
PVC-C (Based on average temperature of 80°C)		12 to 25	0.5	1.0
		32 to 63	0.8	2.2
Unplasticized polyvinyl chloride (PVC-U) to BS EN 1452-2 and chlorinated polyvinyl chloride (PVC-C) to BS 7291	$\frac{3}{8}$	16	0.8	1.2
	$\frac{1}{2}$	20	0.9	1.35
	$\frac{3}{4}$	25	1.0	1.5
	1	32	1.1	1.65
	$1\frac{1}{4}$	40	1.2	1.8
	$1\frac{1}{2}$	50	1.3	1.95
	2	63	1.4	2.1
	3	90	1.6	2.4
	4	110	1.8	2.7
	6	160	2.1	3.15

Jointing methods for PVC-U pipes

For below ground use and in confined locations above ground, it is recommended that mechanical joints are used in preference to solvent welding, particularly where conditions are wet and muddy. Any metal fittings should be immune to dezincification.

Threaded joints should not be used for PVC-U pipes to BS 3505. However, this method can be used for pipes to BS 3506 class 7 up to 2 inches diameter, providing the pressure does not exceed 9 bar.

Compression type mechanical joints for smaller sizes (see figure 11.32) are similar to those described for polyethylene. Any fittings made of brass and used below ground should be dezincification resistant or immune, e.g. gunmetal. These are suitable for joining tubes of up to 50 mm diameter. For larger diameter pipes a mechanical joint similar to that shown in figure 11.46 may be used.

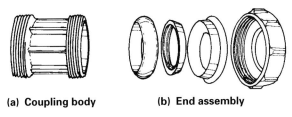

(a) Coupling body   **(b) End assembly**

Similar fitting can be used for polyethylene pipes.

Care should be taken not to overtighten joints.

Figure 11.32   Compression joint for PVC-U pipes

Push-fit mechanical joints for PVC-U pipes use integral elastomeric sealing rings that become compressed when the plain-ended pipes are inserted into the adjoining sockets. The plain pipe ends need to be chamfered and the surfaces cleaned and lubricated prior to insertion. The chamfered pipe end should be inserted fully into the adjoining socket (except where provision is to be made for expansion). An example of a push-fit joint can be seen in figure 11.33.

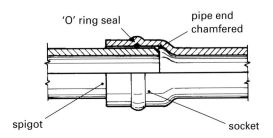

The push-fit joint for PVC-U is the simplest and cheapest form of installation.

Mechanical joints of push-fit type are suitable for use on pipes of 50 mm diameter and above.

Care should be taken to ensure that bends and branches are secure and properly anchored before pressure is applied.

Pipe end should be lubricated with suitable bactericidal, non-toxic lubricant to assist insertion into socket.

Pipe end should be inserted fully into the socket, making any allowance for expansion as recommended by the manufacturer.

Figure 11.33   Mechanical joint for PVC-U pipe (push-fit type)

Push-fit joints are popular because they are quick and easy to install and, more importantly, readily permit thermal movement to take place. They are available in a wide range of sizes and are illustrated in figures 11.24–11.27.

Flanged joints used for connections to valves, fittings and pipes of other materials should use full-face flanges or stub flanges, both with corrosion resistant or immune backing rings and bolts (see figure 11.35). Where appropriate the joint should be wrapped with a suitable protective material.

Solvent-welded joints are commonly used for the jointing of PVC-U pipes. For below ground use it is recommended that mechanical joints are used in preference to PVC-U, particularly where conditions are wet and muddy. Where solvent-welded joints are used on long pipelines, thermal movement should be accommodated by the use of the occasional mechanical or push-fit joint. Failures with solvent cement jointing will nearly always result from operatives not adhering to manufacturers' jointing procedures.

Solvents used should comply with BS 4346-3. Any cleaning fluid used should be from the same manufacturer as the solvent cement.

Figures 11.34 and 11.35 show examples of typical solvent cement joints.

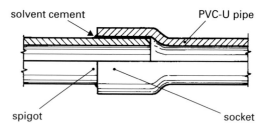

Jointing method:
(1) Cut pipe end square and remove internal burrs.
(2) Slightly chamfer outer edge of pipe (at about 15° to pipe axis).
(3) Roughen joint surfaces, using clean emery cloth or medium glasspaper.
(4) Degrease joint surfaces of pipe and fitting with cleaning fluid, using absorbent paper.
(5) Using a brush, apply an even layer of cement to both fitting and pipe in a lengthwise direction, with a thicker coating on the pipe.
(6) Immediately push the fitting on to the pipe without turning it. Hold for a few seconds, then remove surplus cement.
(7) Leave undisturbed for five minutes, then handle with reasonable care.
(8) Allow 8 hours before applying the full rated pressure, and 24 hours before testing at one-and-a-half times the full rated pressure.
      For lower pressure, allow one hour per bar, e.g. 3 bar would require 3 hours drying time.

Figure 11.34   Solvent cement joint for PVC-U pipe

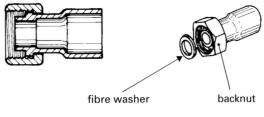

fibre washer   backnut

(a) Straight tap connector

(b) Equal tee 90°

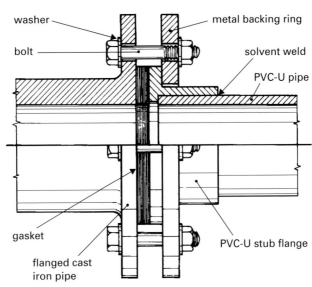

washer —   metal backing ring

bolt —   solvent weld

PVC-U pipe

gasket   PVC-U stub flange

flanged cast
iron pipe

Flange adaptor is solvent welded to pipe end.

(c) Flanged connections for PVC-U

Figure 11.35   Examples of solvent cement joints

Pipes of chlorinated polyvinyl chloride (PVC-C)

PVC-C pipes to BS 7291: Parts 1 and 4 are available in sizes 12 mm to 63 mm and are suitable for both hot and cold water applications. Jointing is similar to that shown for PVC-U. See table 11.13 for fixing distances.

Cross-linked polyethylene (PE-X)

Pipes and fittings of cross-linked polyethylene (PE-X) to BS 7291-1 and 3 are suited to both hot and cold water applications. The material is particularly suitable when resistance to freezing temperatures and abrasion is required. Joints may be push-fit or mechanical types and will be similar to those illustrated for polyethylene. These include fittings made from a plastics material that meets the applicable requirements of BS 7291, and copper and copper alloy compression fittings conforming to BS EN 1254-2 and/or BS EN 1254-3. PE-X cannot be solvent welded.

Polybutylene (PB)

Polybutylene should conform to BS 7291-1 and 2 and can be used for both hot and cold water. It is also suited to use where pipes may be subject to abrasive conditions. Push-fit, mechanical or thermal fusion joints may be used but PB cannot be solvent welded.

Propylene copolymer (PP)

Propylene copolymer pipes and fittings to BS 4991 are suited only to cold water applications up to 20°C. Above these temperatures the material's properties are reduced. Propylene copolymer (PP) cannot be solvent welded.

Acrylonitrile butadiene styrene (ABS)

Pipes and fittings made from acrylonitrile butadiene styrene (ABS) conforming to BS 5391-1 and BS 5392-1, or to BS EN ISO 15493, are suitable only for cold water applications. The use and installation of ABS pipes should follow the recommendations of BS 312-2.

Mechanical joints should conform to CP 312 and may be made of plastics or proprietary metal compression fittings as recommended by the pipe manufacturer. Fittings should include liners to support the bore of the pipe unless the manufacturer of the fitting instructs otherwise.

Compression joints must only be used with ABS piping of nominal diameter up to 63 mm. Joints should be of the non-manipulative type and care must be taken to avoid overtightening.

Solvent cement welded joints in ABS piping should be made using a solvent cement recommended by the manufacturer of the pipe. The dimensions of the pipes, spigots and sockets shall conform to BS 5391-1, BS 5392-1 or BS EN ISO 15493, as applicable.

Flanged joints used for connections to valves, fittings and pipes of other materials for ABS piping should use full-face flanges or stub flanges with corrosion resistant or immune backing rings and bolts. Where appropriate, joints should be wrapped with a suitable protective material.

Table 11.14 gives further general information on the above plastics materials and spacing for fixings are shown in tables 11.15 and 11.16.

Table 11.14   Plastics pipes and their uses

Pipe material	Use of material	Maximum temperature limits	Range of pipe sizes	Jointing methods
Blue polyethylene (PE) to BS 6572	Cold below ground only	20°C max	20–63	Compression Push-fit Thermal fusion
Black polyethylene (PE) to BS 6730	Cold above ground only	20°C max	20–63	Compression Push-fit Thermal fusion
Unplasticized polyvinyl chloride (PVC-U) to BS 3505	Cold	20°C max	3/8 to 24″	Compression Push-fit Solvent cement weld Mechanical
Unplasticized polyvinyl chloride (PVC-U) to BS 3506	Cold Non-drinking	20°C max	3/8 to 24″	Compression Push-fit Solvent cement weld Mechanical
Propylene copolymer (PE) to BS 4991	Cold	20°C max	1/4 to 24″	Compression Push-fit
Polybutylene (PB) to BS 7291	Cold and hot	83°C max	10 to 35 and 10 to 32	Compression Push-fit
Cross-linked polyethylene (PE-X) to BS 7291	Cold and hot	95°C max	10 to 35 and 10 to 32	Compression Push-fit
Chlorinated polyvinyl chloride (PVC-C) to BS 7291	Cold and hot	83°C max	12 to 63	Compression Push-fit Solvent cement weld
Acrylonitrile butadiene styrene (ABS) to BS 5391-1, 5392-1 or BS EN 15493	Cold	20°C max	–	Compression Push-fit Solvent cement weld Mechanical

Table 11.15   Maximum spacing of fixings for internal PB and PE-X pipes

Type of piping	Nominal size of pipe mm	Spacing on horizontal run m	Spacing on vertical run m
Polybutylene (PB) to BS 7291: Parts 1 and 2 and cross-linked polyethylene (PE-X) to BS 7291: Parts 1 and 4	up to 16 18 to 25 28 32 35	0.3 0.5 0.8 0.9 0.9	0.5 0.8 1.0 1.2 1.2

Table 11.16   Maximum spacing of fixings for ABS pipes

Type of piping	Nominal diameter of pipe (DN)		Spacing on the horizontal run	Spacing on the vertical run
	inches	mm	m	m
Acrylonitrile butadiene styrene (ABS) to BS 5391, BS 5392 and BS EN ISO 15493	$\frac{3}{8}$	16	0.8	1.2
	$\frac{1}{2}$	20	0.9	1.35
	$\frac{3}{4}$	25	1.0	1.5
	1	32	1.1	1.65
	$1\frac{1}{4}$	40	1.2	1.8
	$1\frac{1}{2}$	50	1.3	1.95
	2	63	1.4	2.1
	3	90	1.6	2.4
	4	110	1.9	2.85
	6	160	2.3	3.45

11.5   Iron pipes

These are made in three types: vertically cast, spun iron and ductile iron. However, the production of vertically cast and spun iron pipes has now virtually ceased in favour of pipes in ductile iron to BS EN 545 which have much improved mechanical properties.

Since all iron pipes are liable to corrosion they are factory treated inside and out. Also, many water authorities may require pipes to be sheathed in a blue polyethylene sleeve to BS 6076 for further protection from aggressive soils, and for identification.

Ductile iron pipes are available in sizes DN 40 to DN 2000 and in three thickness classes. These are suitable for a range of working pressures depending on the size and pressure rating. For example, Class 40 pipe in the range DN 40 to DN 150 is suitable for working pressures of up to 64 kPa (6.4 bar).

Fixings for cast iron pipes

See figure 11.36 and table 11.17 for examples of fixings and spacings for cast iron pipes.

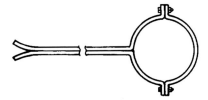

(a) Holderbat build-in type in mild steel

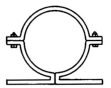

(b) Holderbat screw-to-wall type in mild steel

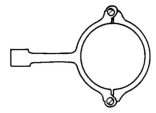

(c) Holderbat hinged build-in type in cast iron

Figure 11.36 Brackets for cast iron pipe

Table 11.17 Maximum spacing of fixings for above ground cast iron pipes

Type of piping	Nominal size of pipe mm	Spacing on horizontal run m	Spacing on vertical run m
Vertically cast or spun iron complying with BS 1211 or BS 2035	51	1.8	1.8
	76	2.7	2.7
	102	2.7	2.7
	152	3.6	3.6
Ductile iron complying with BS EN 545, BS EN 598	80	2.7	2.7
	100	2.7	2.7
	150	2.7	3.6

Jointing methods for cast iron pipes

See figures 11.37–11.42.

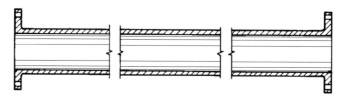

Available in sizes 75 mm to 300 mm.

Figure 11.37   Flanged spun iron and cast iron pipes

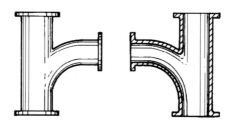

Figure 11.38   Flanged junction

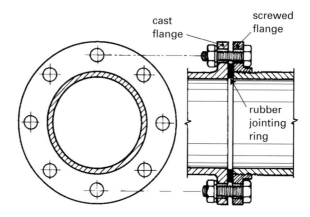

Figure 11.39   Flanged joint detail

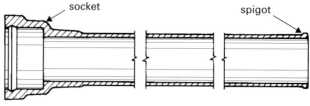

Plain joint for lead.

Figure 11.40   Cast iron pipe with plain socketed joint

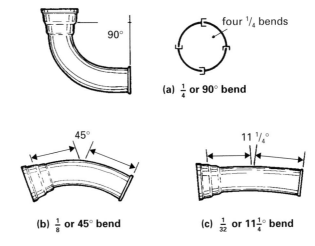

(a) $\frac{1}{4}$ or 90° bend

(b) $\frac{1}{8}$ or 45° bend

(c) $\frac{1}{32}$ or $11\frac{1}{4}$° bend

Figure 11.41   Plain socketed pipe bends

Caulking tool chosen to match pipe size and width of joint.

Caulk when cold to finish about 3 mm inside socket face.

Use synthetic yarn that will not promote the growth of bacteria. Yarn should be caulked tightly to approximately one-third depth of the joint, to prevent direct contact between lead and water, and to centralize pipe in socket.

No longer permitted for use on potable water installations.

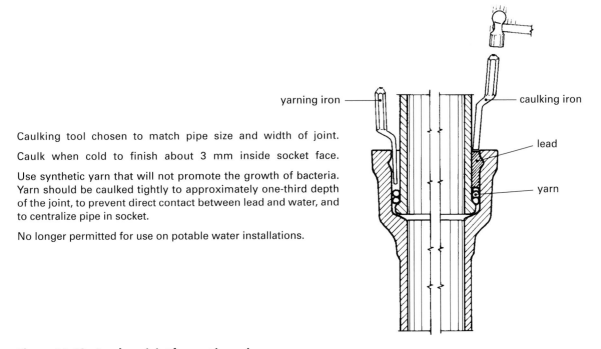

Figure 11.42   Lead run joint for cast iron pipes

Joints for ductile iron pipe will include flanged joints (mainly for above ground use) or more commonly push-fit 'O' ring type mechanical joints for use below ground (see figures 11.43 and 11.44), giving good flexibility of movement without loss of joint seal. Other joints used for cast iron pipes can be seen in figures 11.45 and 11.46.

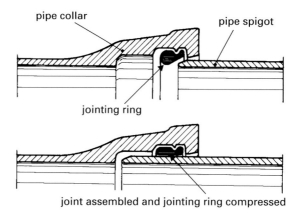

Available in sizes 80 mm to 600 mm.

Suitable for spun iron and ductile iron pipes.

Figure 11.43   'Tyton' slip-fit joint

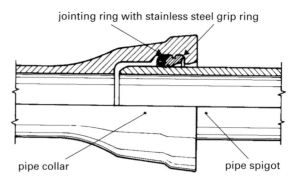

Stainless steel grip ring makes joint resistant to pulling apart.

Figure 11.44   Slip-fit joint with self-anchoring gasket

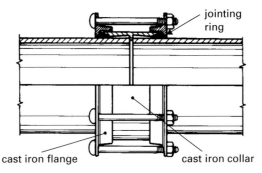

jointing ring

cast iron flange   cast iron collar

Also suitable for PVC-U, steel and asbestos cement pipes.

(a) Straight coupler

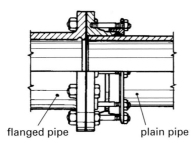

flanged pipe   plain pipe

Used for jointing plain ended pipes to flanged pipes or fittings.

Each adaptor consists of flanged sleeve, end flange, wedge rubber packing ring, bolts, studs and nuts.

(b) Flange adaptor

Figure 11.45   Viking Johnson joints

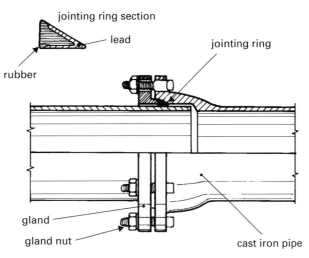

jointing ring section

lead

jointing ring

rubber

gland

gland nut

cast iron pipe

More commonly used with gas.

Figure 11.46   Bolted gland joint

11.6 Asbestos cement pipes

Asbestos cement pipes, in common with many asbestos-based materials, are subject to the requirements of the Asbestos at Work Regulations. These regulations require that any dust liberated is restricted to a low level of concentration. Asbestos cement pipes contain only a small percentage of asbestos and are considered safe to handle. Where a limited amount of cutting and turning by hand tools is done in the open air, the dust level is generally below the minimum set in the regulations. However, if any doubt exists, clarification must be sought from the local office of HM Factory Inspectorate, or from the manufacturer of the material.

Asbestos cement pipe is corrosion resistant in most soils but since it is very brittle it is liable to break where soils, such as clays, move because of seasonal loss or gain of moisture.

The use of asbestos cement pipes is restricted to installations below ground. Pipe is available from 50 mm to 900 mm diameter. See table 11.18 and figures 11.47 and 11.48.

Asbestos cement pipes, formerly to BS 417, now come within the scope of BS EN 512-19, which also covers cement fibre pipes that contain no asbestos material.

Table 11.18　Class and pressure range of asbestos cement pipes

Class	Metric colour coding (pipes, joints and rings)	Works test pressure bar	Maximum working pressure bar	Test pressure
15	Green	15	7.5	One-and-a-half to
20	Blue	20	10	twice the expected
25	Violet	25	12.5	working pressure

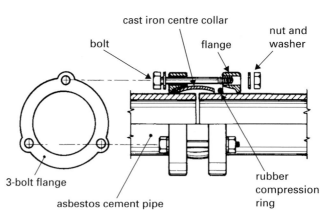

This joint needs careful protection from corrosion otherwise the advantage of the non-corrosive properties of asbestos cement pipes is lost.

Made also for cast iron pipes.

Figure 11.47　Detachable joint made of cast iron

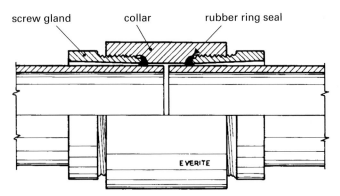

screw gland   collar   rubber ring seal

EVERITE

Gives complete immunity from corrosion.

Figure 11.48   All-asbestos cement screwed gland joint

11.7   Lead pipes

No lead pipe or other water fitting containing lead is permitted to be used in water systems containing wholesome water, even for repair. Where new pipes, e.g. copper, are to be connected to an existing lead pipe, protection against electrolytic (galvanic) action is required. If possible lead pipes should be removed.

Lead solders are not permitted for use on pipelines for wholesome water.

11.8   Connections between pipes of different materials

As far as possible connections between different pipe materials should be avoided, especially when jointing dissimilar metals, as this may lead to electrolytic corrosion. However, when dissimilar pipes have to be used, for example, in the repair or renewal of part systems or connections to existing pipelines, they may be connected by one of the following three methods:

(1)   using 'inert' material between the dissimilar metals (see figure 11.49);
(2)   ensuring that water flows towards a potentially stronger material in the electrochemical series from a weaker one:

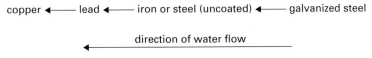

copper ◄—— lead ◄—— iron or steel (uncoated) ◄—— galvanized steel

◄————————— direction of water flow —————————

Water flowing in the opposite direction will carry dissolved particles of potentially stronger material which will adversely affect the weaker pipeline;
(3)   using a sacrificial anode of a potentially weak material (see figure 11.50).

Adaptor couplings are available for the joining of a wide range of differing materials. A number of these have been shown in previous illustrations.

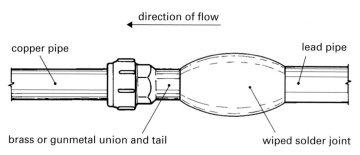

direction of flow

copper pipe

lead pipe

brass or gunmetal union and tail

wiped solder joint

Brass or gunmetal tail used to prevent direct contact between copper pipe and lead pipe. Direct contact is not permitted.

The wiped soldered joint is the traditional means of connecting to existing lead pipes.

An alternative jointing method is to use a 'patent' compression joint.

Figure 11.49   Use of inert material between connections of dissimilar metals

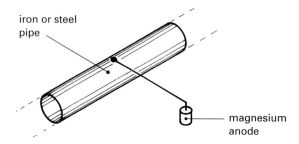

iron or steel pipe

magnesium anode

Anode will corrode and leave the pipe intact.

Sacrificial anodes can be fitted to cisterns, tanks and cylinders.

Figure 11.50   Sacrificial anode used on a pipeline

11.9   Connections to cisterns and tanks

The following general instructions should be followed.

(1)   Provide proper support for cisterns and tanks to avoid undue stress on connections and deformation of cistern or tank walls.
(2)   Use proper tools for hole cutting (see figure 11.51) not flame cutters, not hammer and chisel.
(3)   Holes must be truly circular with clean edges.
(4)   All debris or filings must be removed from inside the tank or cistern.

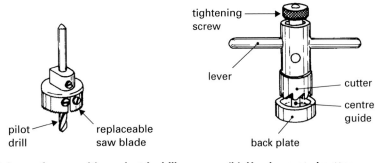

(a) **Hole saw for use with an electric drill**   (b) **Hand-operated cutter**

Figure 11.51   Tools for cutting holes

Figure 11.52 shows examples of cistern connections. Additional considerations for thermoplastics cisterns are as follows:

(1)   Scribing tools should *not* be used to mark position of holes.
(2)   Cistern wall should be supported with wooden strut or similar during cutting.
(3)   Pipes should be carefully fitted and supported to avoid distortion of cistern or tank.
(4)   Corrosion-resisting support washers should be used inside and outside the cistern to strengthen the joint area.
(5)   Float valves should be fitted through a supporting back plate to stabilize the cistern wall against the thrust of the lever arm.
(6)   Linseed oil based sealants must not be used with plastic cisterns or pipes. Where a jointing sealant is used, it should conform to BS 6356-5.

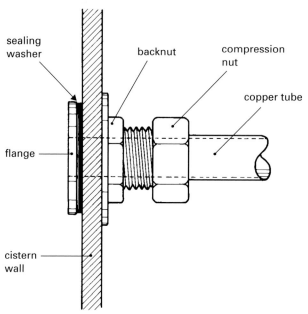

Example shows type A compression joint to copper tube.

(a) Copper connection to cistern

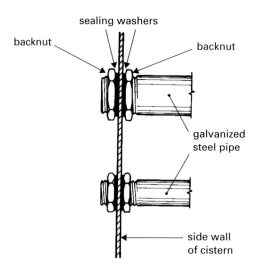

Copper pipe should not be connected to galvanized steel cisterns or tanks.

Pipe sealed with proprietary washers.

(b) Galvanized steel connections to steel cistern

Figure 11.52 Examples of cistern connections

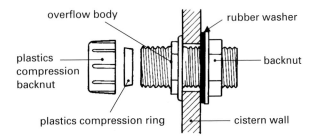

(c) Plastics overflow connection to WC cistern

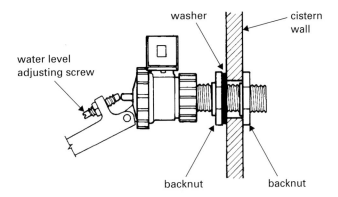

(d) Plastics float valve connection to cistern

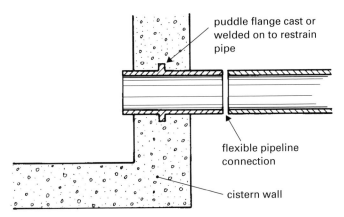

Puddle flange properly aligned before casting into concrete.

Concrete to be compacted to ensure watertight joint.

(e) Connection to concrete tank or cistern

Figure 11.52   continued

11.10   Branch connections for buildings

Service connections to mains are normally made by drilling and tapping the top of the main and screwing in a union ferrule (see figures 11.53 and 11.54).

The ferrule is set on top of the main to avoid drawing sediment into the service pipe.

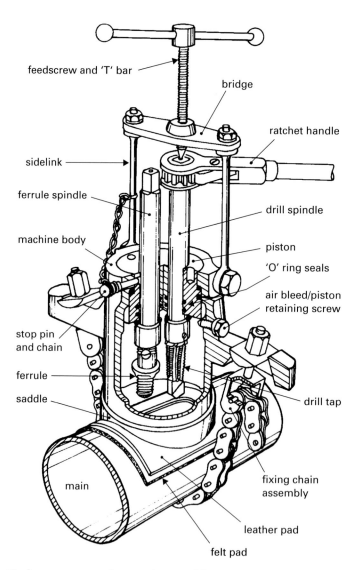

feedscrew and 'T' bar

bridge

ratchet handle

sidelink

ferrule spindle

drill spindle

machine body

piston

'O' ring seals

air bleed/piston retaining screw

stop pin and chain

ferrule

saddle

drill tap

main

fixing chain assembly

leather pad

felt pad

Figure 11.53   Under pressure mains tapping machine

Ferrule screwed directly into top of cast iron water main.

Branch pipe to run parallel to main before taking its final route.

Gooseneck to permit movement in soil without damage to pipe.

Gooseneck bent to ensure that ferrule will not loosen if the service pipe should settle.

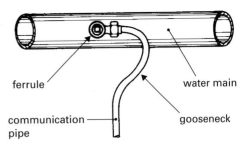

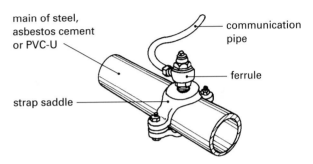

(a) Connection of ferrule to main

Mains pipes of asbestos cement, steel and PVC-U are strengthened at the point of connection by the use of a strap saddle.

Strap saddle of cast iron for asbestos cement and steel mains, and of gunmetal for PVC-U mains.

(b) Use of strap saddle for mains connections

Self-tapping ferrule is shown here.

Strap saddles of PVC-U are available for solvent welding or with mechanical connection to the main.

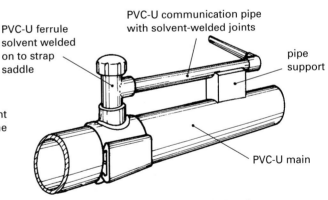

(c) PVC-U connection to PVC-U main

Figure 11.54   Mains connection

Where rigid pipes are used for the branch service pipe connect them using a short length of suitable flexible pipe.

Depending on the size of the service pipe and the main, ferrules may not always be suitable. For larger connections or smaller mains a more suitable method would be the use of a leadless collar or a tee connection (see figures 11.55 and 11.56, and table 11.19).

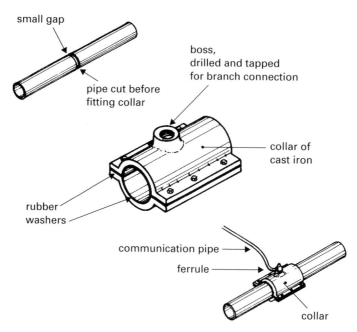

Figure 11.55   Leadless collar

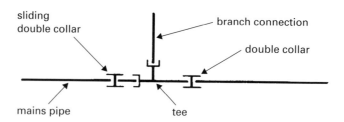

Figure 11.56   Tee connection

Table 11.19   Method of branch pipe connection

Nominal size of branch pipes		Nominal diameter of main pipe				
mm	inches	80 mm	100 mm	150 mm	200 mm	250 mm and over
15	$\frac{1}{2}$	F	F	F	F	F
22	$\frac{3}{4}$	T	F	F	F	F
25	1	T	T	F	F	F
35	$1\frac{1}{4}$	T	T	T	F	F
42	$1\frac{1}{2}$	T	T	T	F	F
54	2	T	T	T	T	F

F: Ferrule; T: Tee or leadless collar.

11.11   Contamination of mains

When cutting into mains to make branch connections, precautions should be taken to avoid contamination.

(1)   Sterilize the trench around the branch connection before cutting.
(2)   Take care to avoid the entry of soil or water from the trench.
(3)   Insert sterilizing tablets into the pipe when making the connection.

11.12   Laying underground pipes

Qualifications and standards

Installations should conform to National or European Regulations and Standards as applicable and particularly to the requirements of the Water Fittings Regulations. Account should also be taken of any specific instruction of the manufacturer of pipeline components.

Installation work should be carried out by competent operatives and employers should ensure that:

• contractors and work operatives are qualified to carry out the work; and
• supervisors are suitably qualified to assess the quality of the work.

Transportation and storage of pipeline components

Pipeline components should be protected against damage during transportation and in storage. Equipment used for the loading, unloading and transportation of materials should be suitable for the purpose. Care should be taken to ensure that materials and components do not come into contact with hazardous substances, such as earth, mud and sewage. Where contamination has occurred or is unavoidable, the pipeline components should be cleaned and sterilized before they are installed. Manufacturers' guidance on materials handling, and the avoidance of damage, degradation and contamination, should be followed.

Health and safety

The requirements of the Health and Safety at Work Act 1974 and relevant safety regulations should be strictly observed. Operatives should be aware of any health or safety hazards, and relevant health and safety regulations should be observed (e.g. wearing protective clothing and using safe working methods).

Sites should be equipped with appropriate alarm devices and other emergency equipment for use in an accident. These should be regularly checked and maintained in good order, and defective equipment removed from the site and replaced.

Site operations, including movement of materials, must not cause any undue hazard to operatives or to other persons or property in and around the site (e.g. barriers, lighting, etc. should be used). Trench excavations should be supported from collapse as appropriate to soil conditions. Access ladders should be provided where necessary and secured in position when in use. Construction operations should not cause damage to, or undermine, existing structures.

When working with asbestos cement components precautions must be taken when cutting, machining or carrying out other operations likely to create dust.

Pipe trenches

Care should be taken to ensure that excavations do not affect the stability of other installations.

Pipe trenches should be large enough to provide a working space for operatives as well as for the pipeline and its surrounding material. The trench should provide a frost-free depth of cover over the pipe and, where required, space below the pipe for a suitable bedding material. The trench should be wide enough to allow pipes to be properly bedded, jointed and backfilled. The depth, gradient, width and condition of the trench bottom should be checked before the pipe is laid.

The trench bottom should be carefully prepared to give a firm even surface that will give adequate support throughout the whole length of each pipe (see figure 11.57). Mud, rock projections, boulders, hard spots and local soft spots should be removed and replaced with selected fill material consolidated to the required level. Where rock is encountered, the trench should be cut at least 150 mm below pipe level and made up with well rammed bedding material.

Where pipes are to be laid in loose or non-load bearing soil, or where the soil contains flints, stones or rock, the trench should be deep enough to insert a selected bedding material. The bedding should be levelled and compacted to provide adequate support under and around the full length of each pipe. Where soil conditions are suitable, an additional bedding material will not be needed.

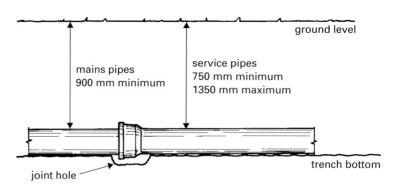

Pipeline to follow contours of ground level whilst ensuring adequate depth of cover.

Pipe to lie flat on trench bottom if soil conditions permit.

Additional excavation needed to leave space for collar.

Figure 11.57   Support of pipes in trench

Installation of pipes and components

As far as possible, pipes should be laid in straight lines to permit easy location later. However, copper or polyethylene pipes should be snaked within the trench to allow for settlement and moisture movement in the soil (see figure 11.58).

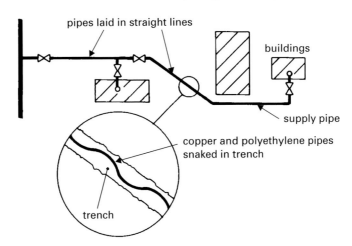

Large diameter PVC-U pipes must not be bent within their length. However, some deflection is permissible at flexible joints.

Figure 11.58   Location of pipelines

Pipes should be laid on a firm even base, evenly supported throughout their length, and must not rest on their sockets, bricks or other makeshift supports. Plastics pipes should be laid on a bed free from sharp stones.

To guard against contamination, pipelines and joints should be kept clean internally whilst being laid. After laying, or when work is interrupted (or completed), all openings in the pipeline should be sealed and temporarily plugged until jointing takes place. Precautions must be taken to prevent the plugged pipes floating if the trench becomes flooded.

Measures should be taken to make sure that pipes are not obstructed by, or do not obstruct other services or structures. Where possible, pipelines should be laid at least 400 mm horizontally from foundations or other similar structures. Where water pipes run parallel to other pipelines or cables, the horizontal distance between them should be at least 400 mm, but where there is congestion, a minimum distance of 200 mm should be maintained. In all cases direct contact between services should be avoided.

Where cables and pipelines cross each other, a clearance of at least 200 mm should be maintained, but if this is not possible, measures should be taken to prevent direct contact (see figure 11.59).

Joints below ground

Joints below ground (and in other inaccessible places) should be kept to a minimum and avoided where possible. It is preferable that smaller diameter pipes of copper and polyethylene are laid in one length.

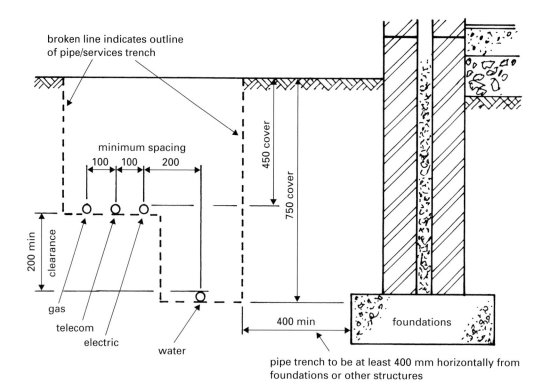

broken line indicates outline
of pipe/services trench

minimum spacing
100   100   200

450 cover

750 cover

200 min
clearance

gas

telecom

electric

water

400 min

foundations

pipe trench to be at least 400 mm horizontally from
foundations or other structures

Space requirements around pipe provide access for repair and prevent
damage to other pipelines when excavations are carried out.

Figure 11.59   Space requirements between services and structures

Pipelines with unrestrained joints should be securely anchored at blank ends, tees, bends, tapers and valves to resist thrust movement due to internal pressure. Anchors and thrust blocks should be constructed to withstand any forces resulting from internal pressure, including site testing. The size and weight of anchors and thrust blocks will depend on the bearing capacity of the soil. Restrained joints should be installed in accordance with the manufacturer's instructions.

Welded joints (e.g. on steel or polyethylene pipes) should be carried out by suitably trained and qualified operatives and the work should conform with national or European standards, using welding equipment and methods approved by the pipe and fitting manufacturer.

Any jointing lubricants that might come into contact with potable water should be approved to show they comply with the Water Fittings Regulations and relevant British or European Standards and are not a contamination risk.

Protection against corrosion

Pipes and pipelines should be resistant to, or protected against, any corrosive effects of the soil. Examples of pipe protection may include the use of plastics sleeving, bitumen coatings, protective tape or anti-corrosive blankets. Where pipes have plastics coatings or loose plastics sleeving, care should be taken to avoid damage by large sharp-edged stones, shale, flints or other harmful substances. Any damage to coatings during installation should be repaired before backfilling the trench.

Repairs and additions to the pipe coatings, at faults or at pipe joints, should follow the pipe manufacturer's instructions.

Embedment and backfilling of trenches

When backfilling trenches, the pipes must be surrounded by a suitable bedding and surround material, carefully applied so that any backfill load will be distributed evenly along and around the pipe and so that the pipe is not unduly deflected. The bedding and embedment material may be native soil (if suitable) or selected imported material such as pea shingle. Embedment, properly done, is essential for the proper support of the pipe or pipeline (see figures 11.60 and 11.61).

The designer should advise the operative on how much compaction is needed after referring to the structural design calculations. Compaction will vary with different pipe materials depending on the nature of the soil, and the size and quality of the embedment material.

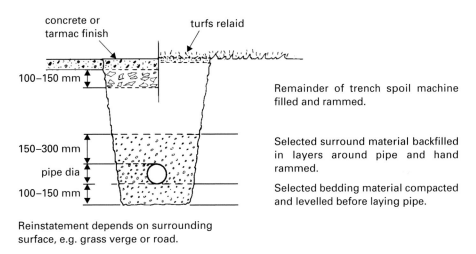

Figure 11.60 Trench backfilling

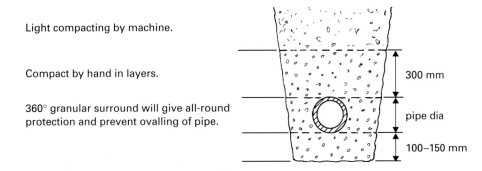

Figure 11.61 Backfilling trenches for PVC-U and PE pipes

Particular attention should be given to compaction under the haunches of the pipe and any voids left by the removal of any temporary trench wall supports, which should be filled.

Material for embedment (e.g. native soil, imported soil, pea shingle, etc.) should have the following properties:

- It should be capable of being compacted to the required density to support the pipeline in the correct position both during and after laying, and enable the installed pipe to accommodate internal and external loads.
- It should not be corrosive or otherwise react with the pipe, components or protective coatings. Nor should it react adversely with the soil or groundwater.
- It should not include debris, organic materials, frozen soil, large stones, rocks, tree roots and similar large objects.

Where used in fine-grained soil, such as clay, silt or sand, and where the embedment is partially or totally below the water table, bedding material used should prevent the soil from mixing with the embedment material. If mixing is likely to occur, a filter fabric should be used.

The main backfill and final surface of the trench should be completed in accordance with the designer's specification for the reinstatement of the trench.

Where specified, tracer tapes for detection, warning and identification purposes should be installed, normally just above the pipe.

Pipes of PVC-U need extra care when being backfilled; otherwise they might become distorted and weakened. With these pipes, granular bedding and surrounds are essential (see figure 11.61).

Restraint of pipes

With most methods of pipe laying and trench backfill for large diameter pipes, joints at changes in direction are liable to move and push apart due to internal thrust pressure (see figure 11.62). To guard against these risks, which will vary according to the internal pressure and how much the pipe direction changes, pipes must be securely anchored at bends, branches and pipe ends. The amount of anchorage required will also depend upon the soil and its bearing capacity. See also tables 11.20–11.22.

Calculation of the thrusts which act in the direction of arrows shown in figure 11.62 may be calculated using the formulae below:

- End thrust (kN) in capped ends and branches $= 100\ AP$
- Radial thrust at bends (kN) $\qquad = 100\ AP \times 2 \sin \theta/2$

where
  A is the cross sectional area of the socket (m²)
  P is the test pressure (bar)
  θ is the angle of deviation of the bend.

Alternatively, when standard fittings are used, the thrusts may be calculated by multiplying the values given in table 11.20 by the test pressure in bars.

Thrust blocks used to restrain pipelines must have sufficient bearing area to resist the thrust under test pressure. This can be calculated by using data for soil-bearing capacities given in table 11.22.

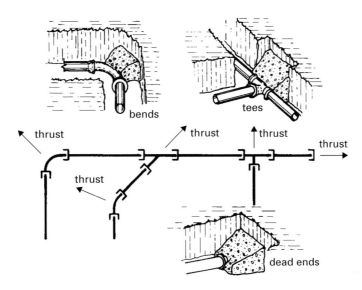

Figure 11.62   Horizontal thrust on buried mains

Table 11.20   Thrust per bar internal pressure

Nominal internal diameter of pipe	End thrust	Radial thrust on bends of angle			
		90°	45°	$22\frac{1}{2}°$	$11\frac{1}{4}°$
mm	kN	kN			
50	0.38	0.53	0.29	0.15	0.07
75	0.72	1.02	0.55	0.28	0.15
100	1.17	1.66	0.90	0.46	0.24
125	1.76	2.49	1.35	0.69	0.35
150	2.47	3.50	1.89	0.96	0.49
175	3.29	4.66	2.52	1.29	0.65
200	4.24	5.99	3.24	1.66	0.84
225	5.27	7.46	4.04	2.06	1.04
250	6.43	9.09	4.92	2.51	1.26
300	9.38	13.26	7.18	3.66	1.84
350	12.53	17.71	9.59	4.89	2.46

Table 11.21   Gradient thrusts on buried or exposed mains

Gradient	Spacing of anchor block		
	m	ft	
1 in 2	5.5	18	
1 in 3	11.0	36	
1 in 4	11.0	36	
1 in 5	16.5	54	gradient 1 in 6
1 in 6	22.0	72	or steeper

Table 11.22   Bearing capacity of soils

Soil type	Safe bearing load kN/m^2
Soft clay	24
Sand	48
Sandstone and gravel	72
Sand and gravel bonded with clay	96
Shale	240

Valve chambers and surface boxes

Surface boxes must be provided to allow access to valves and hydrants, and must be supported on concrete or brickwork which, after making an allowance for settlement, must not rest on the pipes and transmit loads to them. See figure 11.63.

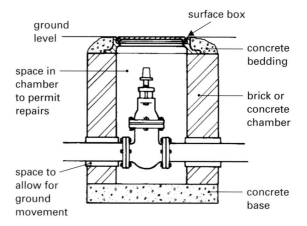

Figure 11.63   Valve chamber

Brick or concrete hydrant chambers (see figure 11.64) must be large enough to permit repairs to be carried out to the fittings.

An alternative is to provide vertical guard pipes or precast concrete sections to enclose the spindles of valves as shown in figure 11.65.

All valves and hydrants should be positioned where they can easily be found and used.

Surface boxes must be positioned and marked to indicate the pipe service, the size of mains, and position and depth below the surface (see figure 11.66).

Surface boxes should be of sufficient strength to withstand likely traffic loads, i.e.:

BS 750   Class A   for use in carriageways where loads may be heavy;
           Class B   for use where vehicles would have only occasional access such as footpaths and verges.

Indicator plates, illustrated in figure 11.67, can be screwed to walls or marker posts, but must be visible.

Drawings should show all pipe runs, valves and hydrants. Working drawings should be amended to show variations from the original design.

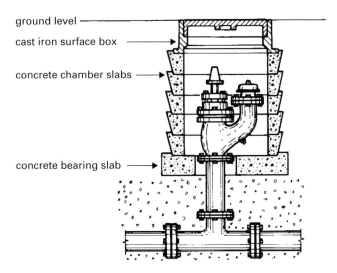

Concrete chamber slabs provide quick and easy method of building chamber plus easy removal for access for repairs.

Hydrant installations to comply with BS 3251.

Internal dimensions 430 mm × 280 mm.

Figure 11.64   Typical hydrant arrangement

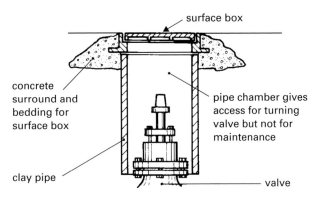

Commonly used because it is cheap, but valve must be dug out for repairs.

Figure 11.65   Valve access

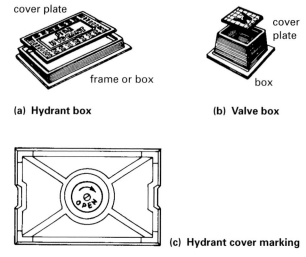

(a) **Hydrant box**   (b) **Valve box**

(c) **Hydrant cover marking**

Cover plates marked to identify type of valve enclosed in chamber.

Hydrant cover must be marked underneath to indicate direction of valve turning.

Figure 11.66   Surface boxes for valves and hydrants

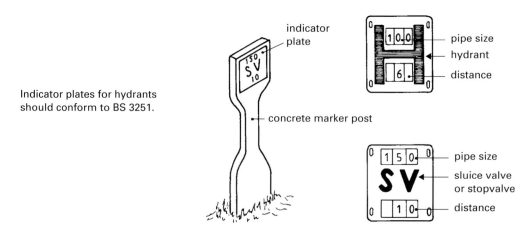

Indicator plates for hydrants should conform to BS 3251.

Figure 11.67   Valve and hydrant marker plates

11.13   Pipework in buildings

Fixings and allowance for thermal movement

Allowance must be made for expansion and contraction of pipes (see figure 11.68), especially where pipe runs are long and where temperature changes are considerable (hot distributing pipes) and where the pipe material has a relatively high coefficient of thermal expansion, e.g. plastics.

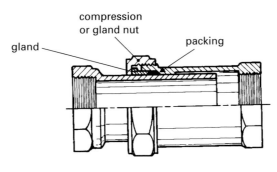
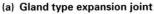

(a) Gland type expansion joint

(b) Expansion loop

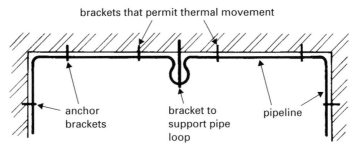

(c) Use of horseshoe expansion loop

Figure 11.68   Expansion joints

Fixing insulated piping

Sufficient space should be allowed behind pipes for insulation to be properly installed, as shown in figure 11.69. Pipelines can lose heat by conduction through metal brackets. Where cold pipes are below ambient temperatures this may lead to increased risk of freezing or condensation. In hot pipelines it can significantly add to heat energy losses. To guard against this, it is preferable that metal brackets are not fixed directly to the pipe; instead they can be fixed to the outside of load-bearing insulation or hardwood support inserts as shown in figure 11.69(c).

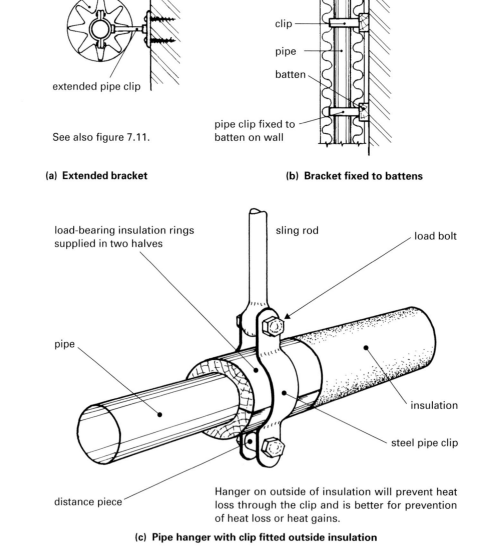

(a) Extended bracket

(b) Bracket fixed to battens

(c) Pipe hanger with clip fitted outside insulation

Figure 11.69   Fixings for insulated piping

Concealed pipes

These must be housed in properly constructed ducts or chases that provide access for maintenance and inspection. See chapter 10.

Pipes passing through structural timbers

Structural timbers must not be weakened by indiscriminate notching and boring. Notches and holes should be as small as practicable. Whenever possible, notches should be U-shaped and formed by parallel cuts to previously bored holes (see figure 11.70). The positions of notches and holes are as important as their sizes, and in timber beams and joists should only be made within the hatched areas shown in figure 11.71.

Before making notches or holes, agreement should be reached on their size and position with the architect, structural engineer or building supervisor. For further information see BS 5268.

A visual inspection should be carried out before running pipework below suspended floors. The position of any electrical cables, junction boxes and ancillary equipment should be noted so as to avoid accidental damage or injury when inserting pipework.

Care should be taken when re-fixing flooring to prevent damage to pipes or other services by nails or screws. Where possible, the flooring should be appropriately marked to indicate for others the position of pipes.

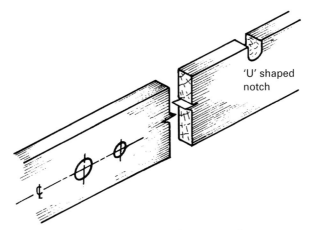

'U' shaped notch

Holes and notches to be as small as practicable but large enough to permit pipes to expand and contract.

To avoid weakening the joist, holes and notches should be restricted to the hatched areas shown in figure 11.71.

Figure 11.70   Pipes passing through structural timbers

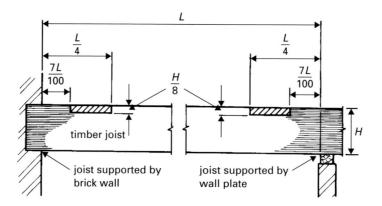

(a) Notches

Note Notches must be restricted to the hatched areas shown if the joists are not to be weakened.

L is the length of the structural member
H is the height of the structural member
If H exceeds 250 mm then for calculation purposes it is deemed to be 250 mm

Example For a joist 6 m long and 300 mm deep, notches must be:
(a) not more than $H/8$ deep, i.e.:

$$\frac{250}{8} = 31 \text{ mm}$$

(b) (i) at least 7 $L/100$ from bearing, i.e.:

$$\frac{7 \times 6000}{100} = \frac{42000}{100} = 420 \text{ mm}$$

(ii) not more than $L/4$ from bearing, i.e.:

$$\frac{6000}{4} = 1500 \text{ mm} = 1.5 \text{ m}$$

Figure 11.71 Dimensions for notches and holes

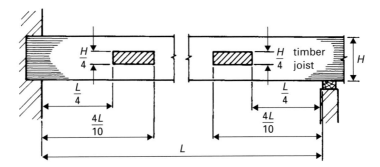

(b) Holes

Note   Holes must be restricted to the hatched areas shown if the joists are not to be weakened.

L is the length of the structural member

H is the height of the structural member

If *H* exceeds 250 mm then use *H* = 250 mm for calculation purposes.

Example   For a joist 4 m long and 200 mm deep, holes must be:

(a)   not more than *H*/4 in diameter, i.e.

$$\frac{200}{4} = 50 \text{ mm}$$

(b)   (i)   at least *L*/4 from bearing, i.e.

$$\frac{4000}{4} = 1000 \text{ mm} = 1 \text{ m}$$

     (ii)   not more than 4*L*/10 from bearing, i.e.

$$\frac{4 \times 4000}{10} = 1600 = 1.6 \text{ m}$$

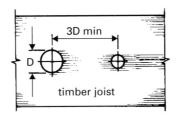

(c) Spacing of holes

The distance between adjacent holes in joists should be at least three times the diameter *D* of the larger hole.

Holes should pass through the centre of the joist and be parallel.

Figure 11.71   continued

Penetration of fire walls and floors

As required by the current Building Regulations, penetration of compartment walls and floors and fire barriers must be fire-stopped to prevent the passage of smoke and flame.

Only pipes of non-combustible materials such as cast iron, steel or copper are permitted to pass through compartment walls, a non-combustible material being one that will not soften or melt when exposed to a temperature of 800°C. Additionally, these pipes must be no greater in diameter than 160 mm.

Where pipes of PVC or other plastics are taken through compartment walls they must pass through a sleeve of non-combustible material. For more information on fire protection, see chapter 13.

11.14   Electrical earthing and bonding

Water pipes should not be used as an electrode for earthing purposes, but all metal pipes entering buildings should be bonded to the electrical installation as near as possible to the point of entry to the building.

Where pipes, fitting or appliances are to be replaced, the earth continuity and equipotential bonding should be maintained. Meter installations, for example, that are designed to permit replacement should have a suitable conductor permanently fitted between inlet and outlet pipework. where no permanent conductor is in place a temporary conductor should be fitted for the duration of the replacement work. Any electrical earthing or bonding installation that has been disturbed should be properly tested to ensure continuity is maintained.

Electrical installations, including earthing and bonding arrangements, should be carried out by a person who is competent to install, inspect and test the work, and qualified to complete an electrical installation certificate. Electrical installations should comply with Building Regulations (Part P) and conform to the recommendations of BS 7671 *Requirements for electrical installations*. Earthing is covered by BS 7430.

Supplementary equipotential bonding may be required in special locations, such as bathrooms.

11.15   Jointing of pipework for potable water

Research and the testing of water samples from mains and services have shown that some traditional jointing materials harbour or promote the growth of bacteria. As a result, some of these materials are now banned and others discouraged.

For example, bacterial contamination may be caused by linseed oil-based compounds commonly used with screwed joints. New compounds now available do not promote bacterial growth.

Again, if soldered joints are badly made, lead can be leached into solution and consumed. So, the Water Regulations now permit only lead-free solder in capillary joints for hot and cold water supplies.

However, there are problems when choosing jointing materials because for some applications no suitable alternatives are available for the traditional materials.

Table 11.23, adapted from BS 6700, lists permitted jointing materials and gives guidance on their use.

Table 11.23   Jointing of potable water pipework

This table lists jointing methods and materials that should be used for jointing potable water pipework. Materials and products that have been assessed under the Water Regulations Advisory Scheme and listed by them are considered to meet the requirements of this table.

Type of joint	Method of connection	Jointing material	Precautions/limitations
Lead to brass, gunmetal, or copper, pipe or fitting	Plumber's wiped soldered joint	Tallow Flux	Of very limited application See note
Copper to copper pipe (pipe to pipe or fitting, or fitting to fitting)	Capillary ring or end feed, soldered joint	Flux	No solder containing lead to be used See note
Copper to copper (pipe to fitting or fitting to fitting)	Non-manipulative compression fittings	—	Above ground only
	Manipulative compression fittings	Lubricant on pipe end when required	See note
Copper to copper (pipe to fitting or fitting to fitting)	Bronze welding or hard solder	Flux	See note
Galvanized steel (pipe to pipe or fitting), including copper alloy fittings	Screwed joint, where seal is made on the threads	PTFE tape or proprietary sealants	PTFE tape only up to 40 mm ($1\frac{1}{2}$ in diameter) See note
Galvanized or copper (pipe to pipe or fitting)	Flanges	Elastomeric joint rings complying with BS EN 681, or corrugated metal. Vulcanized fibre rings complying with BS 5292 or BS 6091: Parts 1 and 2	See note
Long screw connector	Screwed pipework with BS 2779 thread	Grummet made of proprietary paste and hemp	Must not promote growth of bacteria
Shouldered screw connector	Seal made on shoulder with BS 2779 thread	Elastomeric joint rings complying with BS EN 681 and plastics materials	—
Unplasticized PVC (pipe to fitting)	Solvent welded in sockets	Solvent cement complying with BS EN 3452	Follow manufacturers' recommendations
	Spigot and socket with ring seal. Flanges. Union connectors	Elastomeric seal complying with BS EN 681. Lubricants	Lubricant should be compatible with the unplasticized PVC and elastomeric seal
Cast iron (pipe to fitting)	Caulked lead	Sterilized gaskin yarn/blue lead	See note
	Bolted or screwed gland joints	Elastomeric ring complying with BS EN 681	—
	Spigot and socket with ring seal	Elastomeric seal and lubricant	—

Table 11.23   continued

Type of joint	Method of connection	Jointing material	Precautions/limitations
Copper or plastic (pipe to tap or float-operated valve)	Union connector	Elastomeric or fibre washer	—
Stainless steel (pipe to fitting, including copper alloy fittings)	Non-manipulative compression fittings	Lubricant when required	See note
	Manipulative fittings	Elastomeric seals when required	—
Pipework connections to storage cisterns (galvanized steel, reinforced plastics, polypropylene, polyethylene)	Tank connector/union with flanger backnut	Washers: elastomeric, polyethylene, fibre	—
Polyethylene (pipe to fitting)	Non-manipulative fittings	—	Do not use lubricant
	Thermal fusion fittings	—	Follow manufacturers' directions
Polybutylene (pipe to pipe or fitting)	Non-manipulative fittings	Lubricant on pipe end when required	Lubricant if used should be listed and compatible with plastics
	Thermal fusion fittings	—	Follow manufacturers' directions
Polypropylene (pipe to pipe)	Non-manipulative fittings	Lubricant on pipe end when required	Lubricant if used should be listed and compatible with plastics
	Thermal fusion fittings	—	Follow manufacturers' directions
Cross-linked polyethylene (pipe to fitting)	Non-manipulative fittings	Lubricant on pipe end when required	Lubricant if used should be listed and compatible with plastics
Chlorinated PVC (pipe to fitting)	Solvent welded in sockets	Solvent cement to BS EN 1452	Follow manufacturers' recommendations

Note. Where non-listed materials are to be used, due to there being no alternative, the procedure used should be consistent with the manufacturer's instructions taking particular note of the following precautions:

(1) use least quantity of material to produce good quality joints;
(2) keep jointing materials clean and free from contamination;
(3) remove cutting oils and protective coatings, and clean surfaces;
(4) prevent entry of surplus materials to waterways;
(5) remove excess materials on completion of the joint.

11.16   Testing

Smaller services should be tested by filling up the system at normal working pressure and inspecting all joints and fittings for leakage.

Any pipes below ground, buried under screeds or in other inaccessible places should be tested before being covered. Hot water pipes should be similarly checked after heat has been applied. See chapter 12 for further information on testing.

11.17   Identification of valves and pipes

Below ground

The position of underground pipes and the location of valves should be recorded on a plan of the premises. Valve surface boxes should be marked to indicate the service below them. On mains and larger service pipes, indicator plates should be set up to show the size and position of valves and hydrants. These are illustrated in figures 11.66 and 11.67.

Water pipes below ground should be coloured blue to distinguish them from other services.

Above ground

Valves on hot and cold pipes should be fitted with an identification label of non-corrodible and non-combustible material, as shown in figure 11.72. The label should describe the size and function of the valve.

Alternatively, the label may be marked with a reference number which relates to a durable diagram of the water system showing valve reference numbers, and fixed in a prominent position. See figure 11.72.

Pipes within buildings should be colour banded as shown in figure 11.73, to identify the pipe and the service for which it is used.

Pipes supplying water solely for firefighting purposes should be clearly marked to distinguish them from each other and from other pipes in the building.

Any pipe or fitting used to supply rainwater, recycled water or any fluid that is not drinking water or not supplied by a water undertaker is required to be clearly identified to easily distinguish it from any supply pipe or distributing pipe.

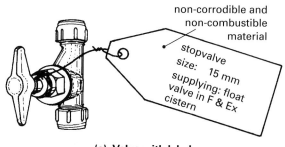

(a) Valve with label

Key

Control valve	Pipe run	
SV1 SV2 SV3	sp3 sp2 sp4 } supply pipe	
SV4 } Servicing SV5 } valves	h1 h1 } distributing header	
SV6	dp1	distributing pipe
SV7	cf1	cold feed pipe
SV8	dp2	distributing pipe
Cisterns		
CWSC No.1 } CWSC No. 2 }	Combined feed and storage cistern (linked)	
F & Ex C	Feed and expansion cistern	

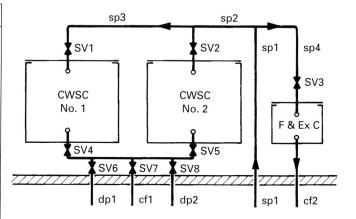

(b) Services and components diagram

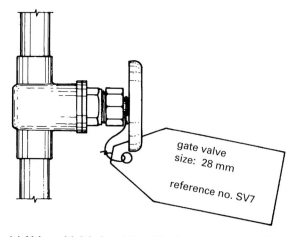

(c) Valve with label and identification reference number

Figure 11.72   Identification of pipes and services

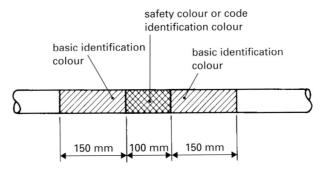

Colour code identification may be one or more colours.
Dimensions show minimum length for each colour.

(a) Colour coding

pipe contents	basic colour	colour code indication			basic colour
drinking water	green	auxiliary blue			green
boiler feed	green	crimson	white	crimson	green
central heating	green	blue	crimson	blue	green
cold down service	green	white	blue	white	green
hot water supply	green	white	crimson	white	green
fire extinguishing	green	red			green
grey water	green	green	black	green	green

(b) Example of colours used for water pipelines

Figure 11.73   Identification of pipelines above ground to BS 1710

Chapter 12
Commissioning and maintenance of pipelines, services and installations

Commissioning includes inspection, testing and cleansing of pipes and installations, and is an integral part of the work. It should be allowed for when estimating costs and should be undertaken at appropriate times as the work proceeds.

Inspection, testing and cleansing should ensure that:

(1) materials and equipment conform to British Standards or other forms of approval;
(2) installations are in accordance with the specification;
(3) all relevant laws and regulations are complied with;
(4) water drawn from pipes and fittings is of good quality and fit for human consumption;
(5) records of tests are kept and copies handed over to the client on completion of the contract.

Additionally, tests should ensure that end fittings such as draw-off taps, shower fittings and float-operated valves are operating correctly and flow rates match up with specified requirements.

The water undertaker should be notified and given the opportunity to witness any tests that are undertaken. On successful completion, the water supplier shall be informed that the system is available for permanent connection to the supply.

12.1 Inspections

Inspection of below ground installations

During the visual inspection of pipelines, particular attention must be paid to the pipe bed, the line and level of the pipe, irregularities at joints, the correct fitting of air valves, washout valves, sluice valves and other valves together with any other mains equipment specified. This should include the correct installation of thrust blocks, where required, and ensure that protective coatings are undamaged.

Trenches must be inspected to ensure that excavation is to the correct depth to guard against frost and mechanical damage due to traffic, ploughing or other agricultural activities on open land.

Trenches should not be backfilled until these conditions have been satisfied and the installation is seen to conform to the drawings, specifications and appropriate regulations.

Valve and hydrant boxes should be properly aligned, and suitable valve-operating keys provided before a pipeline is accepted.

Inspection of installations within buildings

All internal pipework must be inspected to ensure that it has been securely fixed.

Before testing takes place all cisterns, tanks, hot water cylinders and water heaters must be inspected to ensure that they are properly supported and secured, that they are clean and free from swarf, and that cisterns are provided with correctly fitting covers.

Water undertakers expect to be notified of certain water installations before the work is carried out, and particularly those with a high contamination risk. Proper notification as in Regulation 5 allows inspectors to discuss situations prior to installation and so avoid problems later. This also enables them to ensure that completed installations comply with relevant requirements.

An approved person registered with a water industry-approved Competent Persons Scheme may be permitted to self-certify that the installation conforms to the Water Fittings Regulations.

Visual inspection is an essential part of both interim and final tests and will detect many faults that the formal test will not pick up and which might lead to failure at a later date. Visual inspections should be made before any work is concealed. A careful record should be kept of such inspections and notes should be made to help with the preparation of 'as installed' drawings.

12.2   Testing for soundness

All water supply installations are required by Water Regulations to be tested for soundness. The guidance document is quite specific and states the following:

> 'Installations, including all supply and distributing pipes, fittings and connections to appliances, must be tested hydraulically (water pressure test) for a test period of one hour, at a test pressure of 1.5 times the maximum operating pressure, or the maximum operating pressure plus an allowance for any expected surge, whichever is the greatest. During this, there should be no visible leakage, and no loss of pressure. Normal working pressure for pumping mains should take account of any likely surge pressures.'

BS 6700 gives advice on the testing of installations above and below ground and sets out test procedures for rigid and elastomeric pipes, which are given in section 12.3.

Timing of tests

Interim tests should be applied to every pipeline as soon as practicable after completion of that particular section, with particular attention to all work that will be concealed. For buried pipelines these should be carried out before backfilling is placed over joints. However, some backfilling will be necessary to hold the pipes in place and prevent any movement of them during the test period. Long pipelines should be tested in sections as the work proceeds.

Final tests should be carried out on completion of all relevant work. Completion of buried pipelines includes backfilling, compacting and surface finishes. Final tests are generally carried out immediately before the hand-over date, but where the installation area is not likely to be affected by site works, the test may be done as work is completed.

Retests. Items failing any test should be corrected immediately and retested before further work proceeds.

12.3   Testing methods

Mains should be pressure tested at 1.5 times the normal working pressure (see figure 12.1).

Normal working pressures in pumping mains should take account of any likely surge pressure. Where surge is likely, a higher test pressure may be needed, but no higher than the design pressure of the pipeline and for no longer than 1 hour.

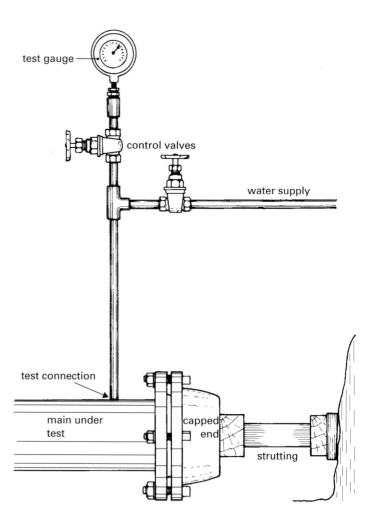

Pressure gauge fixed at lowest point on main.

Check and calibrate pressure gauge before beginning test.

Air valves positioned at high points to release trapped air.

Capped ends strutted securely against solid ground. All bends and branches securely anchored.

Place sufficient backfill before testing to prevent movement of pipes under test.

Allow 30 minutes before test period for water absorption and for water temperatures to warm/cool to ambient temperature.

Bring main up to test pressure slowly and release carefully after testing.

Using mains water, fill main slowly to expel air and disconnect from main before beginning test.

Allow time for water absorption and for stabilization of water temperatures before starting test.

Figure 12.1   Testing underground pipelines

Before testing

On above ground installations, all jointing should be completed with pipes and components properly secured before commencement of the test. Pipes below ground should be fully installed, ends capped and pipes fully anchored to prevent movement under test. Pipes and components should be inspected for obvious signs of leakage or other irregularities. Any valve within a test section should be fully open to ensure the whole section is tested.

To avoid the risk of contamination, water used for testing should be obtained from a drinking supply and the supply protected from risk of backflow. Water systems should be tested before a connection is made to the water supplier's main.

Testing procedure for rigid pipe installations

After installation:

(1)   Fill the installation slowly with drinking water allowing all the air to escape. (It is extremely difficult to pressurize a system containing air pockets.)
(2)   Allow to stand for 30 minutes to allow water temperature to stabilize.
(3)   Inspect the whole system and its joints visually for leakage.
(4)   Using mains water, pump up to test pressure (1.5 times normal working pressure).
(5)   Leave to stand for 1 hour.
(6)   Check for visible leakage and for loss of pressure. If neither occurs, the test is deemed to be satisfactory.
(7)   Otherwise repeat test after locating and repairing any leakage.

Testing elastomeric pipes

Two procedures are shown in BS 6700 for the testing of elastomeric pipes. These methods can also be used where the installation includes both rigid and plastics pipes.

Test procedure A (figure 12.2)

(1)   Fill the installation slowly, allowing the air to escape, and raise (or lower) the pressure in the system to 100 kpa (1 bar).
(2)   Allow to stabilize for 45 minutes.
(3)   Inspect system visually for leaks.
(4)   Apply test pressure (1.5 times maximum working pressure) by pumping for a period of 30 minutes and inspect visually for leakage.
(5)   Reduce pressure by bleeding water from the system to one third maximum working pressure. Close the bleed valve.
(6)   Visually check and monitor for 45 minutes. If the pressure remains at or above 0.5 times maximum working pressure, the system can be regarded as satisfactory.

Note: test pressure is 1.5 times maximum working pressure.

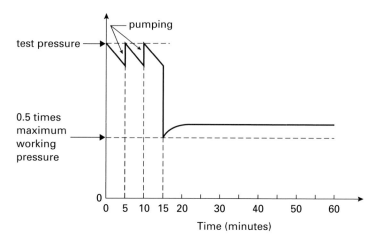

Figure 12.2   Testing elastomeric pipe systems – test procedure A

Test procedure B (figure 12.3)

(1) Fill the installation slowly, allowing the air to escape, and raise (or lower) the pressure in the system to 1 bar.
(2) Allow temperature to stabilize for 30 minutes.
(3) Inspect system visually for leaks.
(4) Apply test pressure (1.5 times maximum working pressure) by pumping for a period of 30 minutes. Note the pressure and inspect visually for leakage.
(5) Note the pressure after a further 30 minutes. If the pressure drop is less than 60 kPa (0.6 bar) the system can be considered to have no obvious leakage.
(6) Visually check and monitor for 120 minutes. If the pressure drop is less than 20 kPa (0.2 bar) the system can be regarded as satisfactory.

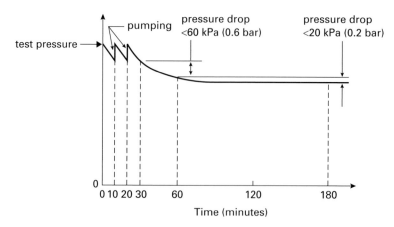

Figure 12.3   Testing elastomeric pipe systems – test procedure B

12.4   Flushing and disinfection

It is a requirement of Water Regulations that 'water fittings shall be . . . flushed and where necessary, disinfected prior to use'. Flushing and disinfection are not a substitute for cleanliness during installation.

Flushing

All pipes, fittings, appliances, storage vessels and other components should be thoroughly flushed out with drinking water before being brought into use. Additionally, if a system is not used immediately before commissioning, it should be flushed at regular intervals and disinfected before being brought into use.

Flushing is important to prevent stagnation both in new and existing installations that may be filled with water but are temporarily out of use. This advice is particularly important to combat the effects of 'blue water' in new copper installations where the water is left to stagnate before the system is put into use. It is thought that excess soldering flux left inside the pipe may lead to corrosion that releases copper particulate to be absorbed into the water giving it a blue, cloudy appearance.

To minimize stagnation, new systems should either be:

- flushed and drained down until needed, then flushed again before putting the pipe(s) into use; or
- flushed twice weekly to keep the system 'sweet' and avoid stagnation until the system is put into use.

Additionally, in newly occupied premises, occupants should be advised that for two weeks after occupation, they should run drinking water taps briefly until noticeably cooler, to clear standing water before drawing water off for use.

Excess flux can be difficult to remove. Flushing will usually clear the problem although in some cases it may be necessary to consider warm water and/or dynamic flushing.

Larger diameter pipes may also need to be swabbed to remove debris in addition to simple flushing. This applies particularly to pipework below ground.

After flushing, all systems other than single private dwellings should be disinfected. For single dwellings, thorough flushing is considered to be sufficient.

Further advice on flushing is given in the WRAS advisory leaflet *Commissioning Plumbing Systems*.

Disinfection

Disinfection should always be carried out in the following instances:

- in all new hot and cold installations, appliances and components, except single dwellings and other similar small installations;
- in all systems that have been modified or altered in any way;
- on pipes laid below ground, except for minor localized repairs or the insertion of tee junctions;

- junctions and fittings used for minor localized repairs should be washed or immersed in disinfectant before insertion into the pipeline;
- on any new system that has been standing unused after completion, for up to 30 days depending on the characteristics of the water (contact water supplier for advice);
- on any system that has been taken out of use or not used regularly;
- on any installation (new or existing) where any form of contamination is suspected.

When disinfection processes are carried out on mains pressure installations and there is no backflow device fitted to prevent backflow to the mains, the water undertaker should be informed. Where water for disinfection is to be discharged to a sewer or water course, the appropriate authority should be consulted.

BS 6700 shows two disinfection methods:

(1)   using chlorine as the disinfectant; and
(2)   using approved disinfectants other than chlorine. In these cases manufacturers' recommendations should be strictly followed.

Any chemicals used should be chosen from those listed in the Drinking Water Inspectorate's *List of Approved Substances* and published in the *Water Fittings and Materials Directory*.

Before commencing the disinfection process, checks should be made to ensure the disinfection liquid will not adversely affect the materials or any protective coatings used in the system.

Safety

It is important that no other chemicals, such as toilet cleansers, are added to any system until cleansing and disinfection are complete as the mixing of chemicals in this way could lead to the generation of toxic fumes.

All users of the building, including temporary and part-time users, e.g. cleaners and security staff, should be notified before the disinfection process is commenced, and notices should be displayed to show that the system and its appliances are out of use. Where possible, affected areas should be closed off.

Safety Notice

DISINFECTION IN PROGRESS

DO NOT USE

The disinfection sequence should be: water mains, supply pipes, cisterns and finally, the distributing system.

For supply pipes, disinfection liquid should be injected through a properly installed injection point.

Procedures for disinfection of both supply pipes and distributing systems are set out in figures 12.4 and 12.5.

Mains and large diameter supply pipes

Piping must be effectively cleansed and disinfected before first being used and after being opened up for repair or alteration (see figure 12.4).

At the time of laying, large pipes should be brushed clean and sprayed internally with a strong solution of sodium hypochlorite. For small diameter pipes insert a polyurethane foam plug soaked in sodium hypochlorite solution at a strength of about 10% chlorine and pass it through the bore.

In pipework under pressure, chlorine should be pumped in through a properly installed injection point at the beginning of the pipeline, until the residual at the end of the pipeline is in excess of 2 ppm (parts per million).

If the pipework is under mains pressure inform the water supplier of the intention to carry out chlorination. All work must be carried out in accordance with the water supplier's requirements.

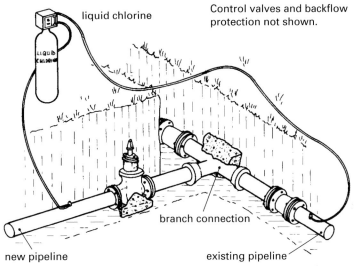

Before connection:
- brush clean and disinfect pipes and fittings before insertion into pipeline;
- disinfect trench around connection before cutting into existing pipeline;
- put chlorine tablet or similar inside pipe before final connection.

Pipe from existing main can be used to fill new main for pressure test and for chlorination purposes, provided precautions are taken against backflow and the water undertaker agrees.

After connection:
- flush out new main using water from existing pipeline;
- fill with chlorinated water (50 ppm);
- allow chlorinated main to stand for 24 hours;
- if free residual chlorine is less than 30 ppm repeat the process; otherwise
- flush out and take samples for chemical and bacteriological analysis. Free residual chlorine to be at level of drinking water supplied.

Figure 12.4   Disinfection of mains connection

Disinfecting supply pipe

(1) Thoroughly flush and empty at least twice.

(2) Fill the system with clean water.

(3) Close all draw-off taps, and valves.

(4) Inject chlorine at rate of 50 ppm.

(5) Open draw-offs in turn starting at the bottom and check for residual chlorine at each draw-off point.

(6) Leave for contact period of 1 hour.

(7) After contact period, check residual chlorine at each draw-off point. If less than 30 ppm the disinfection process should be repeated.

(8) Drain and flush system until residual chlorine is at same level as incoming drinking water.

(9) Take sample for bacteriological analysis.

Disinfecting distributing system

(1) Thoroughly flush and empty at least twice.

(2) Fill the system with clean water. Cistern to be full to overflowing level.

(3) Close all draw-off taps and valves and shut servicing valve (A).

(4) Add chlorine to cistern at rate of 50 ppm.

(5) Open draw-offs in turn starting at the top (nearest the cistern) and check for residual chlorine at each draw-off point.

Note: Cistern may need to be topped up as work proceeds.

(6) Leave for contact period of 1 hour.

(7) After contact period, check residual chlorine at each draw-off point. If less than 30 ppm the disinfection process should be repeated.

(8) Drain and flush system until residual chlorine is at same level as incoming drinking water.

(9) Take sample for bacteriological analysis.

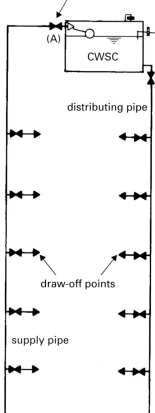

servicing valve to remain closed during test, except when filling or topping up cistern

(A)

CWSC

distributing pipe

draw-off points

supply pipe

temporary connection from mains supply with backflow prevention device

injection point

If direct connection from a main is to be made, backflow protection should be provided and the water supplier should be notified.

The water supplier may require an interposed cistern to be used.

Figure 12.5 Disinfection of water services

Installations within buildings

Storage cisterns and distributing pipes must be disinfected using the following method.

Prior to disinfection any users of the building or water supply should be informed and the system, or part of it, put out of use.

The cistern and pipe must first be filled with water and thoroughly flushed out. The cistern must then be filled with water again and a disinfecting chemical containing chlorine added gradually whilst the cistern is filling, to ensure thorough mixing. Sufficient chlorine should be used to give a concentration of at least 50 ppm.

When the cistern is full, the supply should be stopped and all the taps on the distributing pipes must be opened successively, working progressively away from the cistern. Each tap should be closed when the water discharged begins to smell of chlorine. The cistern should then be topped up with water from the supply pipe and with more disinfecting chemical in the recommended proportions. The cistern and pipes should then remain charged for at least 1 hour, after which a test must be made for residual chlorine. If none is found, the disinfecting process must be repeated.

Finally, the cistern and pipes should be thoroughly flushed out with clean water before any water is used for domestic purposes.

If ordinary 'bleaching powder' is used, the proportions must be 150 g of powder to 1000 l of water; the powder should be mixed with water in a separate clean vessel to a creamy consistency before being added to the water in the cistern. For proprietary brands of chemicals, use the proportions recommended by the manufacturer.

Where cisterns have internal coatings, these must be thoroughly cured before disinfection takes place.

For single dwellings and small pipework alterations, flushing is all that is required unless contamination is suspected, in which case the system should be disinfected.

12.5   Commissioning hot water and heating systems

System flushing and chemical treatment

Primary circuits to hot water and central heating systems should be thoroughly cleaned and flushed on completion and before a new boiler is installed. Once cleaned, the system should be filled with clean water and a chemical water treatment formulation added to control corrosion and the formation of scale and sludge. Guidance on the use of chemical inhibitors is given in BS 7593. The boiler manufacturer's installation instructions should be followed for appropriate treatment products and special requirements for individual boiler models and materials.

Where the mains water hardness exceeds 200 parts per million, the feed water to instantaneous water heaters and the hot water circuit of combination boilers should be treated to reduce the formulation of lime scale.

Commissioning

After installation of a hot water and/or heating system, the system and its components, e.g. boiler, pipes, pumps and controls, should be commissioned in accordance with the manufacturer's instructions.

An important part of the commissioning procedure is the completion of a commissioning checklist. The checklist, when issued by the competent person who did the work, will provide proof that all relevant checks and tests have been made, the system/components are working efficiently and the installation conforms to all relevant Building Regulations. On completion, the commissioning checklist, along with any installation manuals, should be given to the builder or householder for use when servicing or repairs are carried out.

Additionally, the competent person should complete a Building Regulations Compliance Certificate, a copy of which should go to the householder and another copy to the local building control body. In some cases the Building Regulations Compliance Certificate may be sent to the relevant approved certification body, who may in turn notify the local building control body on their behalf.

Where an installer is not registered with a competent person self-certification scheme, he/she, or the person commissioning the work, should notify the building control body before commencement of the work.

12.6 Maintenance

To keep the performance of any installation at the original specified design standard, it will be necessary to carry out some maintenance work. The amount of maintenance work will depend upon the type and size of system, and the risk or effects of breakdown balanced against the frequency and cost of the inspection and maintenance programme.

Planned preventative maintenance, regularly carried out, will help to ensure that systems perform correctly and avoid most breakdowns and the risk of costly damage to components, equipment and buildings. Table 12.1 shows a typical maintenance schedule.

Building owners should be provided with a maintenance schedule and instructions along with detailed and accurate drawings of the installation including all pipe runs, concealed or otherwise.

Maintenance and repairs should be carried out by a competent person, which means one who has the skills, knowledge and experience of water services installations, and of relevant statutory requirements.

Table 12.1 Maintenance schedule

Component	Maximum time interval	Remarks
Inspections	12 months	At frequent intervals in addition to any statutory inspection.
Water analysis	6 months or as required	Larger buildings (not individual dwellings) particularly where drinking water is stored. Gives useful guide to condition of an installation. Check cisterns, hot and cold water outlets.
Water temperatures	6 months or as required	Check during periods of most severe conditions, e.g. high occupancy, extreme heat or cold.
Cleaning and disinfection	12 months or as required	Following alterations to system and when contamination is suspected. Attention should be given to risk of *Legionella*.
Meters	6 months	Read meters for consumption and early warning of wastage. Check that meters are working.
Meter and stopvalve chambers	12 months or as required	Inspect to ascertain state of chamber construction. Clean out as necessary, check that box and lid are in working order and undamaged, and grease hinges.
Stopvalves and servicing valves	12 months	Operate to check that they close tightly and operate smoothly. Repair or renew as necessary. Valve keys should be available for emergency use. Valves should be labelled clearly to show what they do.
Pressure relief valves and temperature relief valves	6 to 12 months	Ease at regular intervals to ensure they are not stuck or that outlet is blocked. Remedy faults immediately. Easing valves may sometimes cause them to leak but this is preferable to an explosion. Check discharge pipe is unobstructed.
Pressure reducing valves	6 to 12 months	Check pressures downstream of valve and investigate any changes from normal.
Vessels under pressure	6 months	Inspect water storage vessels and expansion vessels for signs of deterioration. Measure gas pressures and adjust if not within manufacturer's recommended limits.
Boilers and water heaters	12 months or as manufacturer's recommendations	Inspect and test and adjust. Follow manufacturer's maintenance instructions.
Electrical work and control systems	12 months or as manufacturer's recommendations	Inspect and test and adjust. Check operation. Follow manufacturer's maintenance instructions.
Pressure gauges	6 months	Investigate any change from normal pressure.
Storage cisterns	6 months	Inspect for cleanliness and clean out as required. Check lid is securely in place. Look for signs of leakage or corrosion. Check linked cisterns for stagnant water (taste, odour or dust on surface). Check condition of bearers, safes and safe outlets. Check overflow pipes for obstruction and correct fall. Check insulation before winter.
Float-operated valves	6 months	Check operation and closing. Adjust for correct water level. Open float valve in feed and expansion cistern to prevent it sticking in the closed position.

Table 12.1   continued

Component	Maximum time interval	Remarks
Terminal fittings	6 to 12 months or as required	Rewasher, reseat or renew taps as required to prevent leakage. Tighten packing glands and check spindles for wear and efficient action. Check self-closing taps at intervals and remedy any faults. Clean sprayheads on taps, showers and shower mixers at intervals depending on rate of furring.
Pipework	12 months	Tighten loose fittings and supports and replace any missing ones. Check provision for expansion and contraction (especially plastics pipework). Check and tighten joints, remake or renew as necessary. Before commencing any work check compatibility of pipes and fittings, i.e.: ○ sizes, and whether imperial or metric; ○ metals for corrosion; ○ plastics for jointing method; ○ materials are available for repairs.
Disused fittings	1 month	Sterilize every 6 months or disconnect at branch connection.
Corrosion	12 months	Inspect for outward signs of corrosion. Reduced flow rates may indicate that corrosion products are causing obstruction. Replace corroded or seriously scaled lead pipe with pipe of some other suitable material. Internal corrosion of galvanized steel pipe is usually localized, although in some waters the zinc coating can break up, looking very like sand deposits in cisterns and tanks. In this case replace the complete pipe length. Pitting corrosion of copper pipe may be due to carbon film in pipe bore or cathodic scale. Rapid water velocities may lead to erosion corrosion. Pipes in damp conditions and areas where the air is acidic need regular inspections, and pipes should be protected against corrosive effects. Where pipes enter buildings they are particularly vulnerable at floor level, especially in the back of sink units.
Earth continuity bonding	12 months or as required	Checks to be carried out by competent electrician after pipes, fittings or appliances have been removed or replaced, or additions have been made to installation.
Insulation and fire stopping	12 months	Inspect and make good any damage. A good time would be before the start of winter.
Ducting	6 months	Accessibility is essential. Check that ducting is clear of extraneous matter and free from vermin. Check that access is not obstructed and entry is readily possible. Check crawlways and subways for leakage from pipework, entry of ground or surface water, and accumulation of flammable materials.

12.7   Locating leaks

BS 6700 deals with leak detection to premises supplied by meter only. As most premises in this country are unmetered the author has also included notes relating to these.

The first signs of leakage in any premises are usually:

- Visual.
- Noise. The sound of water escaping through a small crack or hole in a pipe is similar to that of a cistern filling, but of course it will be continuous, and will probably be noticed more at night when the pressures are higher and there are fewer extraneous noises.
- In metered supplies the high meter reading or account is often the first indication, which leads to leak detection taking place.

Water authorities periodically carry out their own waste water detection exercises and many leaks are located in this way. Water undertakers have a duty to keep wastage to a minimum, and if asked will usually send an inspector along to advise, and possibly to assist in locating both the service pipe and the leak. However, there are fairly simple ways to establish first that there is a leak, and secondly its approximate location.

Unmetered supplies

Procedure
(1)   Look for visible signs of leakage, i.e.:
- wet or soggy patches on ground;
- areas of grass that are greener or growing stronger and faster than remainder;
- water in stopvalve chambers.
(2)   Make sure no water is being drawn off or used.
(3)   Listen at taps and stopvalves to see where the noise appears loudest. There are excellent pocket stethoscopes available which can be used directly on a tap or on a stopvalve key which in turn is resting on an underground stopvalve. This will often give a general indication of the whereabouts of the leak.
(4)   Turn off at the main stopvalve and listen again. If the noise has stopped, it usually means the leak is on the supply pipe. If the noise continues the leak is probably on the communication pipe or main.
Note   It is important that the stopvalve turns off effectively, otherwise the leak may not be where it appears to be.
(5)   If there are branch supplies with stopvalves fitted, then the procedure can be repeated by closing each branch supply in turn, to establish if the leak is on the common pipe or one of its branches (see figure 12.6).
(6)   It is not always possible to locate the leak exactly, and in long pipelines the best approach may be to cut and plug the pipe about halfway along its length (or fit an intermediate stopvalve) and retest. This can be repeated a number of times to establish which length of pipe needs to be dug up (see figure 12.7).
(7)   It is also possible to drive a pointed steel bar into the ground at intervals along a pipeline, pulling it out again to see how wet it is. Usually it is wetter nearer the leak.
Note   The occurrence of a leak is often an indication of the state of the pipe, especially with steel pipelines.

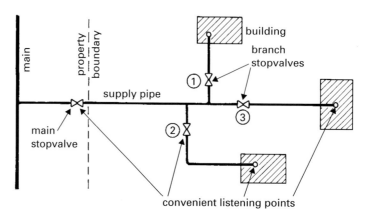

Figure 12.6   Locating leaks in branch supply pipes

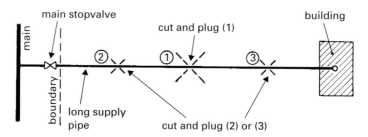

Figure 12.7   Locating leaks in long pipelines

Metered supplies

The method shown in BS 6700 follows a simple logical procedure using the meter and a watch, and requires first that the supply pipe and its stopvalves are located and recorded on a diagram. The isolating stopvalves should be numbered. The following example (see figure 12.8), based on that shown in BS 6700, describes the procedure.

Before the test:

- check that all stopvalves are in working order and will stop the supply when closed;
- make sure that no part of the supply is in use and that all water fittings except isolating valves are closed;
- check that the meter is capable of recording low flows.

Procedure
(1) Record rate of flow using the meter and a watch; if zero, there is no detectable leakage.
(2) If there is a rate of flow, shut off isolating valves in sequence starting at the valve furthest from the meter, i.e. no. 1 in figure 12.8. After each valve has been shut any change in the rate of flow should be noted.
(3) Leakage should be sought in any sections of the network where closure of the isolating valve has reduced the rate of flow.
(4) As a check, repeat the test procedure when the leakages have been detected and repaired.

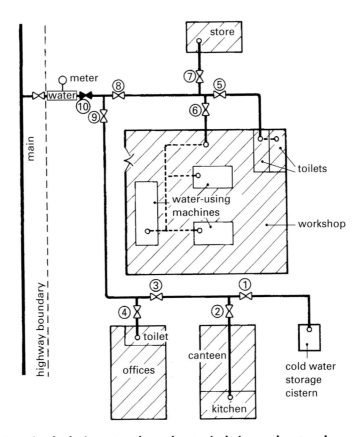

Figure 12.8   Locating leaks in metered supply – typical pipework network

Practical example (see figure 12.8)

Item	Rate of flow l/min	Change in rate of flow l/min	Remarks
Start of test	90	—	
Shut valve no. 1	90	Nil	
Shut valve no. 2	45	45	Leakage at 45 l/min (in canteen section)
Shut valve no. 3	45	Nil	
Shut valve no. 4	45	Nil	
Shut valve no. 5	30	15	Leakage at 15 l/min (in toilets section)
Shut valve no. 6	30	Nil	
Shut valve no. 7	Zero	30	Leakage at 30 l/min (in supply to store)
Shut valve no. 8	Zero	Nil	
Shut valve no. 9	Zero	Nil	

12.8   Disconnection of unused pipes and fittings

If an installation or part of an installation becomes redundant, or if any appliance or fitting is unused (except for repair, maintenance or renewal) it is important that it be disconnected. The whole of the pipework supplying water to the unused appliance or fitting should be cut off at the source of the connection so that no dead legs of unused pipework are left to stagnate (see figure 12.9).

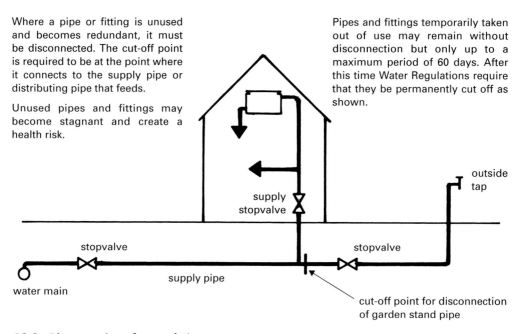

Where a pipe or fitting is unused and becomes redundant, it must be disconnected. The cut-off point is required to be at the point where it connects to the supply pipe or distributing pipe that feeds.

Unused pipes and fittings may become stagnant and create a health risk.

Pipes and fittings temporarily taken out of use may remain without disconnection but only up to a maximum period of 60 days. After this time Water Regulations require that they be permanently cut off as shown.

outside tap

supply stopvalve

stopvalve

stopvalve

water main

supply pipe

cut-off point for disconnection of garden stand pipe

Figure 12.9   Disconnection of unused pipes

12.9   Occupier information

Occupiers and owners of property have an interest in keeping their systems in good order and are in the position, as users, to note the first signs of faults appearing.

Occupiers should be informed of the need for maintenance and be provided with maintenance instructions along with record drawings upon which pipe runs and valves are accurately marked and a full explanation of the system and its operation.

On completion of work the installer should, where applicable, leave with the householder or occupier:

- the manufacturer's user manual;
- the installation manual and maintenance instructions;
- a copy of the commissioning checklist; and
- a copy of the Building Regulations Compliance Certificate.

Chapter 13
Firefighting systems

13.1   Fire safety

Buildings are required under Building Regulations Part B to be constructed in a manner that will ensure the safety of occupants in the event of a fire. Building materials and methods are required to provide adequate means of escape, and minimize the spread of fire and smoke within and between buildings (see figure 13.1).

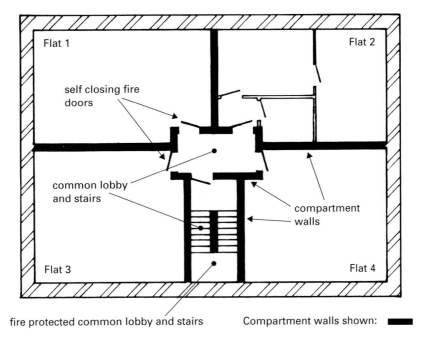

Compartment walls form a fire-resisting barrier penetration to give occupants time to escape in the event of a fire.

Floors/ceilings between upper and lower flats also to be of fire resisting construction to resist fire spread.

Fire doors to have at least 30 minutes fire resistance and be self closing.

Any pipes passing through fire walls should be arranged to resist fire penetration between compartments.

Figure 13.1   Compartmentation in flats and maisonettes

Part of this fire protection relies on 'compartmentation' which means that buildings must be divided into 'compartments', each of which is separated one from the other by walls and floors of fire-resisting materials. Compartment walls and floors should form a complete barrier that will prevent fire penetration long enough to give occupants of the building sufficient time to escape.

Openings in compartment walls are not generally permitted but there are a few limited exceptions including the following:

- a door provided as a means of escape in case of fire. The door must be properly fitted so as to give the same fire resistance as that required for the compartment wall;
- pipes, ventilation ducts, chimneys, appliance ventilation ducts, or ducts encasing flue pipes that need to pass from one compartment to another;
- shafts or ducts used to accommodate pipes, which should also have barriers against the passage of fire where they pass through and between compartments.

There is a risk that the fire resistance in compartment walls or floors could be breached where pipes or ducts pass through them. To prevent this, the following points should be considered:

- the size of the pipe and material from which pipe is made (see table 13.1); and
- the material and method used to seal the hole made to allow the pipe to pass through the wall or floor.

Table 13.1   Maximum nominal internal diameters of pipes passing through a compartment wall or floor

Situation	Pipe material and maximum nominal internal diameter (mm)		
	(a)	(b)	(c)
	Non-combustible materials (1)	Lead, aluminium, aluminium alloy, PVC-U (2), fibre cement	Any other material
Wall separating dwelling houses	160	160 (stack pipe) (3) 110 (branch pipe) (3)	40
Wall or floor separating a dwelling house from an attached garage	160	110	40
Any other situation (see note 4)	160	40	40

Notes
(1)   Any non-combustible material (such as cast iron, copper, or steel) which, if exposed to a temperature of 800°C, will not soften or fracture to the extent that flame or hot gas will pass through the wall of the pipe.
(2)   PVC-U pipes complying with BS4514 and PVC-U pipes complying with BS 5255.
(3)   These diameters are only in relation to pipes forming part of an above ground drainage system.
(4)   Situation 3 will apply to hot and cold pipes and fire installations.

13.2 Openings for pipes

All openings for pipes and ducts passing through a compartment wall or floor or other fire-separating element should be kept as small as practicable; as few in number as possible; and sealed or fire stopped to meet one of the following provisions:

1. **A proprietary seal.** Should be used to provide a fire-resistant seal between the pipe and the hole through which it passes. Any sealing system used should have been tested to show that its fire resistance is at least equal to that of the wall or floor that is being sealed.
2. **Fire stopping.** Pipes with a restricted diameter as shown in table 13.1 may be sealed using a fire stopping around the pipe. The wall opening should be kept as small as possible. Fire stopping around a pipe should be flexible enough to permit thermal movement. Materials used for fire stopping may include the following: cement mortar, gypsum-based plaster, cement- or gypsum-based vermiculite/perlite mixes, glass fibre, crushed rock, blast furnace slag or ceramic-based products (with or without resin binders), and intumescent mastics. For further guidance the local building control office or the local fire brigade should be consulted.
3. **Sleeving.** Where a pipe of lead, aluminium, aluminium alloy, fibre-cement or uPVC (of maximum internal diameter of 160 mm) is installed it may be sleeved through the wall or floor using a non-combustible sleeve pipe as shown in figure 13.2. The pipe opening should be kept as small as possible to accommodate the pipe and its sleeving. The sleeving should fit closely around the pipe. It should be continuous in length and extend to a distance of at least 1 m from either side of the fire-resisting structure.

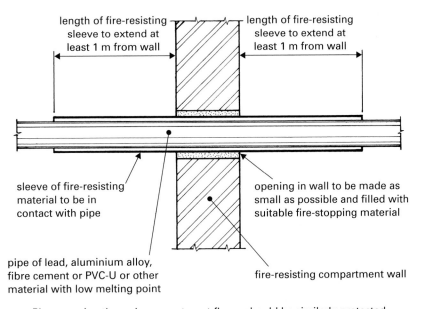

length of fire-resisting sleeve to extend at least 1 m from wall

length of fire-resisting sleeve to extend at least 1 m from wall

sleeve of fire-resisting material to be in contact with pipe

opening in wall to be made as small as possible and filled with suitable fire-stopping material

pipe of lead, aluminium alloy, fibre cement or PVC-U or other material with low melting point

fire-resisting compartment wall

Pipes passing through compartment floors should be similarly protected.

Figure 13.2 Pipes passing through compartment wall

13.3   Fire mains within buildings

Fire mains and fire valves should be readily accessible to the fire service for connecting water hoses to fight fires within the building.

Access for a pump appliance should be provided to within 45 m of all points within a dwelling house. Alternatively, for other dwellings such as flats, a fire main could be installed.

Buildings with a compartment area of 280 m² plus that are more than 100 m from a highway should have fire hydrants installed. Where an assembly building has any floor area exceeding 900 m² and is over 7.5 m above ground level, provision should be made for firefighting shafts. In buildings that are unsprinklered, every part of every storey over 18 m in height should be within 45 m of a fire main outlet.

Fire mains may be of two types:

1.   **The 'dry' system** (see figure 13.3) is normally empty and is supplied when needed, through hose from a fire service pumping appliance.
2.   **The 'wet' system** is kept permanently full of water and under pressure from tanks and pumps within the building. It should have connections that will allow the system to be replenished from a pumping appliance in an emergency.

Fire mains are generally installed in buildings where normal firefighting methods cannot reach, e.g. floors above or below ground level. Fire mains are required in buildings that are provided with firefighting shafts and should be installed within those shafts for access by the fire authority when needed.

In buildings containing a floor that is more than 60 m above the fire service vehicle access level, only the wet system is permitted. Otherwise either wet or dry systems may be used.

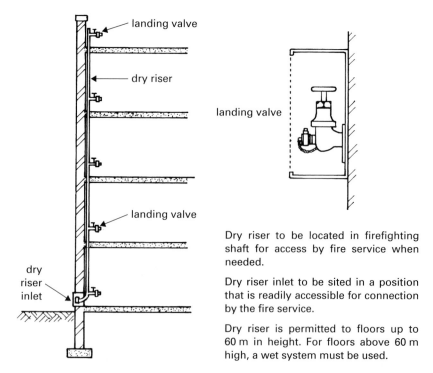

landing valve

dry riser

landing valve

landing valve

dry riser inlet

landing valve

Dry riser to be located in firefighting shaft for access by fire service when needed.

Dry riser inlet to be sited in a position that is readily accessible for connection by the fire service.

Dry riser is permitted to floors up to 60 m in height. For floors above 60 m high, a wet system must be used.

Figure 13.3   Dry riser installation

There should be one fire main in every firefighting shaft with an outlet from the fire main sited in each firefighting lobby giving firefighting facilities to the accommodation at each floor level. A firefighting shaft is illustrated in figure 13.4.

Further information and guidance on the design and construction of fire mains can be obtained from BS 5306: Part 1 *Fire extinguishing installations and equipment on premises, hydrant systems, hose reels and foam inlets.*

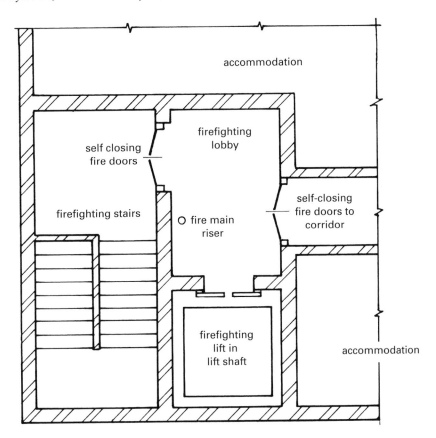

Outlets from fire main to be located in firefighting lobby.

Firefighting lift required for floors over 18 m high.

Figure 13.4   Components of a fire shaft

13.4   Certification and accreditation of fire protection installations

The fire performance of materials and components used in buildings depends upon satisfactory site installation and maintenance. Independent schemes of certification and accreditation of installers and maintenance firms aim to provide confidence in the standard of workmanship provided and may be accepted by building control bodies as evidence of compliance with the relevant standard. However, a building control body may wish to establish, before the work is commenced, that the work carried out under a certification scheme will meet the requirements of Building Regulations.

Chapter 14
Sprinkler systems for domestic and residential premises

'If a building is fitted throughout with a properly installed sprinkler system, it is reasonable to assume that the intensity and extent of a fire will be reduced' (Building Regulations Approved Document B4).

A National Fire Protection Association report states that *'people with smoke alarms in their home have a 50% better chance of surviving a fire. Adding sprinklers and fire alarms increases the chance of surviving a fire by over 97%'*. This suggests that of the 600 lives lost each year through domestic and residential fires, 580 of the victims may have lived if sprinkler systems and smoke detectors were fitted to all dwellings. In addition to the loss of life and many thousands of fire-related injuries each year, a great deal of damage is caused to buildings, partly by the fire itself and partly by the large volumes of water used to extinguish the fires.

A sprinkler system, properly designed and installed, can control a fire and activate its alarm at an early stage of the fire's development. Using a sprinkler will also help to reduce the rate of production of heat and smoke and allow more time for occupants of the building to escape or be rescued.

Sprinkler systems provide a quick response because strategically positioned sprinkler heads, activated by heat of the fire, will automatically tackle the fire very soon after it begins. Because only sprinkler heads in the vicinity of the fire will operate, less water will be needed to control the fire than normal firefighting methods. Without a sprinkler the fire is often very advanced by the time the fire service arrives at the scene, and vast quantities of water are then needed to control the fire. To summarize, sprinkler systems installed in domestic and residential buildings provide the following benefits:

- a quick response that limits the fire to a more localized area of the building;
- less fire damage to the building;
- less water damage to the building; and
- more importantly, greatly reduced risk of injury or loss of life as the result of a fire.

14.1 Scope of BS 9251

Where a sprinkler system is used in residential and domestic properties Building Regulations recommend that the system be designed and installed to conform with BS 9251:2005 *Sprinkler systems for residential and domestic occupancies – Code of practice.*

The aim of this chapter is to provide an outline of the recommendations of BS 9251. It will cover systems suitable for use in buildings ranging from small individual dwellings to those used for multiple occupancies up to 20 m in height, such as apartments, blocks of flats, residential homes, boarding homes, care homes, etc. It does not cover the use of sprinkler systems used in industrial, commercial or other larger buildings that fall within the scope of BS 5306-2 and BS EN 12845.

14.2 Terms and definitions

Definitions from BS 9251 relevant to this chapter include:

alarm device   electrical or mechanical device for detecting water flow into the system and sounding an alarm

alarm system   electrical or mechanical system audible internally and externally, with a built-in precaution to avoid spurious alarms. NOTE: *An electrical system should be mains powered and have a back-up battery of adequate capacity*

alarm test valve   valve through which water may be discharged to test the operation of alarm system

concealed sprinkler   recessed sprinkler with a cover plate that disengages when heat is applied

domestic occupancy   individual dwelling for occupation as a single family unit used or constructed or adapted to be used wholly or principally for human habitation, such as individual dwelling houses, individual flats, maisonettes and transportable homes, with a maximum individual room size of 40 m^2

experienced sprinkler contractor   contractor who is suitably qualified and experienced and has independent documentation providing evidence of this

fire pump   pump that is automatically operated in the event of a fire which supplies water to a sprinkler system from a water storage facility or from a mains supply

fusible link sprinkler   sprinkler which opens when an element provided for that purpose melts

glass bulb sprinkler   sprinkler which opens when a liquid-filled glass bulb bursts

pendent sprinkler   sprinkler in which the nozzle directs the water downwards

priority demand valve   valve for isolating the supply to the domestic service in the event of sprinkler operation

quick response sprinkler   sprinkler with quick response temperature-sensing element which operates to allow water to discharge in accordance with BS EN 12259-1

recessed sprinkler   sprinkler in which all or part of the heat sensing element is above the lower plane of the ceiling

residential occupancy   occupancy for multiple occupation not exceeding 20 m in height, with a maximum individual room size of 180 m^2, such as apartments, residential homes, houses of multiple occupancy (HMOs), blocks of flats, boarding houses, aged persons homes, nursing homes, residential rehabilitation accommodation and dormitories.
Note 1 Where multiple occupation buildings exceed 20 m in height, special circumstances need to be considered and the authority having jurisdiction should be consulted. This matter is receiving the attention of the relevant BSI committee with a view to issuing an amendment.

Note 2   This occupancy classification is not suitable for secure accommodation, asylum centres or large, open, communal dormitories or equivalent hazards.

residential pattern sprinkler   sprinkler which gives an outward and downward water discharge, suitable for use in domestic and residential occupancy

room   area, enclosed by walls and a ceiling, which may have openings to an adjoining room or adjoining rooms provided such openings have a lintel depth of at least 200 mm

service pipe   pipe supplying water from a water supply to any premises that are subject to water pressure from that water supply

sidewall pattern sprinkler   sprinkler which gives an outward half paraboloid pattern of water discharge

subsidiary alternate system   portion of a sprinkler system which is capable of being charged with air or water

upright sprinkler   sprinkler in which the nozzle directs the water upwards

wet pipe system   sprinkler system which is designed to be permanently charged with water

14.3   Consultation

Where a sprinkler system is to be installed, extended or altered, whether for use in new or existing buildings, the following authorities should be consulted and, where appropriate, their approval sought at an early stage:

- the local water undertaker;
- the building control body;
- the fire authority;
- the insurer(s) of the dwelling and dwelling contents.

14.4   Sprinkler systems and water supply methods

The system should be designed and installed by an experienced sprinkler contractor. This means a person who can prove by independent documented evidence that he is suitably qualified and experienced in this type of work.

A sprinkler system should be a wet pipe system, i.e. one that is permanently charged with water. The method of supply should be chosen after investigation of the site and by taking account of local supply conditions. Whichever method of supply is used, it is essential to establish that an adequate water flow and pressure can be maintained, particularly at peak supply periods.

Systems may be chosen from one of the following supply methods:

1. direct mains water connection using one supply pipe to serve both the sprinkler system and the domestic water services (see figure 14.1(a));
2. direct mains water connection using a dedicated water supply pipe (preferred method) (see figure 14.1(b));
3. indirect connection using stored water supplied by gravity from a high level storage cistern;
4. direct or indirect supply using a pressure vessel;
5. automatic fire pump drawing water directly from a mains water supply or indirectly from an elevated storage cistern;
6. automatic fire pump drawing water from a low level break cistern (see figure 14.1(c)).

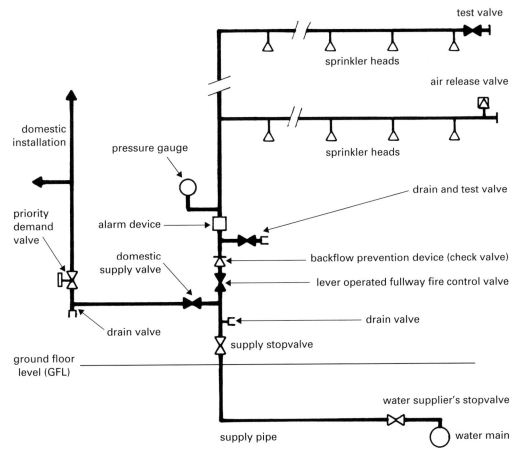

Alarm device is activated by flow of water to set off both internal and external alarms and close priority demand valve on domestic supply.

Priority demand valve is not needed where dedicated water supply is used.

(a) Direct system using one supply pipe to serve both sprinkler system and domestic water

Figure 14.1   Sprinkler systems

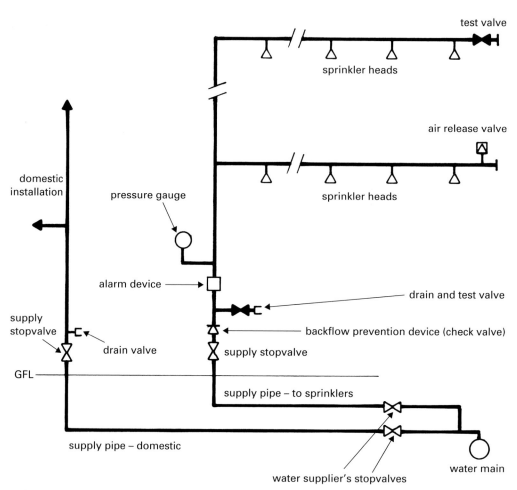

Supply stopvalve to sprinkler system also designated 'lever operated fullway fire control valve'.

(b) Direct system using dedicated water supplies

Figure 14.1   continued

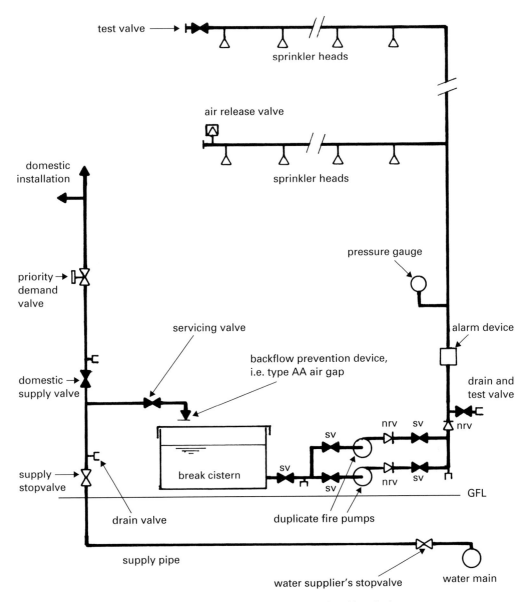

(c) System supplying sprinklers from low level break cistern and domestic water directly under mains pressure

Figure 14.1   continued

Where water is to be supplied directly from a mains water supply the water undertaker should be notified of the installation and consulted at the planning stage regarding minimum mains pressures and capacity. When connecting to an existing service pipe the pressure and flow rate should be checked at the point of entry to the building.

Whichever system is selected, sprinkler protection should be provided in ALL parts of the dwelling, with the permitted exception of:

- bathrooms with a floor area of less than 5 m²;
- cupboards and pantries with a floor area of less than 2 m² where the least dimension does not exceed 1 m and the walls and ceilings are covered with non-combustible or limited-combustible materials;
- non-communicating, attached buildings such as garages, boiler houses, etc., separated from the protected premises by a 30 or 60 minutes fire-resisting construction in accordance with BS 476 or equivalent European Standard depending on the local authority's requirements for the building;
- crawl spaces.

Provision for stored water

A storage cistern designed and installed to supply a sprinkler system should be dedicated to that purpose only. Whilst BS 9251 mentions storage cisterns that are used for both sprinklers and water for domestic purposes, this is not generally recommended and may not be permitted under Water Regulations.

Water stored for use in fire sprinkler systems is likely to present a higher risk of contamination than water for domestic purposes and appropriate backflow protection is required. Wet pipe sprinkler systems without additives are within fluid risk category 2 whilst those that contain additives, e.g. anti-freeze, or systems where the distributing pipe is pumped are considered to be fluid risk category 4. Water stored for domestic purposes is required to be of drinking quality (fluid category 1), and should be stored in a separate 'protected' cistern. Further information on protection against contamination in fire systems is given in chapter 6.

The volume of water stored for the sprinkler system should be calculated to give at least 10 minutes' duration of operating flow. For domestic occupancies, the amount of water stored should be sufficient to maintain pressures and flows for 10 minutes to whichever is the greater of a single operating sprinkler, or a pair of operating sprinklers, in a single room situated in the hydraulically most favourable position. For residential occupancies, the amount of water stored should be sufficient to maintain pressures and flows for a period of 30 minutes to any combination of sprinklers (not more than four) operating in a single room and situated in the hydraulically most favourable position.

14.5   Pressure requirements and system flow rates

Minimum operating pressure

The minimum operating pressure at any sprinkler should be not less than 0.5 bar. Tests should ensure that this pressure is maintained at times of peak demand when pressures are at their lowest.

System flow rates

Where used in domestic occupancies, the system should be capable of providing adequate flow rates for a single sprinkler, or two sprinklers operating simultaneously at the values given in table 14.1 plus any flow for alarm purposes. For residential occupancies, the system should provide sufficient flow rates to allow up to four sprinklers to operate simultaneously at not less than the flow rates given in table 14.1 plus any flow for alarm purposes.

Note   Where the mains water supply serves both the sprinkler system and the domestic or residential occupancy supply, the sprinkler system should be capable of providing additional flow rates at the sprinkler heads as follows:

- plus at least 25 l/min for domestic occupancies;
- plus at least 50 l/min for residential occupancies or the design demand for the residency, whichever is the greater.

An additional flow should also be added to the flow rate for water needed to operate alarm equipment.

Hydraulic calculations should be carried out to determine the pipe sizes required to meet the performance recommendations of BS 9251. The water supplier should be consulted about the size of the service pipe that will supply the sprinkler system. This is particularly important when the system is to be supplied directly under mains pressure.

Table 14.1   Flow rate requirements

For domestic occupancies	60 l/min	through any single sprinkler	plus any flow for alarm purposes	Add at least 25 l/min where mains supply serves both sprinkler and domestic occupancies
	42 l/min	through each of two sprinklers operating simultaneously in a single room	plus any flow for alarm purposes	Add at least 25 l/min where mains supply serves both sprinkler and domestic occupancies
For residential occupancies	60 l/min	through any single sprinkler	plus any flow for alarm purposes	Add at least 50 l/min where mains supply serves both sprinkler and residential occupancies
	42 l/min	for each sprinkler operating simultaneously up to a maximum of four sprinklers in a single room	plus any flow for alarm purposes	Add at least 50 l/min where mains supply serves both sprinkler and residential occupancies

14.6 System components

Fire pump. Where a pump is used, its selection should be based on the pressure and flow rate calculated for the system and:

- should start automatically on demand and shut down manually;
- be located where it is unlikely to be affected by a fire and where the temperature will be maintained above freezing;
- have electrical protection using suitable fusing; and
- be designed and manufactured for annual testing, but not more frequently.

Sprinklers should be suitable for use in residential and domestic applications and should have a quick response thermal sensitivity rating. They may be selected from the following patterns: pendent, upright, sidewall, recessed or concealed types (see figure 14.2).

Only new equipment should be used. Any sprinkler head removed from a system should be discarded.

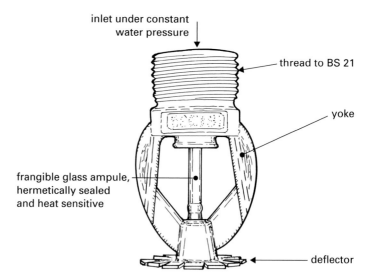

inlet under constant water pressure

thread to BS 21

yoke

frangible glass ampule, hermetically sealed and heat sensitive

deflector

The ampule contains a precise amount of heat-sensitive liquid. When heat is absorbed, the liquid in the bulb expands to increase the pressure within the ampule.

At the prescribed temperature, the increasing pressure will burst the glass, allowing water to pass through the inlet.

The water then sprays on to the deflector to be distributed in the 'approved' pattern to extinguish the fire.

(a) Pendent type

Figure 14.2   Residential pattern sprinklers

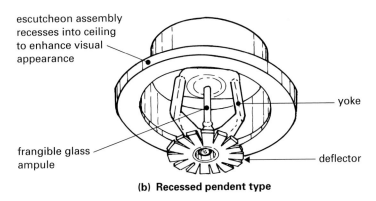

escutcheon assembly
recesses into ceiling
to enhance visual
appearance

yoke

frangible glass
ampule

deflector

(b) Recessed pendent type

Figure 14.2   continued

Size of sprinklers. Sprinklers should be suitable for use with fittings threaded in accordance with ISO 7-1, ISO 65 and BS 21 (see table 14.2 which is based on a similar table in BS EN 12257-1).

Temperature rating of sprinklers. Fusible link sprinklers should be colour coded on the frame or sprinkler body. Glass bulb sprinklers should be colour coded by the bulb liquid in accordance with BS EN 12259-1. The temperature rating of residential type sprinklers should be at least 20°C greater than the highest anticipated ambient temperature of their location and within the range of 79°C to 100°C when installed under glazed roofs (see table 14.3). For normal conditions in the United Kingdom, the sprinkler temperature ratings will be 57°C or 68°C.

Table 14.2   Nominal thread and orifice size of sprinkler heads

Nominal thread size of pipe inches	Nominal diameter of orifice mm
$\frac{3}{8}$	10
$\frac{1}{2}$	15
$\frac{3}{4}$	20

Table 14.3   Sprinkler head temperature ratings

Sprinkler type	Temperature rating °C	Colour code
Fusible link	55/57 80/107	Uncoloured White
Glass bulb	57 68 79	Orange Red Yellow

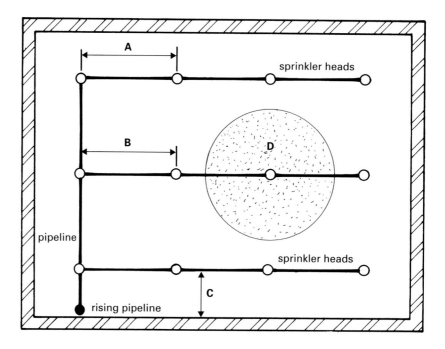

Notes to figure 14.4:

A	maximum distance between sprinklers	4 m
B	minimum space between sprinklers	2 m
C	maximum spacing between sprinkler and wall	2 m
D	maximum area to be protected by any one sprinkler	15 m²

Figure 14.3   Sprinkler positioning

Sprinkler positioning. See figure 14.3. When positioning sprinkler heads, the following recommendations should be followed:

- Pendent and upright sprinklers for residential and domestic applications should be arranged so that their heat-sensitive elements are close to, but not more than 100 mm below the ceiling.
- Heat-sensitive elements of sidewall sprinklers should positioned within 100 mm to 150 mm below the ceiling.
- When recessed and concealed sprinklers are permitted, the water supplier should be consulted and heat-sensitive elements positioned to meet their recommendations. *Note* Before installing concealed and recessed sprinklers, approval of the local/fire authority should be sought.
- When the sprinklers are operated, the whole of the floor area and the walls up to a level 0.7 m below the ceiling should be wetted.
- Where sprinklers are fitted in sloping ceilings they should be positioned to follow the recommendations of the supplier.
- Care should be taken so that the sensitivity and discharge pattern of sprinklers are not adversely affected by obstructions such as constructional beams or light fittings or other sprinkler heads.
- The potential for a shielded fire to develop should be taken into account.

Alarm devices. The alarm system should be fitted with one of the following alarm devices, which should be set off by the flow of water to at least one sprinkler:

1. a mechanically driven alarm for which the flow should be taken into account in the hydraulic calculations; or
2. an electrically operated flow switch connected to an audible alarm; or
3. at least one internal audible alarm which can be easily heard in all parts of the building or dwelling, and an audio-visual alarm positioned externally in a prominent position and clearly labelled 'FIRE ALARM'.

Control valves

Every sprinkler system should be fitted with the following valves, the relative positions of which are illustrated in figure 14. 4.

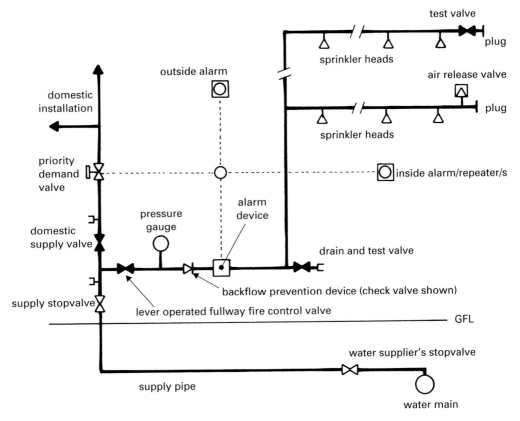

Alarm device may be electrically or mechanically driven and activated by a flow of water in the sprinkler system.

The alarm may be of the audible or strobe type.

The priority demand valve will close off the domestic supply if the alarm is activated.

The type of backflow device used will depend on the degree of risk.

Figure 14.4 Components of a sprinkler system

A **supply stopvalve** should be located in an accessible position inside the premises, above floor level, close to the point of entry of the pipe. It should be situated so that it will shut off the whole of the domestic supply to the premises.

A **further stop valve** or **servicing valve** of the full bore lever type should be used to isolate the sprinkler pipework from mains water supply. The valve should be locked in the open position and labelled 'FIRE CONTROL VALVE' to prevent accidental interruption of the water supply to the sprinkler system.

A **backflow prevention device** should be fitted to prevent mains water contamination:

- In mains fed systems, without additives (fluid category 2), the backflow device will usually consist of a single check valve.
- Where fluids are added for frost protection the water becomes a fluid category 4 risk and requires the use of an RPZ valve.
- Where a system is pumped or a pressure vessel is used, it too will require an RPZ valve for fluid category 4 risk.
- A system supplied by storage where the water is likely to become stagnant is a category 5 risk and requires the use of an air break device such as a type AA or AB air gap.
- Where a high level cistern also supplies water for domestic purposes and the water is continually replenished, a type AG air gap with circular overflow (fluid category 3) may be permitted.

Alarm test valves. Two are needed:

1. A **quick-acting test valve** should be located at the end of the hydraulically most remote range pipe on the system. The pipe should be at least 22 mm nominal diameter with an outlet nozzle equivalent in size to the smallest sprinkler in the system.
2. A **quick-acting drain and test valve** located at the lowest point of the sprinkler pipework. This will allow functional testing and complete draining of the sprinkler system. It should be suitably sized to check the maximum flow rate specified for the system subject to a minimum nominal pipe diameter of 22 mm.

A **priority demand valve** (not needed on dedicated systems) is recommended for use where the supply serves both domestic use and the sprinkler system. It should be fitted on the supply pipe downstream of the supply stopvalve and will shut off the domestic supply when water flow is detected in the sprinkler system.

Electrically operated devices. Electrically operated devices, e.g. pumps, flow switches or alarms, should be capable of carrying out their function in the event of a failure of the mains electrical supply. The electrical supply to fire pumps should be installed so as to minimize the risk of electrical supply failure. It should have a separately fused connection taken off after the meter from the supply side of the domestic fuse box. Fire-resisting cable should be used.

14.7   Installation

Sprinkler systems should be installed to comply with current Water and Building Regulations and any local fire regulations. They should also conform to BS 9251 and take account of guidelines issued by the Fire Protection Association and the Loss Prevention Council.

Before installation begins, the water supply pipeline should be tested to ensure that recommended flow rates and pressure can be maintained. Tests should be taken at peak periods when flows and pressures are likely to be at their lowest.

If tests show that recommended pressures and flow rates cannot be achieved, the designer should be consulted before any work is commenced.

Pipes and fittings for sprinkler installations should comply with relevant British or European Standards and pipe sizes determined by hydraulic calculation. Precautions should be taken to accommodate thermal movement. Capillary fittings for copper tube should be jointed using soldering or brazing alloys that have a melting point of not less than 230°C. Otherwise copper pipework should be installed to the standards described in other parts of this book. Bending of copper piping should only be carried out by an approved method.

Plastics and other pipes and fittings should be of a type recognized by the authority having jurisdiction over sprinkler systems as suitable for the application and should be installed in accordance with manufacturer's instructions.

Where pipes pass through structural timbers, the timbers should not be notched or bored in such a way that the structural strength is weakened (see chapter 11).

Pipework supports should be of non-corrosive metallic materials. Battens and lock type clips should be fitted in close proximity to the sprinkler heads to ensure that heads do not move from the design position.

Frost protection. Any water-filled pipework that may be subjected to low temperatures should be protected against freezing. Frost precautions may include insulation based on the recommendations of BS 5422. Electrical trace heating may also be used. Where anti-freeze is used, precautions should be taken to prevent contamination of the mains water supply. Plastics pipe and fittings may be protected using glycerin-based anti-freeze solutions but NOT glycol-based solutions. Frost precautions are dealt with more fully in chapter 7.

14.8   Commissioning sprinkler systems

Testing for soundness

The system should be flushed, purged of air and tested in accordance with the requirements of the Water Fittings Regulations. The water supply to the system should be isolated and the system tested to a minimum of 1.5 times working pressure. If the system fails to maintain pressure any leak/s should be located and repaired prior to retesting. A fuller account of testing procedures that meet the recommendations of BS 6700 can be seen in chapter 12.

Performance testing

The sprinkler system should be tested to ensure that the specified flow rates can be achieved and pressures maintained at the required level. Tests should be taken at both alarm test valves. If flow rates or pressures cannot be achieved, the system should not be approved for use until the system has been corrected and retested.

An alarm test should ensure that the flow rate through the alarm device is at least equal to the design flow rate. The test will be deemed to be satisfactory if the alarm (and/or repeaters) operates and can be heard in all parts of the property.

Compliance

On satisfactory completion of the commissioning tests, the experienced sprinkler contractor should issue a completion certificate showing that the installation has been passed and complies with the design criteria and all relevant rules and recommendations. Details of tests should be entered on a log book which should be left in the care of the property owner or occupier.

14.9   Maintenance

It is important that the sprinkler system is inspected and tested annually by a suitably qualified and experienced sprinkler contractor, who should make sure that:

- the sprinklers' heat sensing capacity and spray pattern are not impeded;
- minimum recommended flow rates are achieved at the drain and test valve(s);
- the alarm is effective and can be heard in all parts of the building;
- the system has not been modified except as recommended in BS 9251.

Sprinkler system tests

- A visual inspection should be undertaken to check for leakage. Where a leak is suspected the pipework should be tested to the standards required by Water Regulations.
- Both internal and external alarms should be left active during tests so that their satisfactory operation can be audibly verified.
- The sprinkler system should be tested to ensure that specified flow rates and pressures can be maintained at the required level. Tests should be taken at both alarm test valves.
- Stopvalves should be exercised to ensure free movement.
- Any insulation should be checked to make sure it is properly fitted and trace heating checked for satisfactory operation.
- The log book should be completed and signed by the qualified and experienced sprinkler contractor who carried out the test.

14.10   Documentation

For new or extended systems all drawings and other documents should include details of the system including:

- the address and location of the premises;
- the name and address of the approved or experienced contractor;
- the name of the designer; and
- the date of installation.

The approved contractor should provide the following information to the owner or occupier of the building in which the sprinkler system is installed:

- details of the authorities consulted and any response received;
- a signed Certificate of Compliance giving a general description of the system and a statement of compliance with BS 9251, along with any deviations agreed with the authority having jurisdiction for the deviation;
- a layout drawing of the premises and the installation, as fitted, along with a set of the hydraulic calculations;
- details of the water supply including pressure and flow rate data for the commissioned installation and stating the specific location of the test and the time and date of the test;
- a maintenance and inspection programme which should include instructions for any remedial action, should it be needed;
- a list of components used in the installation along with supplier's name/s and parts reference number/s;
- a 24 hour emergency telephone number for use in the event that assistance is needed;
- a log book containing inspection, testing and maintenance documents, and detailing work or tests carried out by an approved contractor;
- essential information for the user, e.g. 'sprinkler heads not to be painted, covered or otherwise' or 'modifications not to be made to sprinkler equipment except in compliance with BS 9251'.

Spare sprinkler heads of the same design as those used in the system should be supplied to the owner or occupier of the premises. These should be kept by the owner/occupier along with a suitable tool for fitting them. However, replacement of spare heads should not be carried out by the owner or occupier but by a suitably qualified and experienced sprinkler contractor.

Log book

The log book should be completed giving details of:

- the date of the inspection;
- details of all tests conducted along with their results;
- confirmation or otherwise of the operational status of the sprinkler system and its alarm system;
- details of any comments or recommendations.

14.11   Hydraulic calculations for sprinkler systems

Hydraulic calculations should be carried out to establish pipe sizes to meet the performance recommendations of BS 9251. BS 9251 gives the Hazen-Williams formula as the basis for friction loss calculations and also provides pressure loss tables based on this formula for pipes of copper, PVC-U and steel. The author prefers the method shown earlier in chapter 5; a pipe sizing example for a domestic property using the tabular method is included below (see figures 14. 5 and 14.6).

		pipe lengths	flow rates
Pipe A	from main to first branch	40 m	1.81 l/s
Pipe B	from first branch to furthest two sprinklers	30 m	1.40 l/s

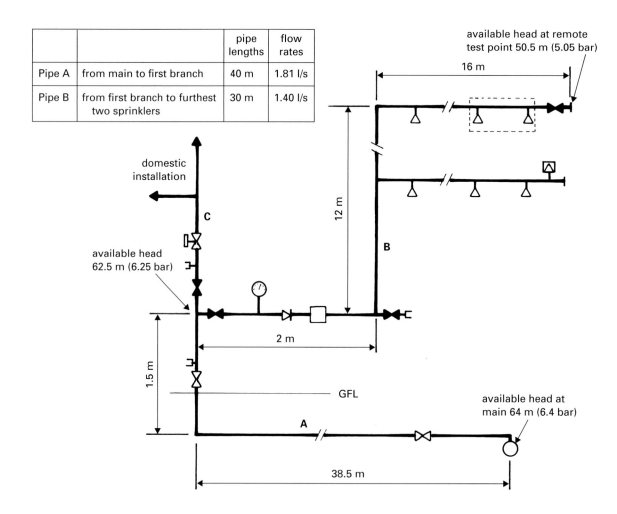

Figure 14.5   Diagram for sizing a sprinkler system

Essential information for calculation

Firstly, the pipe size is assumed, in this case to be 25 mm diameter. The calculation will prove whether this assumption is correct.

Available pressure (see figure 14.5)   – at main   64 m head
                                           – at first branch   62.5 m head
                                           – at remote test point   50.5 m head

Flow rates required (see table 14.1)

Through pipe A (2 sprinklers and domestic) (42 + 42 + 25)   = 109 l/min (1.81 l/s)
Through pipe B (2 sprinklers only) (42 + 42)   = 84 l/min (1.4 l/s)

Note   In this example it is assumed that the alarm does not need to be accounted for.

Measured pipe runs (see figure 14.5)   – pipe A   40 m
                                            – pipe B   30 m

Equivalent pipe lengths for fittings (refer to table 5.2)

Pipe A (2 stopvalves and 1 elbow) (10 + 10 + 1 + 1)   = 22 m head
Pipe B (2 tees, 1 elbow and 1 check valve) (1 + 4.5 + 5.6)   = 11.1 m head

Note   Assume valves on pipe A to be of the screwdown type. Valves on pipe B are assumed to be fullway valves that offer negligible resistance to flow and need not be accounted for.

The resulting figures can be inserted into the calculation sheet, figure 14.6.

Note to calculations   Figure 14.6 shows that a 25 mm supply pipe will provide a flow rate of 1.4 l/s to the most remote sprinklers with a remaining head of 12.33 m (1.23 bar). As the minimum head over sprinklers is required to be 0.5 bar, the calculated pipe size is adequate. If the progressive head exceeds the available head (or is very close) the calculation should be repeated using a larger diameter pipe.

1 pipe reference	2 loading units	3 flow rate (l/s)	4 assumed pipe size (mm)	5 loss of head (m/m run)	6 flow velocity (m/s)	7 measured pipe run (m)	8 equivalent pipe length	9 effective pipe length (m)	10 head consumed (m)	11 progressive head (m)	12 available head (m)	13 final pipe size	14 head remaining
A	n/a	1.81	25.00	0.45	3.50	40.00	22.00	62.00	27.90	27.90	62.50	**25.00**	34.60
B	n/a	1.40	25.00	0.25	2.60	30.00	11.10	41.10	10.28	38.18	50.50	**25.00**	12.33

Notes
(1) Follow method shown in Chapter 5 (figure 5.11).
(2) If progressive head exceeds available head (or is very close) repeat the calculation using a larger diameter pipe.
(3) This figure shows that a 25 mm supply pipe will provide a flow of 1.4 l/s to the most remote sprinklers with a remaining head of 12.33 m (1.23 bar). As the minimum head over sprinklers is required to be 0.5 bar, the calculated pipe size is adequate.

Figure 14.6   Calculation sheet for sizing a sprinkler system

British Standards relevant to this book

Note Standards are continually changing, being revised, renewed, replaced or withdrawn. Many are replaced or partially replaced by EN standards; often one standard is replaced by a series of standards. For some standards it is difficult to find a replacement to follow instead. Even when a standard is withdrawn, materials often remain in use many years after its withdrawal. For these reasons, this list includes a number of seemingly outdated standards but which still have some relevance.

British Standards

BS 21	Pipe threads for tubes and fittings where pressure-tight joints are made on the threads (metric dimensions)
BS 417-2	Specification for galvanized low carbon steel cisterns, cistern lids, tanks and cylinders (metric units)
BS 534	Specification for steel pipes, joints and specials for water and sewage
BS 750	Specification for underground fire hydrants and surface box frames and covers
BS 853-1	Specification for vessels for use in heating systems – Part 1: Calorifiers and storage vessels for central heating and hot water supply
BS 1010-2	Specification for draw-off taps and stopvalves for water services (screw-down pattern) – Draw-off taps and above-ground stopvalves
BS 1192	Construction drawing practice
	Part 1 Recommendations for general principles (withdrawn)
	Part 3 Recommendations for symbols and other graphic conventions (withdrawn)
BS 1211	Specification for centrifugally cast (spun) iron pressure pipes for water, gas and sewage
BS 1212-1	Float-operated valves – Part 1: Specification for piston type float-operated valves (copper alloy body) (excluding floats)
BS 1212-2	Float-operated valves – Part 2: Specification for diaphragm type float-operated valves (copper alloy body) (excluding floats)
BS 1212-3	Float-operated valves – Part 3: Specification for diaphragm type float-operated valves (plastics bodied) for cold water services only (excluding floats)
BS 1212-4	Float-operated valves – Part 4: Specification for compact type float-operated valves for WC flushing cisterns (including floats)
BS 1387	Specification for screwed and socketed steel tubes and tubulars and plain end steel tubes suitable for welding or for screwing to BS 21 pipe threads (withdrawn)
BS 1453	Specification for filler materials for gas
BS 1563	Cast iron sectional tanks (rectangular) (current, obsolescent)
BS 1564	Pressed steel sectional rectangular tanks
BS 1566-1	Open vented copper cylinders. Requirements and test methods

BS 1566-2	Open vented copper cylinders. Specification for single feed indirect cylinders
BS 1710	Specification for identification of pipelines and services
BS 1845	Specification for filler metals for brazing (withdrawn)
BS 1968	Specification for floats for ballvalves (copper)
BS 1972	Specification for polythene pipe (Type 32) for above ground use for cold water services (withdrawn)
BS 2035	Specification for cast iron flanged pipes and flanged fittings
BS 2456	Specification for floats (plastics) for float-operated valves for cold water services
BS 2580	Specification for underground plug cocks for cold water services
BS 2879	Specification for draining taps (screw-down pattern)
BS 3198	Specification for copper hot water storage combination units for domestic purposes
BS 3251	Specification for indicator plates for fire hydrants and emergency water supplies
BS 3284	Specification for polythene pipe (Type 50) for cold water services (withdrawn)
BS 3505	Specification for unplasticized polyvinyl chloride (PVC-U) pressure pipes for cold potable water
BS 3506	Specification for unplasticized PVC pipe for industrial uses
BS 4213	Specification for cold water storage and combined feed and expansion cisterns (polyolefin or olefin copolymer) up to 500 1 capacity used for domestic purposes
BS 4346-1	Joints and fittings for use with unplasticized PVC pressure pipes – Part 1: Injection moulded unplasticised PVC fittings for solvent welding for use with pressure pipes, including potable water
BS 4346-2	Joints and fittings for use with unplasticized PVC pressure pipes – Part 2: Mechanical joints and fittings, principally of unplasticized PVC
BS 4346-3	Joints and fittings for use with unplasticized PVC pressure pipes – Part 3: Specification for solvent
BS 4622	Specification for grey iron pipes and fittings
BS 4772	Specification for ductile iron pipes and fittings
BS 4814	Specification for expansion vessels using an internal diaphragm, for sealed hot water heating systems
BS 4991	Specification for propylene copolymer pressure pipe
BS 5114	Specification for performance requirements for joints and compression fittings for use with polyethylene pipes
BS 5154	Specification for copper alloy globe, globe stop and check, check and gate valves
BS 5163-1	Valves for waterworks purposes. Predominately key-operated cast iron gate valves. Code of practice
BS 5163-2	Valves for waterworks purposes. Stem caps for use on isolating valves and associated water control apparatus
BS 5268-2	Structural use of timber – Part 2: Code of practice for permissible stress design, materials and workmanship
BS 5292	Specification for jointing materials and compounds for installations using water, low-pressure steam or 1st, 2nd and 3rd family gases
BS 5306-1	Code of practice for fire extinguishing installations and equipment on premises. Hose reels and foam inlets
BS 5306-2	Fire extinguishing installations and equipment on premises. Specification for sprinkler systems

BS 5391-1	Specification for acrylonitrile-butadiene-styrene (ABS) pressure pipe – Part 1: Pipe for industrial uses
BS 5392-1	Specification for acrylonitrile-butadiene-styrene (ABS) fittings for use with ABS pressure pipe – Part 1: Fittings for use with pipe for industrial uses
BS 5412	Specification for low resistance single taps and combination tap assemblies (nominal size $\frac{1}{2}$ and $\frac{3}{4}$) suitable for operation at PN 10 max and a minimum flow pressure of 0.01 MPa (0.1 Bar) Parts 1–5
BS 5422	Method of specifying thermal insulating materials on pipes, ductwork and equipment (in the temperature range $-40°C$ to $+70°C$)
BS 5433	Specification for underground stopvalves for water services
BS 5440-1	Installation and maintenance of flues and ventilation for gas appliances of rated heat input not exceeding 70 kW net (1st, 2nd and 3rd family gases) – Part 1: Specification for installation and maintenance of flues
BS 5440-2	Installation and maintenance of flues and ventilation for gas appliances of rated input not exceeding 70 kW net (1st, 2nd and 3rd family gases) – Part 2: Specification for installation and maintenance of 'ventilation for gas appliances'
BS 5449	Specification for forced circulation hot water central heating systems for domestic premises
BS 5493	Code of practice for protective coating of iron and steel structures against corrosion
BS 5546	Specification for installation of gas hot water supplies for domestic purposes (1st, 2nd and 3rd family gases)
BS 5615	Specification for insulating jackets for domestic hot water storage cylinders
BS 5834-2	Surface boxes, guards and underground chambers for gas and waterworks purposes – Part 2: Specification for small surface boxes
BS 5970	Code of practice for thermal insulation of pipework and equipment in the temperature range $-100°C$ to $+870°C$
BS 6076	Specification for polymeric film for use as protective sleeving for buried iron pipes and fittings (for site and factory application)
BS 6144	Specification for expansion vessels using an internal diaphragm, for unvented hot water supply systems
BS 6280	Method of vacuum (backsiphonage) test for water-using appliances
BS 6282-1	Devices with moving parts for the prevention of contamination of water by backflow – Part 1: Specification for check valves of nominal size up to and including DN 54
BS 6282-2	Devices with moving parts for the prevention of contamination of water by backflow – Part 2: Specification for terminal anti-vacuum valves of nominal size up to and including DN 54
BS 6282-3	Devices with moving parts for the prevention of contamination of water by backflow – Part 3: Specification for in-line anti-vacuum valves of nominal size up to and including DN 42
BS 6282-4	Devices with moving parts for the prevention of contamination of water by backflow – Part 4: Specification for combined check and anti-vacuum valves of nominal size up to and including DN 42
BS 6283-1	Safety devices for use in hot water systems – Part 1: Specification for expansion valves for pressures up to and including 10 bar
BS 6283-2	Safety and control devices for use in hot water systems – Part 2: Specifications for temperature relief valves for pressures from 1 bar to 10 bar
BS 6283-3	Safety devices for use in hot water systems – Part 3: Specification for combined temperature and pressure relief valves for pressures from 1 bar to 10 bar

BS 6283-4	Safety and control devices for use in hot water systems – Part 4: Specification for drop-tight pressure reducing valves of nominal size up to and including DN 50 for supply pressures up to and including 12 bar
BS 6351	Electric surface heating. Specification for electric heating surfaces devices
BS 6340-2	Shower units – Part 2: Specification for the installation of shower units
BS 6340-4	Shower units – Part 4: Specification for shower heads and related equipment
BS 6437	Specification for polyethylene pipes (type 50) in metric diameters for general purposes
BS 6465-1	Sanitary installations – Part 1: Code of practice for scale of provision, selection and installation of sanitary appliances
BS 6465-2	Sanitary installations – Part 2: Code of practice for space requirements for sanitary appliances
BS 6465-3	Sanitary installations – Part 3: Code of practice for space requirements for the selection, installation and maintenance of sanitary appliances
BS 6572	Specification for blue polyethylene pipes up to nominal size 63 for below ground use for potable water
BS 6675	Specification for servicing valves (copper alloy) for water services
BS 6700	Design, installation, testing and maintenance of services supplying water for domestic use within buildings and their curtilages. Specification
BS 6730	Specification for black polyethylene pipes up to nominal size 63 for above ground use for cold potable water
BS 6891	Installation of low pressure gas pipework of up to 35 mm ($1\frac{1}{4}$") in domestic premises (2nd family gas)
BS 6920	Suitability of non-metallic products for use in contact with water intended for human consumption with regard for the effect on the quality of the water. Specification
BS 7206	Specification for unvented hot water storage units and packages
BS 7291-1	Thermoplastics pipes and associated fittings for hot and cold water for domestic purposes and heating installations in buildings. General requirements
BS 7291-2	Thermoplastics pipes and associated fittings for hot and cold water for domestic purposes and heating installations in buildings. Specification for polybutylene (PB) pipes and associated fittings
BS 7291-3	Thermoplastics pipes and associated fittings for hot and cold water for domestic purposes and heating installations in buildings. Specification for cross-linked polyethylene (PE-X) pipes and associated fittings
BS 7291-4	Thermoplastics pipes and associated fittings for hot and cold water for domestic purposes and heating installations in buildings. Specification for chlorinated polyvinyl chloride (PVC-C) pipes and associated fittings and solvent cement
BS 7671	2001 Requirements for Electrical Installations (IEE Wiring Regulations)
BS 7874	Method of test for microbiological deterioration of elastomeric seals for joints in pipework and pipe lines
BS 7977-1	Specification for safety and rational use of energy of domestic gas appliances. Radiant convectors
BS 7977-2	Specification for safety and rational use of energy of domestic gas appliances. Combined appliances. Gas fire/back boilers
BS 8000-15	Workmanship on building sites – Code of practice for hot and cold water services (domestic scale) (limited in use, outdated!)
BS 9251	Sprinklers systems for residential and domestic premises. Code of practice
BS 9990	Code of practice for non-automatic fire-fighting systems in buildings

CP 312-1	Code of practice for plastics pipework (thermoplastics materials) – Part 1: General principles and choice of material
CP 312-2	Code of practice for plastics pipework (thermoplastics materials) – Part 2: Unplasticized PVC pipework for the conveyance of liquids under pressure
CP 312-3	Code of practice for plastics pipework (thermoplastics materials) – Part 3: Polyethylene pipes for the conveyance of liquids under pressure

EN Standards

BS EN 26	Gas-fired instantaneous water heaters for the production of domestic hot water, fitted with atmospheric burners
BS EN 200	Sanitary tapware – General technical specifications for single taps and mixer taps (nominal size Vz PN 10) – minimum flow pressure of 0.05 MPa (0.5 bar)
BS EN 257	Mechanical thermostats for gas-burning appliances
BS EN 297	Gas-fired central heating boilers – Type B_{11} and B_{11BS} boilers fitted with atmospheric burners of nominal heat input not exceeding 70 kW
BS EN 378-1	Specification for refrigerating systems and heat pumps – Safety and environmental requirements – Part 1: Basic requirements, definitions, classification and selection criteria
BS EN 483	Gas-fired central heating boilers – Type C boilers of nominal heat input not exceeding 70 kW
BS EN 512-19	Fibre cement products. Pressure pipes and joints
BS EN 545	Ductile iron pipes, fittings, accessories and their joints for water pipelines. Requirements and test methods
BS EN 598	Ductile iron pipes, fittings, accessories and their joints for sewerage applications. Requirements and test methods
BS EN 625	Gas-fired central heating boilers – Specific requirements for the domestic hot water operation of combination boilers of nominal heat input not exceeding 70 kW
BS EN 681-1	Elastomeric seals. Material requirements for pipe joint seals in water and drainage applications. Vulcanized rubber
BS EN 681-2	Elastomeric seals. Material requirements for pipe joint seals in water and drainage applications. Thermoplastic polymers
BS EN 681- 3	Elastomeric seals. Material requirements for pipe joint seals in water and drainage applications. Vulcanized rubber
BS EN 681- 4	Elastomeric seals. Material requirements for pipe joint seals in water and drainage applications. Cast polyurethane sealing elements
BS EN 805	Water supply – requirements for systems and components outside buildings
BS EN 806-1	Specification for installations inside buildings conveying water for human consumption – Part 1: General
BS EN 806-2	Specification for installations inside buildings conveying water for human consumption – Part 2: Design
BS EN 806-3	Specification for installations inside buildings conveying water for human consumption – Part 3: Pipe sizing. Simplified method
BS EN 816	Sanitary tapware – Automatic shut-off valves PN10
BS EN 997	WC pans and WC suites with integral trap
BS EN 1044	Brazing – Filler metals
BS EN 1057	Copper and copper alloys, seamless, round copper tubes for water and gas in sanitary and heating appliances

BS EN 1151-1	Pumps – Rotodynamic pumps – Circulation pumps having a rated power input not exceeding 200 W for heating installations and domestic hot water installations – Part 1: Non-automatic circulation pumps, requirements, testing, marking
BS EN 1254-1	Copper and copper alloys – Plumbing, fittings. Fittings with ends for capillary soldering or capillary brazing to copper tubes
BS EN 1254-2	Copper and copper alloys – Plumbing, fittings. Fittings with compression ends for use with copper tubes
BS EN 1254-3	Copper and copper alloys – Plumbing fittings. Fittings with compression ends, for use with plastic pipes
BS EN 1254-4	Copper and copper alloys – Plumbing fittings. Fittings combining other end connections with capillary or compression ends
BS EN 1254-5	Copper and copper alloys – Plumbing fittings. Fittings with short ends for capillary brazing to copper tubes
BS EN 1452-1	Plastics piping systems for water supply. Unplasticized poly(vinyl chloride) (PVC-U). General
BS EN 1452-2	Plastics piping systems for water supply. Unplasticized poly(vinyl chloride) (PVC-U). Pipes
BS EN 1452-3	Plastics piping systems for water supply. Unplasticized poly(vinyl chloride) (PVC-U). Fittings
BS EN 1452-4	Plastics piping systems for water supply. Unplasticized poly(vinyl chloride) (PVC-U). Valves and ancillary equipment
BS EN 1452-5	Plastics piping systems for water supply. Unplasticized poly(vinyl chloride) (PVC-U). Fitness for purpose of the system
BS EN 1452-6	Plastics piping systems for water supply. Unplasticized poly(vinyl chloride) (PVC-U). Guidance for installations
BS EN 1452-7	Plastics piping systems for water supply. Unplasticized poly(vinyl chloride) (PVC-U). Guidance for assessment of conformity
BS EN 1490	Building valves. Combined temperature and pressure relief valves
BS EN 1491	Building valves. Expansion valves
BS EN 1717	Protection against pollution of potable water in water installations and general requirements of devices to present pollution by backflow
BS EN 10224	Non-alloy steel tubes and fittings for the conveyance of water and other aqueous liquids – Technical delivery conditions
BS EN 10226-1	Pipe threads where pressure tight joints are made on the threads. Taper external threads and parallel internal threads. Dimensions, tolerances and designation
BS EN 10226-2	Pipe threads where pressure tight joints are made on the threads. Taper external threads and taper internal threads. Dimensions, tolerances and designation
BS EN 10226-3	Pipe threads where pressure tight joints are made on the threads. Verification by means of limit gauges
BS EN 10255	Non-alloy steel tubes suitable for welding or threading – Technical delivery conditions
BS EN 10312	Welded stainless steel tubes for the conveyance of aqueous liquids including water for consumption. Technical delivery conditions
BS EN 12201-1	Plastic water systems for water supply. Polyethylene (PE). General
BS EN 12201-2	Plastic water systems for water supply. Polyethylene (PE). Pipes
BS EN 12201-3	Plastic water systems for water supply. Polyethylene (PE). Fittings
BS EN 12201-4	Plastic water systems for water supply. Polyethylene (PE). Valves
BS EN 12201-5	Plastic water systems for water supply. Polyethylene (PE). Fitness for purpose

BS EN12259-1	Fixed fire fighting systems – Components for sprinkler and water spray systems. Sprinklers
BS EN12259-2	Fixed fire fighting systems – Components for sprinkler and water spray systems. Wet alarm valve assemblies
BS EN12259-3	Fixed fire fighting systems – Components for sprinkler and water spray systems. Dry alarm valve assemblies
BS EN12259-4	Fixed fire fighting systems – Components for sprinkler and water spray systems. Water motor alarms
BS EN12259-5	Fixed fire fighting systems – Components for sprinkler and water spray systems. Water flow detectors
BS EN 12449	Copper and copper alloys. Seamless, round tubes for general purposes
BS EN 12536	Welding consumables. Rods for gas welding non-alloy and creep resisting steels. Classification
BS EN 12828	Heating systems in buildings. Design for water based heating systems
BS EN 12831	Heating systems in buildings. Method of calculation of design heat load
BS EN 12897	Water supply, specification for indirectly heated unvented (closed) hot water storage heaters
BS EN 12976-1	Thermal solar systems and components – Factory made systems – General requirement
DD ENV 12977-1	Thermal solar systems – Custom built systems – Part 1: General requirements
BS EN 13076	Devices to prevent pollution by backflow of of potable water. Unrestricted air gap. Family A. Type A
BS EN 13077	Devices to prevent pollution by backflow of potable water. Air gap with non-circular overflow (unrestricted). Family A. Type B
BS EN 13078	Devices to prevent pollution by backflow of potable water. Air gap with submerged feed incorporating air inlet plus overflow. Family A. Type C
BS EN 13079	Devices to prevent pollution by backflow of potable water. Air gap with injector. Family A. Type D
BS EN 14154-1	Water meters. General requirements
BS EN 14154-2	Water meters. Installation and conditions of use
BS EN 14154-3	Water meters. Test methods and equipment
BS EN 14336	Heating systems in buildings. Installation and commissioning of water based heating systems
BS EN 14339	Underground fire hydrants
BS EN 14451	Devices to prevent pollution by backflow of potable water – In-line anti-vacuum valves DN 8 to DN 80 – Family D, Type A
BS EN 14452	Devices to prevent pollution by backflow of potable water – pipe interuptor with permanent atmospheric vent and moving element DN 10 to DN 20 – Family D, Type B
BS EN 14453	Devices to prevent pollution by backflow of potable water – pipe interuptor with permanent atmospheric vent DN 10 to DN 20 – Family D, Type C
BS EN 14454	Devices to prevent pollution by backflow of potable water – hose union backflow preventor DN 15 to DN 32 inclusive – Family H, Type A
BS EN 14455	Devices to prevent pollution by backflow of potable water. Pressurized air inlet valves DN 15 to DN 50 – Family L, Type A and Type B
BS EN 14622	Devices to prevent pollution by backflow of potable water. Air gap with circular overflow (restricted) – Family A, Type F
BS EN 14623	Devices to prevent pollution by backflow of potable water. Air gaps with minimum circular overflow (verified by test of measurement) – Family A, Type G
BS EN 29453/ISO 9453	Soft solder alloys – chemical compositions and forms

BS EN 60335-1	Specification for safety of household and similar electrical appliances. General
BS EN 60335-2-5	Specification for safety of household and similar electrical appliances. Particular requirements for dishwashers
BS EN 60335-2-7	Specification for safety of household and similar electrical appliances. Particular requirements washing machines
BS EN 60335-2-21	Specification for safety of household and similar electrical appliances. Storage water heaters
BS EN 60335-2-35	Specification for safety of household and similar electrical appliances. Particular requirements for instantaneous water heaters
BS EN 60335-2-51	Specification for safety of household and similar electrical appliances. Particular requirements for stationary circulation pumps for heating and service water installations
BS EN 60335-2-73	Specification for safety of household and similar electrical appliances. Particular requirements for fixed immersion heaters
BS EN 60335-2-84	Specification for safety of household and similar electrical appliances. Particular requirements for toilets
BS EN 60730-1	Specification for electrical controls for household and similar use. General requirements
BS EN 60730-7	Specification for electrical controls for household and similar use. General requirements. Timers and time switches
BS EN 60730-8	Specification for electrical controls for household and similar use. Particular requirements. Electrically operated water valves including mechanical devices

BS Codes of practice

CP 312-1	Code of practice for plastics pipework (thermoplastics material) – Part 1: General principles and choice of material
CP 312-2	Code of practice for plastics pipework (thermoplastics material) – Part 2: Unplasticized PVC pipework for the conveyance of liquids under pressure
CP 312-3	Code of practice for plastics pipework (thermoplastics material) – Part 3: Polyethylene pipes for the conveyance of liquids under pressure
CP 342-2	Code of practice for centralized hot water supply – Part 2: Buildings other than individual dwellings

EN ISO Standards

BS EN ISO 15493	Plastics piping systems for industrial applications – acrylonitrile-butadiene-styrene (ABS), unplasticized poly(vinyl chloride) (PVC-U) and chlorinated poly(vinyl chloride) (PVC-C) – Specifications for components and the system – Metric series
BS EN ISO 228-1	Pipe threads where pressure tight joints are not made on the threads. Dimensions, tolerances and designation
BS EN ISO 228-2	Pipe threads where pressure tight joints are not made on the threads. Verification by means of limit gauges
BS EN ISO 3677	Filler metals for soft soldering, brazing and welding
BS EN ISO 9453	Soft solder alloys. Chemical composition and forms
BS EN ISO 9454	Fluxes for soft soldering

References and further reading

Backsiphonage Report, HMSO (1974)

Building Regulations and Approved Documents, Parts B, G, H, J, L, M, and P

Copper Fire Sprinkler Systems for Residential and Domestic Properties, Copper Development Association (1997)

Commissioning Plumbing Services, advisory leaflet, Water Regulations Advisory Scheme.

Domestic Heating Compliance Guide, ODPM. ISBN – 10: 1 85946 255 1, 13: 978 1 85946 225 6

Guidance on the Application and Interpretation of the Model Water Byelaws, Department of the Environment (1986)

Legionnaire's Disease – Good Practice Guide for Plumbers, The Institute of Plumbing (1990). ISBN – 0 9501671 9 3

Model Water Byelaws, HMSO

NJUG Publication No. 6. Service Entries for New Dwellings on Residential Estates

Performance Specifications for Thermal Stores, Water Heater Manufacturers Association (1999)

Plumbing Engineering Services Design Guide, Institute of Plumbing (2002). ISBN – 1 871956 40 4

Preventing Hot Water Scalding in Bathrooms: Using TMVs, BRE (2003). ISBN – 1 86081 651 7

Principles of Laying Water Mains, Water Authorities Association

Recommended Code of Practice for Safe Hot Water Temperatures for All Non-domestic Installations, TMVA (2003)

'Safe' Hot Water and Surface Temperatures, HS (G) 104, Health and Safety Executive (1992). ISBN 0-11-321404-9

Service Cores in High Flats – Cold Water Services, Ministry of Housing and Local Government (1965)

Water Fittings and Materials Directory, Water Regulations Advisory Scheme. ISSN – 0954 3643

Water Regulations Guide, Water Regulations Advisory Scheme. ISBN – 0-9539708-0-9

Water Supply Byelaws Guide, Water Research Centre in association with Ellis Horwood Ltd, Chichester

Water Supply (Water Fittings) Regulations 1999, HMSO

Water Supply (Water Fittings) Regulations 1999, Regulator's Specification for the Performance of WC Suites, HMSO (1999)

Water Supply (Water Fittings) Regulations 1999, Regulator's Specification on the Prevention of Backflow, HMSO (1999)

Acts etc.
  Water Act 1945
  Water Supply (Water Fittings) Regulations 1999 (Statutory instruments 1148 and 1506)
  Water Industry Act 1991
  Water (Scotland) Act 1980
  Water and Sewerage Services (Northern Ireland) Order 1973
  Building Regulations
  Building Act 1984
  Building Standards (Scotland) Regulations
  Building (Northern Ireland) Regulations
  The Health and Safety at Work etc. Act 1974
  The Workplace (Health, Welfare and Safety) Regulations
  The Gas Safety (Installation and Use) Regulations
  Asbestos Regulations 1969

Index

access
    pipes entering buildings, 275–6
    through walls and floors, 275–9
    to cisterns, 281
    to stopvalves, 279–80
access to and use of buildings, 10, 128
air gaps, 203–10
air valves, 58
anti-vacuum valves, 211, 213–214, 216–7
approved contractors' schemes, 4

backfilling trenches, 339–40
backflow prevention (protection), 4, 57, 194,
    201–33
backflow protection, 210–33
    to agricultural and horticultural premises,
      230–31
    to bidets, WCs and urinals, 222–4
    to domestic and other premises, 222–7
    to fire protection systems, 231–3
    to industrial/commercial premises, 227–33
    secondary or zone protection, 220–21
backflow protection devices, 203–17
    application of, 222–33
    mechanical types, 210–17
    non-mechanical types, 203–10
    schedules of, 204–5
backflow risk categories, 216–19
backsiphonage, 201
back pressure, 201
benchmark scheme, 140–41
bidets, 223–4
blue water, 361
boiler efficiency, 260–3
boiler heated systems, 88, 101–14
boilers, 137–41
boosted cold water, 66–74
booster pumps, 71–4, 258
building regulations, 5–11, 16, 79, 92–4, 118,
    120, 124, 128, 147, 162, 197, 198,
    203, 228, 247–57, 258–66, 373

cavitation, 268
central heating combined with hot water,
    108–14, 200

check and anti-vacuum valves, 216–7
check valves, 215–7, 225, 231
circulating head, 105
circulating pumps, 105, 141–2
cisterns
    access, 281
    capacities, 38–9
    connections to, 43–6, 329–31
    control valves, 39–42
    feed (and expansion), 137–8
    float operated valves, 39–42
    interposed, 223–4
    large, 45–6
    linked, 45
    protected, 36–7
    requirements, 36–7
    sizing, 188–9
    storage, 36–50
    support, 38
    warning/overflow pipes, 46–50
cold feed pipes, 105–7, 125–6
cold water storage cisterns, 36–50, 137, 329
cold water systems
    boosted (pumped), 31, 66–74
    drinking water, 30, 34–5, 66–75
    direct/indirect, 31–5
combination boilers, 87–8
combination type storage heaters, 97–8
commissioning, 365–6
common feed and vent arrangements, 128–9
compartmentation, 373–5
competent persons (schemes), 8–9
concealed pipes and fittings, 347–50
connections to mains, 332–4
contamination of water, 57, 193, 196, 335
continuity bonding, 60, 350
corrosion, 108, 327, 338–9
cross connections, 194, 198–201

dead legs, 123–4
definitions, 4–5, 12–16, 379–80
discharge pipes, 162–6
disconnection of unused pipes, 372
dishwashers, 225, 257
disinfection of systems, 361–6

distributing pipe (secondary), 123–5
draining valves, 57, 143, 239–2
drinking water, 30, 34–5
drinking water fountains, 225

electrical earthing and bonding, 60, 350
electrical immersion type storage heaters, 98–102
electrical safety, 10–11
electrical work, 10–11
energy conservation, 9, 258–66
energy control, 155–8
energy loss from pipes, 259
expansion joints, 345
expansion relief valve, 162, 172
expansion vessel, 168, 170–71

feed and expansion cistern, 105–7, 125, 137
fire protection and sprinkler systems, 231–3, 350, 369, 376
flanged gate valve, 52
float operated valves, 39–42
float valve oscillation, 272
flow noise, 267–8
flow velocities, 57, 184
fluid risk categories, 201–2, 216–19
flushing and disinfecting, 361–5
flushing primary circuits, 365
flushing WCs and urinals, 248–55
frost protection
    draining down, 57, 143, 239–2
    insulation, 243–5
    outside buildings, 236–8
    to cisterns, 239–40
    to pipes below ground, 235–8
    to pipes entering buildings, 236–7
    to pipes in roof spaces, 239–40
    trace heating, 246

garden supplies, 225, 227
gate valves, 52, 56
graphical symbols, 17–20
gravity circulation, 102–5
grey water, 257–8

heat pump systems, 122
heat pumps, 120–23
hose pipes, 225–7
hose union taps, 225–7
hot store vessels (cylinders), 128–36
    capacities, 128
    combination units, 97–8, 134–5
    double feed indirect, 131–4
    energy control, 128, 258

heated by electricity, 136
high performance, 136–7
insulation of, 108, 137, 264–6
single feed indirect, 108–10, 134–5
sizing, 189–93
stratification, 129–30
temperature (thermostatic) control, 128, 147–54
types and grades, 130–7
unvented, 92–5, 117
vented, 117, 131–5
hot water and heating, combined, 108–14
hot water circulation
    gravity, 102–3
    primary, 105–14, 117
    pumped, 108–14
hot water for the less able, 147–54
hot water supply requirements, 128
hot water systems
    boiler heated, 88, 101–4
    choice of, 79–80
    combination boiler, 87–8
    combination type storage heaters, 97–8
    direct, 102–3, 105
    electric, 98–101
    gas fired circulators, 101
    heat pumps, 120–3
    indirect, 102–3
    instantaneous, 81–5
    non-pressure or inlet controlled, 95
    pressure or outlet controlled, 96–7
    sealed, 113–14, 162, 200
    single feed indirect, 108–10
    solar, 113–20
    storage type, 88–123
    unvented, 90–4, 168, 170–1
    vented, 88–9, 168
    water jacketed (primary store), 85–8
hydrants (fire), 58, 342–4

identification of pipes and valves, 353–5
immersion heaters, 98–101, 173, 191
initial procedures, 26–9
inspection of pipes and systems, 356–7
installation of pipes below ground, 335–45
installation of pipes in buildings, 345–50
instantaneous water heayers, 81–5
insulation
    of cisterns, 239, 240
    of hot store vessels, 108, 137, 263–6
    of pipes and fittings, 124, 263–6, 346
    of solar systems, 113–20
interposed cistern, 223–4
isolating valves, 58

joints for pipes
    bolted gland, 325
    brazed, 301–3
    braze welded, 303–4
    capillary solder, 295–6
    compression, 291–4, 315
    detachable, 326
    expansion, 345
    flanged, 310, 318, 322, 325
    lead run, 323
    mechanical, 315
    plain socketed, 322–3
    push-fit (slip-fit) and press-fit, 215, 296–300
    screwed gland, 327
    screwed (threaded), 282–5
    solvent cement, 317–18
    thermal fusion, 310–12
    wiped solder, 328
jointing potable pipework, 350–2

laying underground pipes, 335–44
leakage and leak detection, 369–71
legionella, 193, 197–8
loading units, 174, 176–7
local heating, 246

mains connections, 332–5
maintenance, 73, 366–8, 392
maintenance schedule, 367–8
materials, 6–7, 21–2, 155, 194, 197
materials, in contact with water, 194–7
meters and meter reading, 58–65
mixers (mixing valves), 143–6, 153–4, 200

noise and vibration, 267–74
noise
    cavitation, 268
    flow, 267–8
    in pipes and fittings, 267–8
    in pumps, 272
    in taps and valves, 267–72
    thermal transmission, 273–4
    vapour or air, 273
    water hammer, 270–2
notification and self-assessment, 3–4, 8–9

occupier information
open safety vent (vent pipe), 105–7, 126–7
outside taps, 225–8
overflow pipes, 46–50

pipe disinfection, 361–6
pipe fixings, 245, 284, 289–90, 307, 313–4, 319–21, 326, 346

pipe interrupter, 210, 223, 225–7
pipe joints (see jointing)
pipe sizes, 22–5
pipe sizing
    continuous flow, 177
    design flow rates, 174, 177
    flow rate, 176–7
    flow velocities, 57, 184
    head available, 179, 182
    head loss, 178–82
        permissible, 178–82
        through fittings, 179–82
        through pipes, 179–82
    loading units, 174, 176–7
    pipe length
        effective, 178
        equivalent, 178
        measured, 177
    procedures
    tabular method, 184–7
pipes
    cold feed, 125–6
    concealed, 347–50
    distributing, 123–5
    open safety vent (vent pipe), 126–8
    warning and overflow, 46–50
pipes (and fittings)
    of ABS (acrylonitrile butadiene styrene), 318–19
    of asbestos cement, 326–7
    of copper, 286–303
    of dissimilar materials, 327
    of iron, 320–5
    of lead, 327–8
    of plastic, 305–20
    of PB (polybutylene), 318–19
    of PE (polyethylene), 307–12
    of PE-X (cross linked polyethylene), 318–19
    of PP (propylene copolymer), 318–19
    of PVC-C (chlorinated polyvinyl chloride), 314, 317
    of PVC-U (unplasticised polyvinyl chloride), 312, 317
    of stainless steel, 291, 304–5
    of steel, 282–5
pipes outside buildings, 27–8, 57–8
pipes passing through fire walls, 350, 374–6
pipes passing through structural members, 347–50
plastics pipes, 305–20
plug cock, 51
pneumatic pressure vessel, 66, 70–3
pressure (expansion) relief valve, 162
pressure flushing cisterns, 248

pressure flushing valves, 250–1
pressure reducing (limiting) valve, 167–8
prevention of backflow, 194
primary circulation, 105–14, 172
primary stores, 85–8, 265–6
pump
    booster, 71–3
    circulating, 141
pump room 73–4
pumped systems, 31, 66–73, 108–14

rainwater, 198–9
recycled water, 198–9, 257–8
reduced pressure zone valve (RPZ valve),
    210–13

safe hot water temperatures, 147–54
scale control, 260
sealed system, 113–14, 162, 200
secondary backflow protection, 220–1
secondary circulation, 123–5
self assessment, 8
self-closing taps, 256
servicing valves, 54–6
showers, 143, 145–6, 147, 200, 223–5
sizing
    cold water storage, 188–9
    discharge pipes, 165–6
    hot and cold pipes, 174–88
    hot water storage, 189–93
    sprinkler systems, 394–6
solar heating, 113–20
spherical plug valve, 56
spray taps, and aerators, 256
sprinkler systems, 378–96
stagnation of water, 194, 197–8
sterilization of pipework systems, 361–6
stopvalves, 50–4, 279–80
storage cisterns, 36–50, 384
storage type water heaters, 88–97
stratification, 129–30, 190–3
supplementary water heating, 114
surface boxes, 229, 280, 342–4
systems of cold water, 31–5, 57–8, 376–96
systems of hot water, 79–125

taps and valves, 30, 50–57, 141–53, 256
temperature relief and heat dissipation,
    160–2
temperature relief valves, 161, 162
testing of pipes and systems, 353, 357–60,
    391–2
thermal energy cut-out device, 159–60
thermal insulation, 346

thermostatic control, 145–6, 153–4, 157–60
thrust on mains, 340–6
trace heating, 246
trench excavations, 336–40
type AA air gap, 203, 206–7
type AB air gap, 207
type AG air gap, 207–8
type AUK1 air gap, 208–9

under pressure tapping machine, 332
unvented hot water, 92–5, 117, 170–1
urinals (see also WCs and urinals), 223,
    248–55

valve chamber, 58, 342–4
valves
    air, 58
    anti-vacuum, 211, 213–15
    chambers and surface boxes, 58, 342–4
    check, 215–27, 225
    check and anti-vacuum, 211, 213–14,
        216–17
    combination tap, 143
    double check, 216
    draining, 57, 143, 239–2
    expansion relief, 162, 172
    float operated, 39–42
    for sprinkler systems, 389–90
    for the disabled, 147–53
    gate, 52, 56
    hydrant, 58, 342–4
    identification, 353–4
    isolating, 56, 58
    marker posts and indicator plates, 344
    mixing, 143–6, 153–4
    pipe interrupter, 210, 223, 225–7
    plug cock, 51
    pressure flushing, 250–1
    pressure limiting, 168
    pressure reducing, 167
    reduced pressure zone valve (RPZ valve),
        210–13
    screwdown, 51
    servicing, 54–6
    spherical plug, 56
    stop, 50–4, 57, 279–80
    temperature relief, 160–2
    vacuum breaker (see anti-vacuum valve)
    washout, 58
vented hot water system, 89, 168
vent pipe, 105–7, 126–7

warning pipes and overflow pipes, 46–50
washing machines, 225, 257

water
  conservation, 4
  drinking, 30
  economy, 247–57
  meters, 58–65
  pressure, 57, 167–8
  softeners, 73–7
  supplies, 26–8
water closets (WCs) and urinals, 223, 248–56
water jacketed tube heaters, 85–8
water level control, 172–3

water quality (preservation of), 194
water regulations, 2–5, 14–15, 89, 95, 197–8, 203, 267
water systems, 31–5, 57, 66
water systems outside buildings, 57–8, 335–44
water treatment, 73–8
workmanship, 8

zone or secondary backflow protection, 220–1